T0325740

100 YEARS OF FERROELECTRICITY

1921–2021

100 YEARS OF FERROELECTRICITY

1921–2021

JULIO A GONZALO
Universidad Autónoma de Madrid, Spain

GINES LIFANTE
Universidad Autónoma de Madrid, Spain

FRANCISCO JAQUE
Universidad Autónoma de Madrid, Spain

EW JERSEY • LONDON • SINGAPORE • BEIJING • SHANGHAI • HONG KONG • TAIPEI • CHENNAI • TOKYO

Published by

World Scientific Publishing Co. Pte. Ltd.
5 Toh Tuck Link, Singapore 596224
USA office: 27 Warren Street, Suite 401-402, Hackensack, NJ 07601
UK office: 57 Shelton Street, Covent Garden, London WC2H 9HE

British Library Cataloguing-in-Publication Data
A catalogue record for this book is available from the British Library.

100 YEARS OF FERROELECTRICITY 1921–2021

ISBN 978-981-124-309-7 (hardcover)

For any available supplementary material, please visit
https://www.worldscientific.com/worldscibooks/10.1142/12437#t=suppl

Desk Editor: Rhaimie Wahap

Typeset by Stallion Press
Email: enquiries@stallionpress.com

Printed in Singapore

Preface

Since the discovery in 1921 of ferroelectric properties in Rochelle Salt by Valasek, a new ferroelectric material was found until Busch and Scherrer in Switzerland in 1943. However, during the Second World War, researchers in the US discovered ferroelectric properties in perovskites that led to an explosion of research after the war in Western and Central Europe, Russia and Japan, because of the promising and potential applications.

This book presents an overview of the most relevant experimental and theoretical work in ferroelectricity and is organized into three periods. While the first (1921–1960) discusses the early work by Valasek and ends up with work by Cochran on crystal stability and the soft mode theory of ferroelectricity, the second (1961–2002) shows that the number of publications and scientific meetings related to ferroelectricity that began to increase exponentially. New subfields on ferroelectric liquid crystals, thin films and integrated ferroelectrics, dipolar glasses and relaxors emerged.

The last and final section (2002–2021) includes fundamental structural studies of ferroelectric perovskites, neutron diffraction investigations on PZT, quantum effects in ferroelectric transitions lowering and even suppressing the Curie temperatures at low temperatures, and investigations on the dielectric constant of liquid water showing Curie-Weiss behavior as well as detection of a thermal anomaly in liquid water, and other recent developments.

The book will be a useful compendium for materials scientists around the world.

Editors
Julio A Gonzalo, Gines Lifante and Francisco Jaque
Universidad Autónoma de Madrid, Spain

Contents

Introduction

Introduction: IMF1 to IMF14

After the Discovery of "ferroelectricity" in **Rochelle Salt** by Valasek in 1921 in USA, and the observation of hysteresis loops below T_C displaying a strong nonlinearity above T_C (the transition temperature) named by Valasek as the Curie temperature because the remarkable analogy of Polarization versus Electric field in Rochelle Salt with the behaviour of Magnetization versus Magnetic field in Iron, Nicked or Cobalt, no new ferroelectric material was discovered until Busch and Scherrer, working in Switzerland, discovered ferroelectric behaviour at relatively low temperature in **KDP** (KH_2PO_4) and several isomorphs in 1943.

Then, during the Second World War, Wainer and Solomon (USA), Ogawa (Japan) and Wul and Goldman (Russia) discovered ferroelectric behaviour in ceramic simples of **Barium Titanate** ($BaTiO_3$) and isomorphous perovskites. The number of ferroelectrics discovered by Pepinsky *et al.* in the USA increased rapidly. After the end of the war, a number of researchers in the Western and Central Europe, in Russia and in Japan entered the field and because of promising applications, "ferroelectricity" became an active subfield of compounds undergoing phase transitions within Solid State Physics.

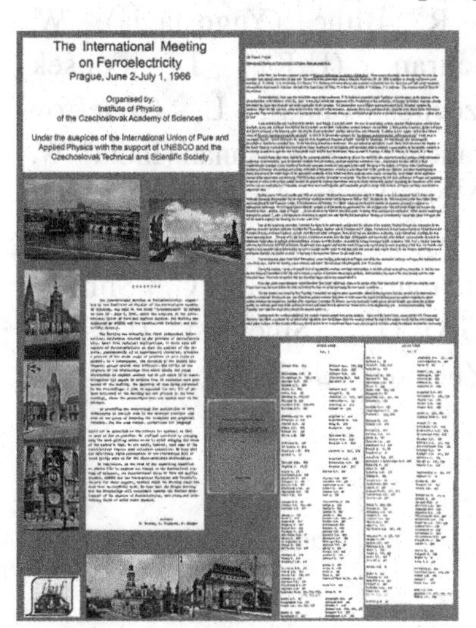

The **First International Meeting on Ferroelectricity** (IMF) was hold in **Prague**, June 2–July 1, 1966. The Editors of the resulting Proceedings were V. Dvorak, A. Fouskova and P. Glogar, No Advisory International Committee was yet elected.

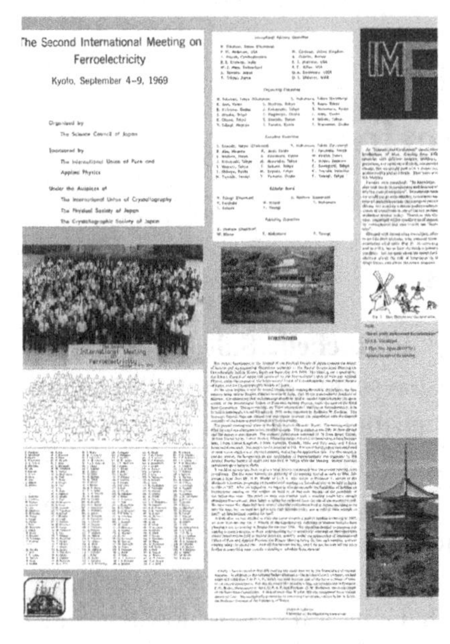

The **Second IMF** was held in **Kyoto**, September 4–9, 1969, with H. Takahasi (Japan) as Chairman and the International Advisory Committee was made up by P. W. Anderson (USA), J. Fousek (Czechoslovakia), S. Sawada (Japan), Y. Takagi (Japan) W. Cochran (U.K), A. Guinier (France), R. C. Miller (US), G. A. Smolenski (URRS) and G. S. Zhudanov (URRS).

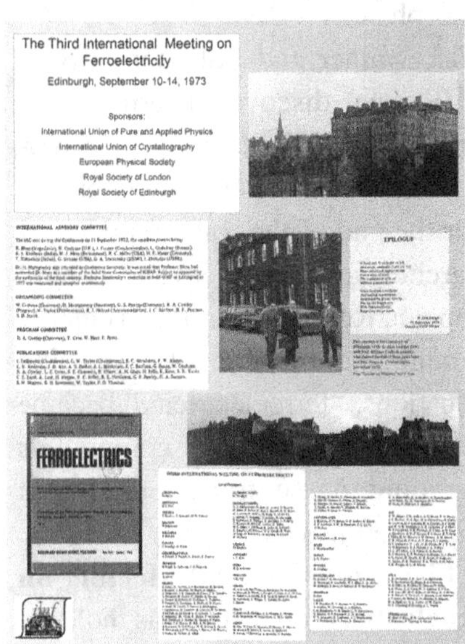

The **Third IMF** was held in **Edinburgh**, September 10–14, 1973. The Chairman of the Organizing Committee was W. Cochran and the International Advisory Commitee was made up by R. Blinc (Yugoslavia), W. Cochran (U.K), J. Fousek (Czechoslovakia), L. Godefroy (France), R. S. Krishnan (India), W. J. Merz (Switzerland), R. C. Miller (USA), H. E. Müser (Germany), T. Nakamura (Japan), G. Shirane (USA), G. A. Smolensky (URRS), I. Zheludev (USSR).

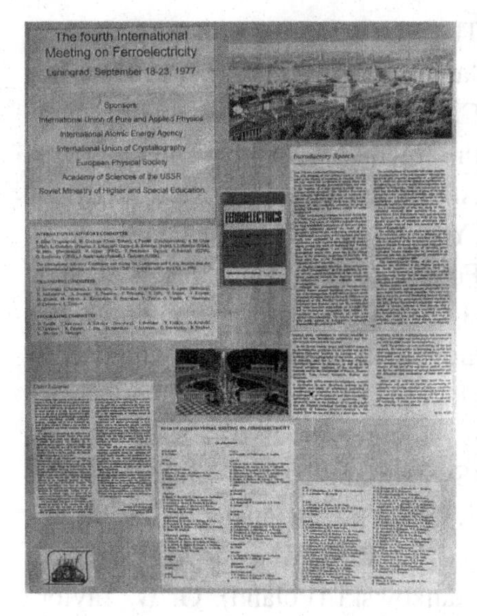

The **Fourth IMF** was held in **Leningrad**, September 18–23, 1977. The Chairman was G. Smolensky (USSR) and the members of the International Advisory Committee were R. Blinc (Yugoslavia), W. Cochran (Great Britain), J. Fousek (Czechoslovakia), R. Krishnan (India), J. Lefkowitz (USA), W. Merz (Switzerland), H. Müsser (FRG), T. Nakamura (Japan), G. Schmidt (GDR), G. Smolensky (USSR) — Chairman, J. Stankowski (Poland), I. Zheludev (USSR).

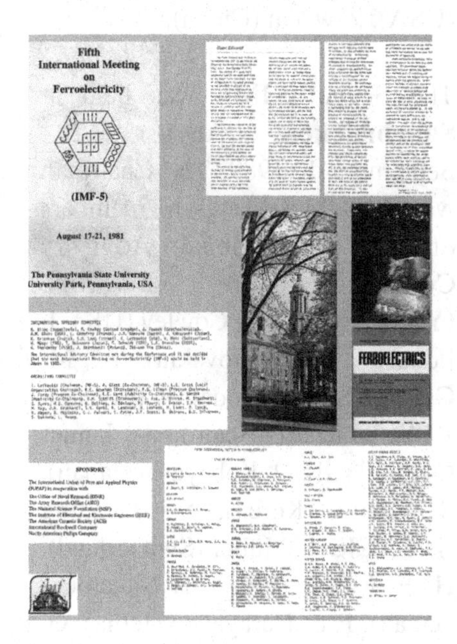

The **Fifth IMF** was held at **The Pennsylvania State University**, University Park, Pennsylvania, USA in August 17–21, 1981. The International Advisory Commitee was made up by R. Blinc (Yugoslavia), R. Cowley (UK), J. Fousek (Czechoslovakia), H.M. Glass (USA), L. Godefroi (France), J. A. Gonzalo (Spain), I. Lefkowithz (USA), W. Merz (Switzerland), H. Müser (FRG), I. Nakamura (Japan), G. Schmidt (GDR), L. A. Shuvalov (USSR), G. Smolensky (USSR), J. Stankowski (Poland), Zhi-wen Yin (China).

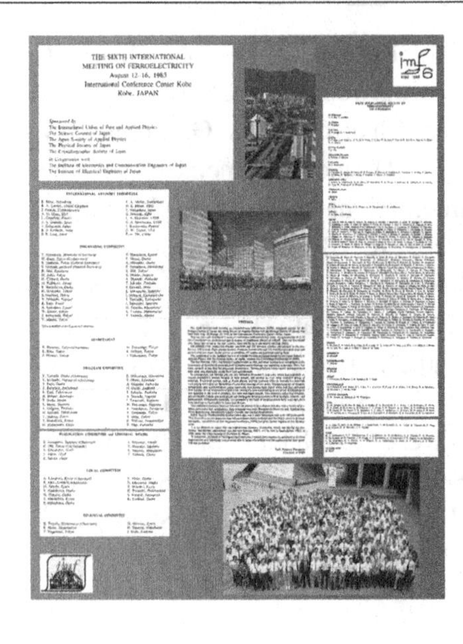

The **Sixth IMF** was held at **Kobe**, Japan, August 12–16, 1985. The Chairman was T. Nakamura, and the members of the International Advisory Committee were R. Blinc (Yugoslavia), R. A. Cowley (UK), J. Fousek (Czechoslovakia), A. Glass (USA), L. Godefroy (France), J. A. Gonzalo (Spain), J. Kobayashi (Japan), R. S. Krishnan (India), S. B. Lang (Israel), K. A, Müller (Switzerland), H. E. Müser (FRG), T. Nakamura (Japan), G. Schmidt (GDR), L. A. Shuvalov (USSR), J. Stankowski (Poland), G. W. Taylor (USA), Z-w Yin (China).

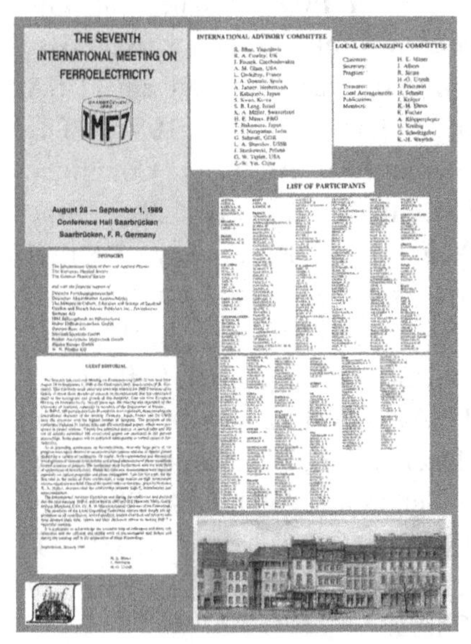

The **Seventh IMF** was held at **Saarbrücken**, August 28–September 1, 1989. The Chairman was H. E. Müser, and the members of the International Advisory Committee were R. Blinc (Yugoslavia), R. A. Cowley (UK), J. Fousek (Czechoslovakia), A. M. Glass (USA), L. Godefroy (France), J. A. Gonzalo (Spain), A. Janner (Netherlands), J. Kobayashi (Japan), S. Kwun (Korea), S. B. Lung (Israel), K. A. Müller (Switzerland), H. E. Müser (FRG), T. Nakamura (Japan), P. S. Narayanan (India), G. Schmidt (GDR), L. A. Shuvalov (USSR), J. Stankowski (Poland), G. W. Taylor (USA), Z-w Yin (China).

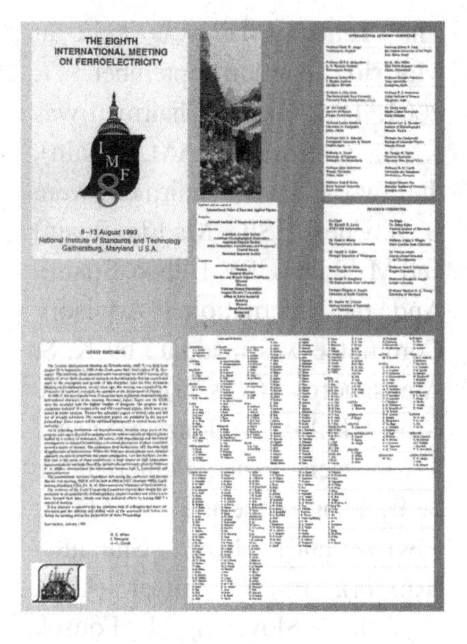

The **Eight IMF** was held at the N. I. of Standards and Technology **Gaithersburg, Maryland** (USA), 8–13 August, 1993. The Advisory International Committee was made up by F. W. Ainger (England) K. S. Aleksandrov (Russia), R. Blinc (Slovenia), I. Eric Cross (USA), J. Fousek (Czechoslovakia), L. Godrefroy (France), J. A. Gonzalo (Spain), A. Janner (Netherlands), J. Kobayashi (Japan), Sook Il Kwun (Korea), S. B. Lang (Israel), K. A. Müller (Switzerland), T. Nakamura (Japan), P. S. Narayanan (India), G. Sorge (Germany), L. A. Shuvalov (USSR), J. Stankowski (Poland), G. W. Taylor (USA), H. G. Unruh (Germany), Z-w Yin (China).

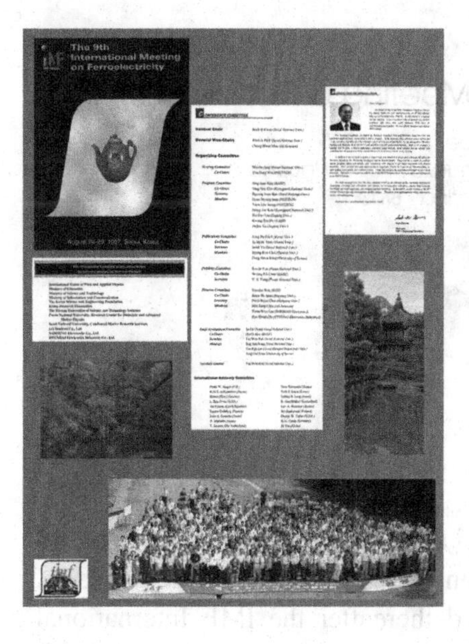

The **Ninth IMF** was held at **Seoul** (Korea), in August 24–29, 1997. The General Chair was held by Suk–Il Kwun and the International Advisory Committee was made up by F. W. Ainger (UK), K. S. Aleksandrov (Russia), R. Blinc (Slovenia), I. Eric Cross (USA), L. Godefroy (France), Julio A. Gonzalo (Spain), Y. Ishibashi (Japan), T. Jansen (The Netherlands), J. Kobayashi (Japan), S. B. Lang (Israel), K. A. Müller (Switzerland), L. A. Shuvalov (Russia), J. Stankowski (Poland), G. W. Taylor (USA), H. G. Unruh (Germany), Xi Yao (China).

The **Tenth IMF** was held at **Madrid** (Spain) in September 1–7, 2001. The General Chairman was J. A. Gonzalo (UAM, Spain), the two Co — Chairmen were B. Jiménez (CSIC, Madrid, Spain) and M. Tello (UPV, Bilbao, Spain). And the International Advisory Committee was made up by K. S. Aleksandrov (Russia), L. Eric Cross (USA), J. A. Gonzalo (Spain), T. Janssen (The Netherlands), S. B. Lang (Israel), K. A. Müller (Switzerland), J. A. Shuvalov (Russia), F. W. Ainger (UK), R. Blinc (Slovenia), J. Fousek (Czech Republic), B. Hilczer (Poland), Y. Ishibashi (Japan), W. Kleemann (Germany), M. Maglione (France), J. Petzelt (Czech Republic), G. W. Taylor (USA).

The **Eleventh IMF** took place in 2005 at **Iguassu,** in the border between **Brazil** and **Argentina**. The Chairmen were J.A. Eiras (Brazil) and A. López García (Argentina) who joined thereafter the IMF International

Advisory Committee formed by K. S. Aleksandrov (Russia), A. S. Bhalla (USA), W. K. Choo (Korea), L. E. Cross (USA), J. A. Gonzalo (Spain), M. Glazer (UK), B. Hikzer (Poland), N. Ichinose (Japan), Y. Ishibashi (Japan), T. Janssen (Holland), W. Kleemann (Germany), S. B. Krupanidhi (India), S. Kwun (Korea), S. B. Lang (Israel), M. Maglione (France), D. Michel (Germany), K. A. Müller (Switzerland), J. Petzelt (Czech Republic), A. Sigov (Russia), A. Stemberg (Baltic States), G. W. Taylor (USA), M. Tello (Spain) and X. Yao (China)

The **Twelfth IMF** was held together with the IEEE–UFFC Ferroelectrics Committee, at **Xian** (China), the ancient capital of China. The Chairperson was Prof. Xi Yao (Xi'an Jiaotong University) the year 2009. The list of members comprised K. S. Aleksandrov (Russia), A. S. Bhalla (USA), R. Blinc (Slovenia), W. K. Choo (Korea), L.E. Cross (USA), J. A. Eiras (Brazil), J. A. Gonzalo (Spain), A. López García (Argentina), M. Glazer (UK), B. Hilczer (Poland), N. Ichinose (Japan), Y. Ishibashi (Japan), W. Kleemann (Germany), S. B. Krupanidhi (India), S. I. Kwun (Korea), J. B. Lang (Israel), M. Maglione (France), D. Michel (Germany), K. A. Müller (Switzerland), J. Petzelt (Czech Republic), A. Sigov (Russia), A. Stemberg (Baltic States), G. W. Taylor (USA), M. Tello (Spain) and X. Yao (China).

At the banquet held at the Tang Yue Palace, Distinguished Contributions Awards were presented by Prof Yao Xi and Prof. J. A. Gonzalo to: R. Blinc, L. Eric Cross, J. Fousek, Y. Ishibashi, V. H. Schmidt and G. Taylor.

13th International Meeting on Ferroelectricity | IMF-13

13th International Meetings on Ferroelectricity

2 września 2013 - 6 września 2013

The August Chełkowski Institute of Physics, The Department of Ferroelectrics Physics , University of Silesia, Katowice, Poland & The Institute of Physics, Jagiellonian University, Kraków, Poland & Polish Physical Society

http://imf-13.us.edu.pl/

Kraków, Poland

The **Thirteenth IMF** was held the year 2009 at **Krakow**, Poland. The Chairman was Prof. Krystian Roleder and The International Advisory Committee was formed by A. S. Bhalla (USA), S. G. Baik (Korea), L. E. Cross (USA), J. A. Eiras (Brazil), M. Glinchuk (Ukraine), J. A. Gonzalo (Spain), A. López–García (Argentina), A. M. Glazer (UK), N. Ichinose (Japan), W. Kleemann (Germany), S. B. Krupanidhi (India), T. W. Noh (Korea), S. B. Lang (Israel), M. Maglione (France), D. Michel (Germany), Y. Noda (Japan), B. Noheda (Netherland), J. Petzelt (Czech Rep.), W. Ren (China), K. Roleder (Poland) and A. Safart (USA), N. Setter (Switzerland), A. Sternbergs (Latvia), G. W. Taylor (USA), M. L. Tello (Spain), I. Ytseng (China), K. S. Alexander (Russia), R. Blinc (Slovenia).

The **Fourteenth IMF** took place the year 2013 at **San Antonio** (TX, USA). The chairman was A. Bhalla (USA) and members of the International Advisory committee were Baik, S. G. (Korea), Banys, J. (Lithuania), Eiras, J. A. (Brazil), Ganzer, M. (Ukhraine), Gonzalo, J. A. (Spain) Gregg J. (North Ireland), Hinka, J. (Czech Rep.), Kleemann, W. (Germany), Kreisel, J. (Luxemburg), Lang, S. (Israel), Maglione, M. (France), Michel, D. (Germany), Migoni, R. (Argentina), Noda, Y. (Japan), Noheda, B. (The Netherlands), Ren. W. (China), Roleder, K. (Poland), Setter, N. (USA), Tello, M. L. (Spain), Tseng, T.-Y. (Taiwan), Tsurumi T. (Japan), Vakhrushev, S. (Russia), Yao Xi (China), Ye, Z. G. (Canada).

1. Ferroelectricity: Early Work, 1921–1961

The discovery of "Ferroelectricity" was reported the year 1921 in Rochelle Salt (sodium potassium tartrate tetrahydrate: $NaKC_4H_4O_6$), a material which had interesting piezoelectric properties capable of detecting ultra-sound waves propagating underwater.

The upper transition temperature between a polar, ferroelectric state and a non-polar, paraelectric state was $T_C \simeq 22°C = 295°K$, and Valasek called it Curie Temperature because of the close analogy between ferroelectricity which implied a strongly nonlinear dependence of P (polarization) and E (electric field), and **ferromagnetism**, discovered by Pierre Curie, which, as noted by Pierre Weiss in 1907, implied such nonlinear dependence of M (magnetization) and H (magnetic field). The maximal spontaneous polarization was relatively low: $P_S = 0.25\ \mu C/cm^2$.

No other ferroelectric material was discovered in decade and a half, and many physicists begun to think that Rochelle Salt was a very strange material. However, in 1935, working in Switzerland, Busch and Scherrer discovered ferroelectricity in **KDP** (Potassium Dihydrogen Phosphate: KH_2PO_4) at relatively low temperature $T_C \simeq -150°C = 123°K$. The saturation spontaneous polarization, at $T_C \simeq -177°C$ was $P_s \simeq 4.75\,\mu C/cm^2$.

About a decade later another ferroelectric material at room temperature, a perovskite, **$BaTiO_3$**, was discovered with a Curie temperature $T_C \simeq 120°C = 393°K$, and very much larger spontaneous polarization at room temperature (23°C) such as $P_S \simeq 26\ \mu C/cm^2$. Very soon ceramic capacitors of this material began to be produced in large scale. And many other ferroelectric and antiferrolectric perovskites were discovered, showing clearly that ferroelectricity in solids was not a rare phenomenon.

TGS (tri-glycine sulphate: $(NH_2CH_3COOH)_3H_2SO_4$, with a $T_C \simeq 49°C = 322°K$, $P_S(RT) \simeq 2.8\ \mu C/cm^2$, easily switchable, was discovered in 1956.

$LiNbO_3$ and $LiTaO_3$, which can be considered twisted perovskites and have very large transition temperatures and spontaneous polarization came out next.

And the field of ferroelectric materials, their solid solutions and the electro-optic, electromechanical, electrothermal devices made up using them was growing steadily.

The following table illustrate the broad field of "Cooperative Phenomena" as it was developed since the 19th century to the early 21st century.

Table of Cooperative Phenomena			
Transition	**Conjugated variables**	**Discovery**	**Theory**
Liquid/Vapor	P – V	T. Andrews, 1872	J. van der Waals, 1874
Ferro/Paramagnet	H – M	P. Curie, 1895	P. Weiss, 1907
Superconductors/ Normal	—	K. Onnes, 1906	Bardeen, Cooper, Schrieffer, 1957
Order/Disorder (in alloys)	—	G. Tammann 1919	W. Bragg & E. Williams
Ferro/Paraelectric	E – P	J. Valasek, 1921	Mason & Matthias, 1948
Superfluid/ Normal	—	W. Keesom	N. Bogoliubov, 1946
Ferro/Paraelastic	X – x	—	K. Aizu, 1969

In "**Ferroelectricity: The Fundamentals Collection**" Wiley-VCH, 2005, J. A. Gonzalo, B. Jiménez **Editors** a set of relevant experimental and theoretical papers published in the **Physical Review**, **Philosophical Magazine** and **Journal of the Physical Society of Japan** appeared in the period 1921 to 1961 some of which are reproduced below.

1.1	J. Valasek	Piezo-electric Activity of Rochelle Salt Under Various Conditions	Physical Review, 1921
1.2	C. B. Sawyer and C. H. Tower	Rochelle Salt as a Dielectric	Physical Review **35**, 1930

(*Continued*)

1.3	C. Busch and P. Scherrer	A New Seignette-Electric Substance	Naturwiss **23**, 1935
1.4	W. P. Mason and B. T. Matthias	Theoretical Model for Explaining the Ferroelectric Effect in Barium Titanate	Physical Review **74**, 1948
1.5	G. Shirane, K. Suzuki, and A. Takeda	Phase Transitions in Solid Solutions of PbZrO3 and PbTiO3 (II) X-ray Study	Journal of the Physical Society of Japan **7**, 1952
1.6	B. T. Matthias, C. E. Muller, and J. P. Remeika	Ferroelectricity of Glycine Sulphate	Physical Review **104**, 1956
1.7	W. Cochran	Crystal Stability and the Theory of Ferroelectricity	Physical Review Letters **3**, 1958

2. Ferroelectrics 1961–2001

2.1	A. M. Glazer	The Classification of Tilted Octahedra in Perovskites	Acta Crystallographica Section B, 1972
2.2	A. M. Glazer	Simple Ways of Determining Perovskite Structures	Acta Crystallographica Section A, 1975
2.3	J. A. Gonzalo	Equation of State for the Cooperative Transition of Triglycine Sulfate near T_C	Physical Review B, 1970
2.4	C. Alemany, J. Mendiola, B. Jimenez, and E. Maurer	X-ray Structural Damage of Triglycine Sulphate (TGS)	Acta Crystallographica, 1973
2.5	J. R. Fernández-del-Castillo, J. Przeslawski, and J. A. Gonzalo	Equation of State for the Pressure- and Temperature-Induced Transitions in Ferroelectric Telluric Acid and Ammonium Phosphate	Physical Review B: Rapid Communications, 1996
2.6	B. Noheda, J. A. Gonzalo, L. E. Gross, R. Guo, S. E. Park, D. E. Cox, and G. Shirane	Tetragonal-to-Monoclinic Phase Transition in a Ferroelectric Perovskite: The Structure of $PbZr_{0.52}Ti_{0.48}O_3$	Physical Review B, 2000
2.7	V. Ya. Shur, E. L. Rumyantsev, E. V. Nikolaeva, and E. I. Shishkin	Formation and Evolution of Charged Domain Walls in Congruent Lithium Niobate	Applied Physics Letters, 2000

(*Continued*)

2.8	V. Ya. Shur, E. L. Rumyantsev, E. V. Nikolaeva, E. I. Shishkin, D. V. Fursov, R. G. Batchko, L. A. Eyres, M. M. Fejer, and R. L. Byer	Nanoscale Backswitched Domain Patterning in Lithium Niobate	Applied Physics Letters, 2000
2.9	C. L. Wang, Z. K. Qin, and D. L. Lin	First-Order Phase Transition in Order-Disorder Ferroelectrics	Physical Review B, 1989
2.10	N. Cereceda, B. Noheda, T. Iglesias, R. Fernández del Castillo, J. A. Gonzalo, N. Duang, Y. L. Wand, D. E. Cox and G. Shirane	O_3 Tilt and Pb/(Zr/Ti) Displacement Order Parameters in Zr-Rich Pb $Zr_{1-x}Ti_xO_3$ from 20 to 500 K	Physical Review B, 1997
2.11	V. Wesphals, W. Kleemann, and M. D. Glinchuk	Diffuse Phase Transitions and Random-Field-Induced Domain States of the “Relaxor” Ferroelectric $PbMg_{1/3}Nb_2/3O_3$	Physical Review Letters, 1992

3. Ferroelectrics 2001–2021

3.1	V. V. Shvartsman, S. Bedanta, P. Borisov, and W. Kleemann	(Sr, Mn)TiO_3: A Magnetolectric Multiglass	Physical Review Letters, 2008
3.2	C. L. Wang, C. Aragó, J. García, and J. A. Gonzalo	Quantum Tunnelling Versus Zero-Point Energy in Double-Well Potential Model for Ferroelectric Phase Transitions	Physica A: Statistical Mechanics and Its Applications, 2002
3.3	C. L. Wang, C. Aragó, and M. I. Marqués	Characteristic Temperatures of First-Order Ferroelectric Phase Transitions: Effective Field Approach	Journal of Advanced Dielectrics, 2012
3.4	J. C del Valle, E. Camarillo, L. Martínez Maeso, J. A. Gonzalo, C. Aragó, M. I. Marqués, D. Jaque, G. Lifante, J. García Solé, K. Santa Cruz, R. Carrillo Torres, and F. Jaque	Dielectric Anomalous Response of Water at 60ºC	Philosophical Magazine, 2015
3.5	R. Jiménez, B. Jiménez, J. Carraud, J. M. Kiat, B. Dkhil, J. Holc, M. Kosec, and M. Alguero	Transition Between the Ferroelectric and Relaxor States in $0.8Pb(Mg_{1/3}Nb_{2/3})O_3$-$0.2PbTiO_3$ Ceramics	Physical Review B, 2006
3.6	W. Kleemann, J. Dec, A. Tkach, and P. M. Vilarinho	$SrTiO_3$ — Glimpses of an Inexhaustible Source of Novel Solid-State Phenomena	Condensed Matter, 2020

1

Early Work 1921–1961

PIEZO-ELECTRIC ACTIVITY OF ROCHELLE SALT UNDER VARIOUS CONDITIONS.

BY J. VALASEK.

SYNOPSIS.

Electrical Properties of Rochelle Salt Crystal are analogous to the magnetic properties of iron, the dielectric displacement D and polarization P varying with the electric field E in the same general manner as B and I vary with H for iron, and showing an *electric hysteresis* with loops distorted by an amount corresponding to the *permanent polarization* P_0, whose value is about 30 e.s.u./cm.^{3} but varies for different crystals. The *dielectric constant* ($\kappa = dD/dE$) was measured *from* $-70°$ *to* $30°$ *C.* and found to be surprisingly large, increasing from about 50 at $-70°$ to a maximum of about 1,000 near 0°. The modulus of *piezo-electric activity for shearing stresses* (δ) *varies with temperature,* $-70°$ *to* $40°$ *C.*, in a very similar manner, increasing from less than 10^{-6} at $-70°$ to a maximum of about 10^{-4} at 0°. The ratio δ/κ *varied with the electrode material,* being greater for tin foil than for mercury electrodes. The difference may be due to the alcohol used in shellacking the tin-foil electrodes on. There are other indications that δ and κ are related. The *variation of* δ *with humidity* is such as can be accounted for by the decrease in the dielectric constant of the outer layer as a result of dehydration. The change of polarization produced by pressure as measured by the change in the hysteresis loop agrees with the value found directly from the piezo-electric response, as required by Lord Kelvin's theory. Also *fatigue effects on* δ produced by temporarily applied fields are traceable to fatigue in the polarization. The *electrical conductivity below* $45°$ is less than 5×10^{-9} mhos/cm.^{3} but *from* $43°$ *to* $57°$ increases rapidly to 5×10^{-4}.

Optical Properties of Rochelle Salt as Calculated from the Natural Polarization.— Assuming only one electron is displaced the *natural period* corresponds to a wavelength of 4.2 μ and the *specific rotation* for sodium light comes out 10°, the observed value being 22°.1.

RECENTLY[1] the writer described some experiments on the dielectric and piezo-electric properties of Rochelle salt, which were made for the purpose of correlating and explaining the effects observed chiefly by Cady and by Anderson. The plates used were cut with faces perpendicular to the $\breve{a}$ axis and with edges at 45° with the $\bar{b}$ and c axes. The present paper is a continuation of the work, the variations in the electrical properties having been studied more extensively. The apparatus and method of observation have been already described in the paper referred to above. The more important results obtained at that time can be summarized as follows:

In the case of Rochelle salt the dielectric displacement D, electric intensity E, and polarization P behave in a manner analogous to B, H, and I in the case of magnetism. Rochelle salt shows an electric

[1] J. Valasek, PHYS. REV. (2), XVII, p. 475.

hysteresis in P analogous to the magnetic hysteresis in the case of iron, the loops however being distorted by an amount corresponding to the permanent polaiization of the crystal in the natural state. This point of view is very effective in accounting for many of the peculiarities observed.

In an electric field the piezo-electric activity has a maximum for a definite value of the field and decreases to a small value in both directions. The position of the maximum corresponds to the greatest rate of change of polarization with electric field in the case of the condenser experiments. In fact if force and electric field are equivalent in changing the piezo-electric polarization then the response for a given force in various applied fields must necessarily give curves of the same general nature as curves of $\partial P/\partial E$ or $\partial D/\partial E$ against E. It is permissible to interchange D and P in most cases because of the large dielectric constant of Rochelle salt.

RELATION BETWEEN POLARIZATION AND PIEZO-ELECTRIC ACTIVITY.

The activity of a piezo-electric crystal is intimately related to the natural polarization. According to Lord Kelvin this natural moment is masked by surface charges so that the crystal appears to be uncharged. This polarization or piezo-electric moment can be measured independently of the charges on the electrodes, through the distortion of the hysteresis loop. The center A of the loop is found by a consideration of symmetry and may be assumed to represent the condition of no polarization. If the natural condition of polarization is assumed to be half way between the two branches of the loop at zero field then the value of the permanent polarization P_0 is proportional to AB, Fig. 1. There being

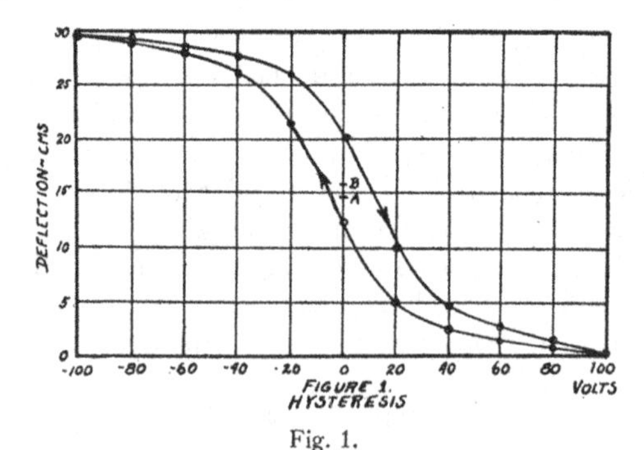

Fig. 1.

no field applied, the equation for the work done per unit charge carried through the condenser is:

$$4\pi\left(\frac{Q_0}{S} - P_0\right)t = 0,$$

so that

$$P_0 = \frac{Q_0}{S},$$

where Q_0 is the apparent average permanent charge at zero field given by AB (Fig. 1) and where S is the area of the plate. Calculation gives the value: 30 e.s.v./cm^2.

According to Lord Kelvin's theory an applied stress will change this polarization so as to create free charges on the electrodes. A force of 250 grams applied to the crystal should consequently shift the loop by an amount equivalent to the piezo-electric response for 250 grams. When this experiment was performed another, but more unsymmetrical loop, was obtained. The change in polarization by the loop method was 114 e.s.u./cm.^{2} while the piezo-electric response amounted to 121 e.s.u./cm.^{2}

The value of P_0 obtained from the hysteresis loops is only approximate because of the assumptions involved in its determination. It cannot, moreover, be fixed definitely enough to be put down as a physical constant of Rochelle salt because it varies with different specimens, besides changing with temperature, pressure and fatigue. The value $P_0 = 30$ e.s.u./cm.^{2} is thought to be a representative value and is checked by other measurements. The writer would not be surprised, however, to find other specimens giving several times this value. The change in polarization due to pressure however is derived by a differential method eliminating much of the uncertainty in measurements on one loop. The result in this case should be fairly definite, as indeed it seems to be.

Piezo-electric activity depends on both the crystalline structure and on the polarization. It is greatest for a polarization somewhat larger than normal and decreases in both directions for changes in this quantity, the polarization being changed by applying an electric field. It has been shown by the writer that this relation between activity and applied field is approximately like that of the derivative $\partial D/\partial E$ of the curve relating the dielectric displacement D and the electric field E of the crystal used as a condenser. Since this latter relation is in the form of a hysteresis loop it follows that the activity is also a double-valued function of the applied field depending on the direction of variation of the field. A curve illustrating this effect is reproduced in Fig. 2. The readings were taken in as short a time as possible to eliminate fatigue. These curves show that the piezo-electric response at zero field depends on the previous electrical treatment of the crystal. The latter fact has also been noted by W. G. Cady in the report previously referred to.

This after-effect does not persist very long but dies off exponentially with the time. The piezo-electric response or ballistic throw of the galvanometer for 250 grams has been observed to return to half value in 1 minute and to normal in over 20 minutes after fields of 150 volts have been applied for 3 minutes previously. There is a much greater after-effect in the direction of increased activity.

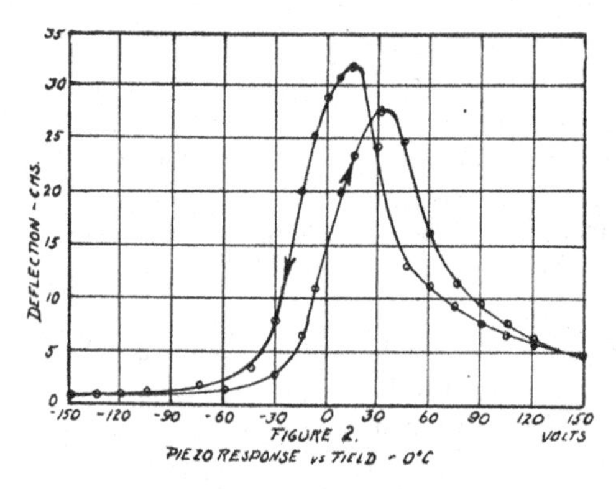

Fig. 2.

A corresponding dielectric effect is indicated by the double value of the condenser charge at zero field in the hysteresis loops. This is clearly due to a fatigue in the polarization and it also dies off exponentially with time. Herein is probably found the explanation of the "storage battery effect" described by W. G. Cady who observed that after applying a field of 100 volts for some time there was, on removal of the field, a small current which decreased gradually and flowed from the crystal as from a miniature storage battery.

The piezo-electric fatigue may well be a direct result of the fatigue in the polarization, as there seems to be a close relation between piezo-electric activity and polarization. It appears that the activity is approximately proportional to the rate of change of polarization with applied field and hence proportional to the dielectric constant. An examination of the temperature variation of the two quantities leads to this conclusion. It is further confirmed as regards field variation by the fact that the relation of activity to applied field is like $\partial D/\partial E$ vs. E where $\partial D/\partial E$ is merely the instantaneous value of the dielectric constant κ. As an approximation we can write the piezo-electric modulus δ proportional to κ:

$$\delta = A \cdot \kappa.$$

If this equation were exact A would be a fundamental piezo-electric constant of the substance, being of the order of 1×10^{-7} between $-20°$ C. and $+20°$ C. At some temperatures and for some exceptional specimens the relation does not seem to be so simple.

Effect of Moisture on Piezo-electric Properties.

In order to investigate the effect of dryness on the activity of Rochelle salt, some phosphorus pentoxide was enclosed in the chamber containing the crystal. The crystal soon started to dehydrate and after a few days was covered by a white coating. The piezo-electric throw for a load of 250 grams continually diminished. When the response was tested at different fields a more interesting fact was observed. Besides the decrease in response, the maxima were displaced along the field axis into a condition of greater polarization. This is shown by Fig. 3, the curves

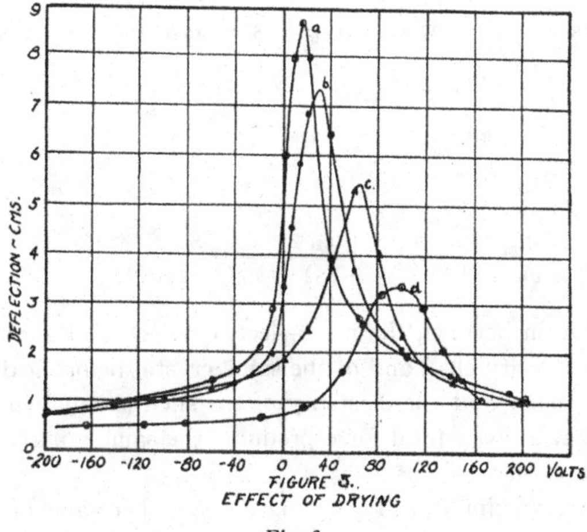

Fig. 3.

being taken after the lapse of the following times: (b) 1 day, (c) 3 days, (d) 12 days.

The decrease in the maxima and also their displacement is in the same direction as, and may be entirely due to, the effect of different dielectric properties of the crystal and of the dehydrated layer. In other words the presence of a layer of inactive dielectric of relatively low specific inductive capacity will diminish the charge on the plates due to the polarization of the central active layer, and thus decrease the piezo-electric response. It will also diminish the effective field across the active layer making it necessary to increase the potential difference between the plates to produce the same field across the inner layer, thus shifting the position of maximum activity. The effects due to uniform layers can be readily calculated. Let P_0 be the polarization produced in the middle layer by pressure, let P_1 and P_2 be the electrically induced polarizations in the dielectrics 1 and 2 respectively (Fig. 4). Since the dielectric displacement is solenoidal we have:

$$D' = E_1 + 4\pi P_1 + 4\pi P_0 = E_2 + 4\pi P_2 = 4\pi\sigma.$$

Since

$$P_1 = (\kappa_1 - 1)E_1 \quad \text{and} \quad P_2 = (\kappa_2 - 1)E_2$$

we can write

$$D' = \kappa_1 E_1 + 4\pi P_0 = \kappa_2 E_2.$$

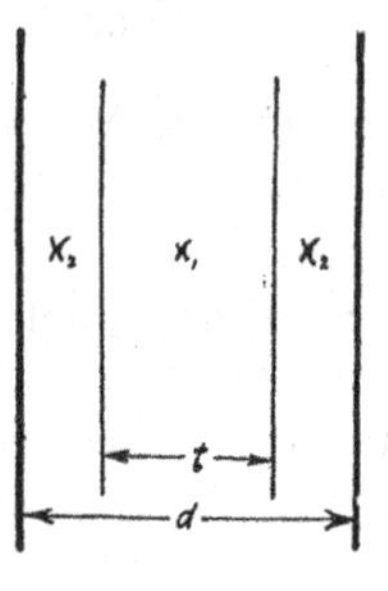

Fig. 4.

The difference of potential between the plates is zero so that, replacing P_0 by σ_0:

$$0 = E_2(d - t) + E_1 t$$
$$= \frac{4\pi\sigma}{\kappa_2}(d - t) + \frac{4\pi}{\kappa_1}(\sigma - \sigma_0)t,$$

giving:

$$\frac{\sigma}{\sigma_0} = \frac{Q'}{Q} = \frac{\kappa_2 t}{t(\kappa_2 - \kappa_1) + \kappa_1 d}.$$

This gives us a relation between the piezo-electric response at zero field of the crystal with the dry shell and of the same crystal before it dried. The assumption is made that the elasticity of the shell is equal to that of the crystal so that a given total force produces the same polarization in the crystalline portion.

The position of the maximum will be changed to another value of total potential difference on the crystal. Let V be the total potential drop and V' be the drop across the crystalline part. When there is no dehydrated layer present

$$V = V' = dE,$$

where E is the field strength in the dielectric. When there is a layer of uniform thickness $(d - t)/2$ on both faces then

$$V = (d - t)E'' + tE',$$

where E'' and E' are the field strengths in the dielectrics 2 and 1 respectively. The dielectric displacement

$$D = \kappa_2 E'' = \kappa_1 E' + 4\pi P_0,$$

is solenoidal, and we can eliminate E'' from equations above and write:

$$V' = dE = (d - t)\frac{\kappa_1}{\kappa_2}E' + tE' + 4\pi P_0\left(\frac{d - t}{\kappa_2}\right).$$

Since the last term is small compared to the rest of the expression, this gives:

$$\frac{E'}{E} = \frac{\kappa_2 d}{t(\kappa_2 - \kappa_1) + d\kappa_1}.$$

The following quantities were measured and substituted in these equations.

$$\kappa_1 = 1000, \qquad d = 0.22 \text{ cm.},$$
$$\kappa_2 = 180, \qquad t = 0.14.$$

The quantity t is an average obtained by breaking the crystal in several places and it is probably not very accurate because of the irregularity of the outer layer. We should, however, get a rough check on the plausibility of the proposed explanation. We find that

$$\frac{Q'}{Q} = 0.24$$

and that

$$\frac{E'}{E} = 0.38.$$

While the values of Q'/Q and E'/E from the maxima of curves a and d of Fig. 3 are respectively 0.39 and 0.33. The agreement is not as good as could be desired even after making allowance for the difficulties in estimating t. Possibly there is a true humidity effect with respect to piezo-electric activity but the above shows, at least, that it is quite small.

Piezo-electric Activity and Temperature.

In order to investigate the variation of activity with temperature, the chamber holding the crystal was immersed in CO_2 snow. After everything was thoroughly cooled and at $-75°$ C., the chamber was allowed to heat up. Above $-35°$ C. an electric heater was used. It was wound on a glass jar and insulated from the crystal chamber by a felt jacket. This jar was immersed in an oil bath to steady the heating rate while the felt eliminated any rapid changes of heating of the crystal. The current was gradually increased so as to keep the rate of heating uniformly between $\frac{1}{2}°$ and $1°$ C. per minute so as to eliminate thermoelastic stresses. The temperature was measured by means of a copper-constantan thermocouple directly soldered to an electrode on the crystal. In this way the actual temperature of the crystal itself was measured.

When the piezo-electric response or galvanometer throw for 250 grams was measured at the various temperatures for the first specimen the curve of Fig. 5 was obtained. This was duplicated to check the second

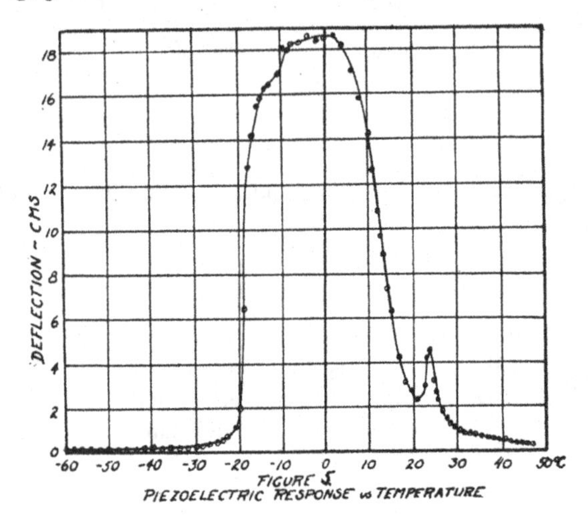

Fig. 5.

maximum. At − 70° C. the piezo-electric activity is comparatively negligible. As the temperature is raised slowly the activity stays small until − 30° C. is reached. At − 20° C. it is rising very rapidly, reaching a maximum at about 0° C. It decreases again but at 23° C. comes to a small but sharp maximum from which it diminishes slowly, becoming very small at + 50° C. The magnitude of the second maximum varies with the temperature at which heating begins. This second maximum was found in the case of two crystal plates provided with tinfoil electrodes attached by shellac.

Three other crystals were prepared with electrodes of mercury held against the crystal by two rectangular cups attached by wax. The thermocouple wires were immersed in the mercury. None of the crystals so prepared gave the second maximum. Moreover, none of them were as active as those used above. The variation of piezo-electric response of these specimens is shown in Fig. 6. The increase at − 30° to − 15° C. and the decrease at + 20° C. to + 30° C. are remarkably consistent. Between − 15° and + 20° C., however, they each show different characteristics. These mercury electrode crystals seemed to give more constant results than the crystals with tinfoil electrodes attached by shellac.

It was then suspected that the increased response of the crystal with

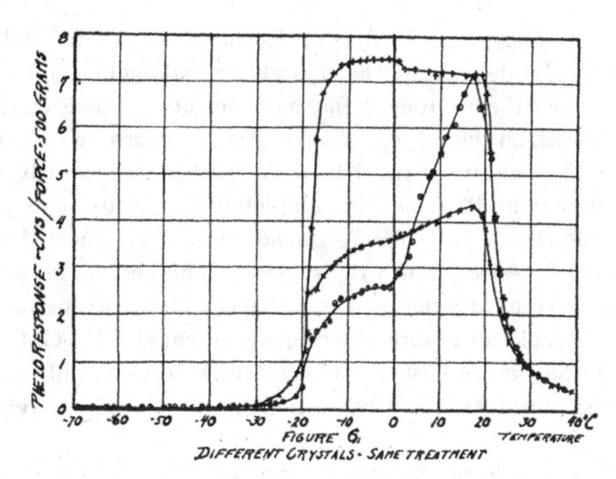

Fig. 6.

tinfoil electrodes and the presence of the second maximum was in some way due to the penetration into the crystal of the alcohol solvent of shellac. Accordingly one of the crystals originally with mercury electrodes was provided with the other type. As soon as the shellac was sufficiently dry, Curve *b*, Fig. 7 was obtained, the response originally

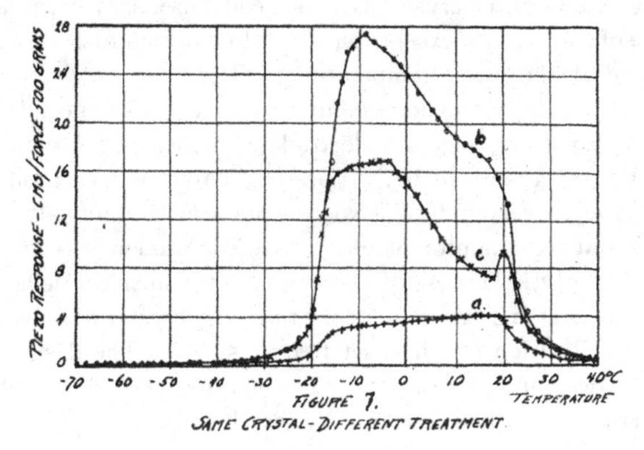

Fig. 7.

having followed Curve *a*. The sensitivity increased seven-fold at some temperatures but there was no second maximum. In two weeks the characteristics had changed to those shown in Curve *c*, seemingly checking the suspicion that alcohol was responsible for the second maximum and for the increased sensitivity. It would be interesting to use alcohol cup electrodes and investigate the continuous effect of alcohol soaking into the crystal. The effect is probably not chemical.

In a paper on the piezo-electric effect on Rochelle salt, A. M. Nicolson[1] describes a method for desiccating the crystals by soaking in alcohol and heating, thus making them stronger and more sensitive. The writer is certain, from his work on the subject, that complete desiccation will make the crystal entirely inactive. In the above method, apparently, the treatment is not prolonged enough to completely dehydrate more than a shell around the crystal and its effectiveness may be connected with the penetration of the alcohol into the crystal. The heating at 40° C. may also be effective in allowing a rearrangement and recrystallization of some of the molecules or groups not properly oriented. W. G. Cady, as well as the writer, has observed that heating the crystal will usually increase its sensitivity permanently, although sometimes the reverse is true.

An interesting side-light on the temperature variation of piezo-electric activity is offered by a study of curves like that of Fig. 2 but at different temperatures. They show that the effect of temperature is not so much to change the piezo-electric activity as to shift the position of the maximum from one value of the field to another. This is probably connected with the variation of the dielectric constant with temperature which will be taken up presently.

The charging throws of the crystal used as a condenser show variations similar to those of the activity except that they do not tend to zero but to a constant value at the lower temperatures. The crystals giving the second maximum on the piezo-electric curve show a similar peculiarity here. The crystals with the mercury electrodes give more regular curves. At 20° to 25° C. the crystals of both types begin to conduct, causing a steady drift of the galvanometer. Experiments seem to indicate that Ohm's law holds at least approximately. The conduction was at first thought to be electrolytic because of the manner in which Rochelle salt melts. Instead of real melting it appears that the crystal dissolves in its water of crystallization which is set free at 55° C. The desiccated crystal, however, decomposes into a tarry product and emits heavy white fumes above 150° C., without melting. The dehydrated crystal also begins to conduct above 20° C., making it probable that electronic and not electrolytic conduction is observed.

Measurements of conductivity were made on the natural crystal at various temperatures up to its liquefying point. The values obtained after the conductivity was sufficiently large to use a Wheatstone bridge are as follows:

[1] A. M. Nicolson, Proc. Am. Inst. Ele. Eng., Vol. 38, p. 1315 (1919).

TABLE I.

Temperature.	Conductivity.
Less than 43° C.	Less than 0.5×10^{-8} mhos/cm³.
43	0.5×10^{-8}
45	1.0×10^{-8}
47	0.3×10^{-7}
49	0.5×10^{-7}
51	0.5×10^{-6}
53	0.6×10^{-4}
54	1.7×10^{-4}
57	5.0×10^{-4}
Greater than 57	5.0×10^{-4}

At temperatures below 20° the dry crystal is a fairly good insulator having a specific conductivity of 5×10^{-12} mhos/cm.³ at 0° C. The conductivity decreases slightly at still lower temperatures. In all these measurements the surfaces were thoroughly dried by the presence of phosphorus pentoxide in the crystal chamber.

TABLE II.

Temp. Cent.	Dielectric Constant.		Piezoelectric Modulus.	
	A	*B*	*A*	*B*
−70	71	42	0.041×10^{-5}	0.017×10^{-5}
−50	85	50	0.068 "	0.017 "
−30	140	146	0.41 "	0.065 "
−20	386	252	5.4 "	1.08 "
−10	943	924	18.9 "	6.07 "
0	1,380	956	22.9 "	6.75 "
10	1,100	928	18.9 "	7.42 "
20	688	645	13.5 "	8.10 "
30	423	146	2.2 "	1.08 "
	Conduction commences		0.41 "	0.41 "

Table II. gives values of the dielectric constants for a field changing from 0 to 880 volts/cm. and of the piezo-electric modulus δ_{14} for shearing stresses of 220 grams/cm.². The modulus δ_{14} is defined by the relation given by Voigt $\sigma_1 = -\delta_{14} Y_z$ where σ_1 is the surface density of charge and Y_z is the shearing stress producing it. The given values are thought to be the most representative in each case. They are subdivided into two classes, according to whether the electrodes were tinfoil attached by shellac (column *A*) or mercury in direct contact with the crystal (column *B*). The former method is the most convenient to use in general practice, but the latter is thought to give more exactly the properties of Rochelle salt in the direction of the ă axis. The dielectric constants are surprisingly large, a fact noticed by Pockels[1] who supposes that this is due to "internal conductivity." The writer however has measured separately the conductivity at these temperatures and is of the opinion that this is a true dielectric constant arising from polarization of the dielectric, and for this reason being so closely related to the piezo-electric effect. Because of the relatively low specific inductive capacity of the desiccated crystal it is thought that the high specific inductive capacity of Rochelle salt is partly due to the water molecule.

Magnetic Analogy.

There seems to be a strong analogy between the behavior of Rochelle salt as a dielectric possessing hysteresis and having an exceptionally large dielectric constant, and the phenomena of ferromagnetism. Accordingly some of the features of Weiss's theory of magnetism may find their counterpart in the phenomena in Rochelle salt. Weiss[2] plots the susceptibility against the reciprocal of the absolute temperature and finds that the curve may be represented by a succession of straight lines. He interprets the sudden changes in slope as due to changes in the number of magnetons. If the data of Figs. 8*a* and 12 are plotted against the reciprocal of the absolute temperature one likewise gets what may be considered to be a succession of straight lines. Actually however there occur rounded corners where the curves suddenly change direction. This may be due to slight non-uniformity in heating which occurred in spite of the precautions taken. It is considered that the straight portions are at least as definite as those shown in Weiss's paper. The most abrupt changes are at $-20°$ C. and at $+20°$ C. These may accordingly be considered as the "Curie points" in Rochelle salt.

Relation to Optical Properties.

Some of the optical properties of Rochelle salt can be at least approximately found from the electrical data given. In the course of this calculation it is of course necessary to introduce some assumptions notably as to the nature of the permanent polarization. If one knew just how the permanent polarization was produced he could find the free period of this mechanism. The data needed are the force per unit displacement f and the mass m of that part of the molecule. The wave-length corresponding to the free period is given by the expression

$$\lambda = 2\pi c\sqrt{\frac{m}{f}}$$

c being the velocity of light.

[1] Pockels, Lehrbuch der Krystaloptik, p. 508.

[2] P. Weiss, J. de Physique, Vol. 1, p. 968 (1911).

Let us assume for simplicity that only one electron is involved in the creation of the permanent moment. The quantity f can be derived from the value of the permanent moment $P_0 = 30$ e.s.u./cm.^{2} and from the displacement of the electron producing it. Since the force on an electron inside a dielectiic of polarization P_0 is roughly equal to $\frac{1}{3}P_0e$[1] the expression for the wave-length may be put in the form:

$$\lambda = 2\pi c\sqrt{\frac{3md}{P_0e}}\,.$$

The value of d, the displacement of the electron, can be found as follows:

Taking 30 e.s.u./cm.^{2} as the natural polarization, the moment per molecule is obtained by dividing by the number of molecules per c.c., the result being 7.1×10^{-21} e.s.u. In each of these molecules there are 140 electrons, this being the sum of the atomic numbers of the constituent elements in Rochelle salt. If we suppose as above that only one of these is effective in producing the piezo-electric moment, its displacement from the center of force of the rest of the molecule will be $d = 2.7 \times 10^{-11}$ cm. It would of course be more reasonable to suppose that at least several of the electrons are displaced by different amounts, and to the extent that we do this the value calculated above becomes smaller.

Using these values of P_0 and d for the simple case treated above we find for the wave-length the value:

$$\lambda = 4.2\ \mu.$$

Coblentz[2] shows the presence of fairly strong absorption in this region of the infra-red. This may, however, be due to the water of crystallization and not to the cause cited above. These two possibilities should be distinguishable experimentally because the character of the absorption due to these electrons should change greatly with the temperature, as the piezo-electric elasticity or force per unit displacement of the electrons changes.

The natural period as found above should be the same as that involved in rotatory dispersion formulæ, since both the piezo-electric effect and optical rotation are due to an unsymmetrical or twisted structure of the molecule. J. J. Thomson[3] gives an approximate formula for the specific rotation, namely:

$$\rho = \frac{e^2dp^2}{c^2Mmn^2}\,,$$

[1] H. A. Lorentz, Theory of Electrons, p. 306.
[2] W. W. Coblentz, Infra Red Spectra, Vol. 2, p. 38.
J. J. Thomson, Phil. Mag., Dec., 1920.

in which e is the charge of the electron, c is the velocity of light and M and m are the masses of the molecule and of the electron, d is the radius of the molecule, p is the free period, and n is the frequency for which ρ is to be calculated. Using the value of p derived from piezo-electric data we find for sodium light the specific rotation of the order of magnitude of 10°, the tables giving 22.1°. Considering the fact that so little is known of the structure of the Rochelle salt molecule, the approximation is fair.

The writer is indebted to Professor W. F. G. Swann, who initiated this research and gave many helpful suggestions, and to Dr. W. R. Whitney, Director of the Research Laboratory of the General Electric Company, whose presentation of two beautiful crystals made the work possible.

Physical Laboratory,
University of Minnesota.

FEBRUARY 1, 1930 *PHYSICAL REVIEW* *VOLUME 35*

ROCHELLE SALT AS A DIELECTRIC

By C. B. Sawyer and C. H. Tower
The Brush Laboratories, Cleveland

(Received November 6, 1929)

Abstract

Both saturation and hysteresis appear in Braun tube oscillograms made at various temperatures with a condenser whose dielectric consists of Rochelle salt slabs cut perpendicular to the a-axis. The dielectric constant for such slabs may reach a value of 18,000. Curves are also given, showing the variation in mechanical and electrical saturation with temperature. These correspond in only a general way to the piezoelectric constant's variation with temperature. Certain marked peculiarities are noted in the resulting mechanical deformation when Rochelle salt is excited with alternating potentials. Clear Rochelle salt half-crystals have been produced up to forty-five centimeters in length.

THE remarkable physical properties of Rochelle salt, the most piezoelectric active of all crystalline substances, have been reported by other authors.[1] Comparatively small plates and few crystals were used in their determinations.

Work at this laboratory has been carried on for a number of years on Rochelle salt with a view towards commercialization. It has, therefore, been necessary to produce large clear crystals in quantity. Clear, flawless half-crystals are grown up to 45 cm in length and 2 kg in weight.

The dielectric strength and insulation value of plates from such crystals is very high. Many hundreds of plates (mostly perpendicular to the a-axis of the crystal) have been produced and their electrical properties measured. Thus a Rochelle salt plate 4.75 mm thick shows a dielectric constant of 18,000 when tested at 15°C at 60 volts 60 cycles alternating current. The highest previously reported value which has come to the attention of the authors is about 1380.[2] An air condenser of area and capacity equal to that of the crystal plate would have a plate separation of only 0.00475 mm, if 1380 be taken as the crystal dielectric constant; and 0.000262 mm (0.0001″) if 18,000 be taken. It is thus evident that a comparatively thin layer of cement or dehydrated Rochelle salt between the body of the crystal and the foil electrode will introduce a very large error in the determination of the dielectric constant. Any adhesive such as balsam in xylol, Japan Gold size, or beeswax dissolved in benzol with a small addition of rosin, may be used in dilute solution for atttaching the foil. It is important, subsequently, to

[1] Frayne, Phys. Rev. **21,** 348 (1923); Isley, Phys. Rev. **24,** 569 (1924); Laurey and Morgan, J. Am. Chem. Soc. **46,** 2192–6, (1924); Pockels, Encyklopadie der Math. Wiss. Vol. 5, Part 2; Valasek, Phys. Rev. **17,** 475 (1921); **19,** 478 (1922); **20,** 639 (1922) and **24,** 560 (1924). Voigt, Lehrbuch der Kristallphysik, Chap. 8, Leipsig (1910).

[2] Valasek, Phys. Rev. **19,** 488 (1922).

rub down the foil very thoroughly to bring it as close to the crystal surface as possible.

A series of Braun tube oscillograms was obtained. For this purpose, and for all other results reported in this paper, a crystal plate was employed measuring about 8.5×5.5×0.5 cm, cut with its plane perpendicular to the

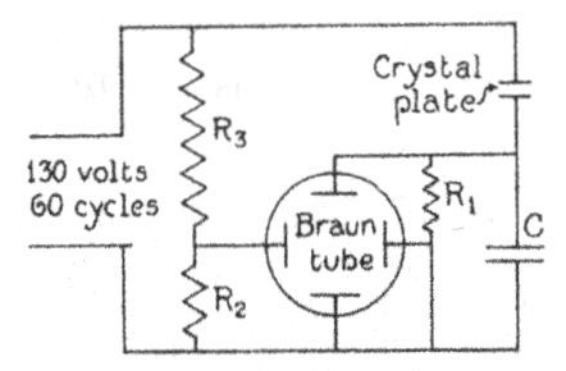

Fig. 1. Schematic connection of Braun tube. Capacity of crystal plate 0.004 to 0.2 Mf; of C 0.7 Mf. R_1=0.45 megohms; R_2=3180 ohms; R_3=31800 ohms.

a-axis, and its long edges parallel to the c-axis. All measurements and oscillograms were carried out with 60 cycle current from the power lines. All vertical deflections are on the same scale as in Fig. 2.

Fig. 1 shows the connections employed with the Braun tube for obtaining crystal oscillograms. At the left is a resistance acting as a voltage divider. To the right is the crystal plate under test, connected in series with a con-

Free

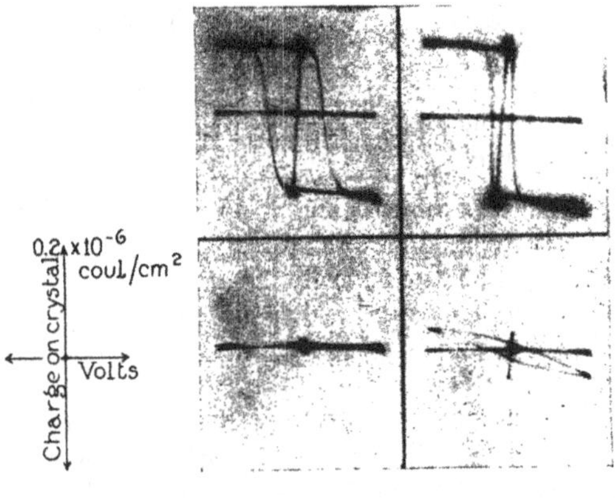

0.2 × 10^{-6} coul/cm^2

Charge on crystal

Volts

387 v. per cm 1161 v. per cm

Restrained

Fig. 2. Oscillograms of crystal plate, free and restrained. Temperature 15° C; frequency 60 cycles per sec.

denser giving voltages proportional to the charge on the crystal. The resulting oscillograms, such as shown in Fig. 2, have ordinates proportional to crystal charge and abscissae proportional to applied voltage.

Fig. 2 shows comparative oscillograms of a plate when entirely free, and of the same plate restrained by cementing it between two thick aluminum plates, thus very largely precluding mechanical motion due to piezo-activity. The left-hand vertical pair is for 387 peak volts per cm; the right-hand vertical pair is for 1161 peak volts per cm. The upper pair is unrestrained; the lower pair is restrained. Dielectric constants calculated from these oscillograms show exceedingly interesting and suggestive values: from the left-hand oscillogram (restrained plate) about 430; from the left-hand oscillogram (free plate) in the saturated range about 330; from the same oscillogram for a complete cycle, excluding saturation range 10,500; and for maximum instantaneous value not less than 200,000.

Such enormous values of the dielectric constant in connection with less efficient foiling of the crystal plates, may account in part for the previously observed storage battery effect. Supplementary tests indicate that little

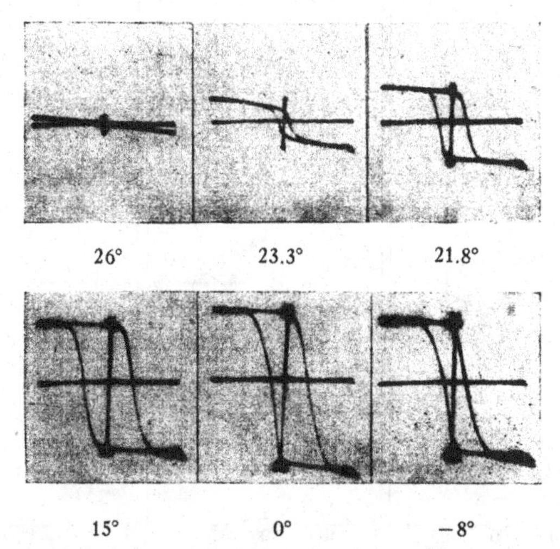

Fig. 3. Hysteresis and saturation of Rochelle salt plate II. Potential gradient in dielectric 387 (peak) volts per cm; frequency 60 cycles per sec.

change in the value of the dielectric constant is to be looked for as a result of improvement in foiling. In these supplementary tests, electrodes of saturated Rochelle salt solution were used and results did not differ significantly from those obtained from a carefully foiled plate. Moreover these large values of the dielectric constant of Rochelle salt have been observed in many hundreds of plates of various dimensions from many different crystals. Determination of the dielectric constant was usually made by applying 112 volts of 60 cycles alternating potential to the free crystal plate and noting the resulting current. In addition, circuit resonance and condenser substitution methods served to check this first method, all three giving results in substantial agreement.

Fig. 3 comprises a series of comparative oscillograms made from the same crystal plate at different temperatures as indicated. Proceeding from top left to bottom right it is evident that as the temperature is decreased both the voltages and charges required for saturation greatly increase. So also do the areas of the hysteresis loop. Here again the method of applying the foil electrodes to the crystal is of great importance as the shape and area of the loop will vary somewhat with this factor.

All of the oscillograms were made with the greatest care. A second crystal plate gave results identical with the first. Two other plates of the same dimensions as before but with their long edges cut at 45° to the c-axis, showed no essential differences in the derived oscillograms. Though no special humidity precautions were observed, the resistance of the plates, at 100 volts constant potential, never fell below many megohms.

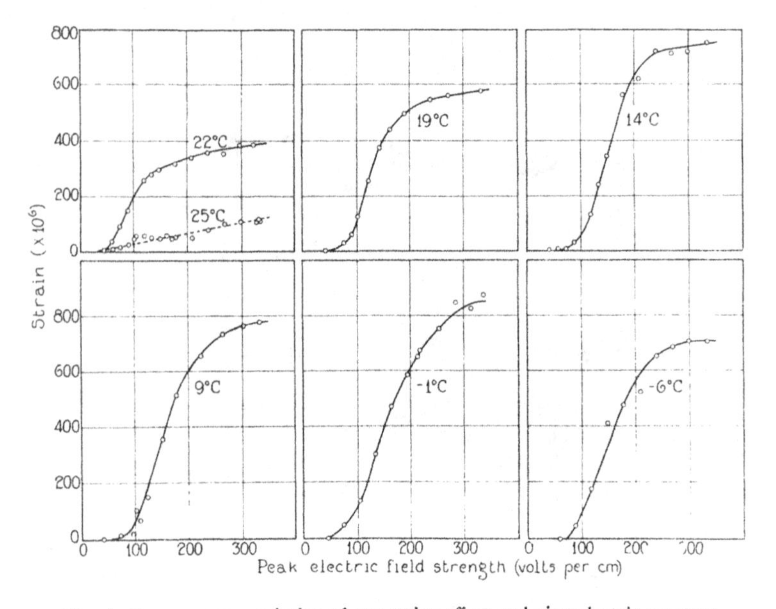

Fig. 4. Temperature variation of saturation effect and piezoelectric constant.

If a standard plate—its long edge being cut parallel to c-axis—is electrified with an alternating potential, it will be deformed and such deformation can be observed and measured conveniently with a microscope. For the results shown in Fig. 4, one short edge of the plate was cemented to a large lead block and various values of 60 cycle potential were applied to the electrodes. The alternating motion produced under these conditions lies in the plane of the plate and is perpendicular to the c-axis. The relation between total deformation and electrification is shown for various temperatures.

Saturation is again in evidence and saturation values again increase greatly with decrease in temperature. Keeping close pace with it is the voltage required to produce saturation. But it is very noteworthy that considerable voltage must be applied before the crystal shows appreciable deformation.

Fig. 5 shows the close relationship existing at different temperatures between: 1st, volts per cm required for mechanical saturation; 2nd, the energy loss per cubic centimeter per cycle; 3rd, the charge per cubic centimeter required for electrical saturation. Though not shown in this figure, these curves are followed closely by those of the voltage required for electrical saturation and of the deformation at mechanical saturation. No determina-

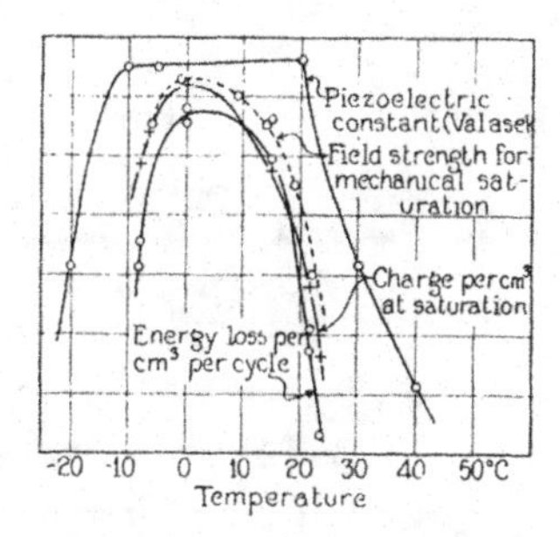

Fig. 5. Properties of Rochelle salt at various temperatures.

tions of the piezoelectric constant were made, but Valasek's[3] most recent curve of the temperature variation of the piezo-electric constant is included for the sake of comparing temperature variation of this property with those of the others.

It has been the great privilege of the authors to carry on this work begun under the very able leadership of the late Charles F. Brush, Jr.

[3] Valasek, Science Vol. LXV No. 1679, p. 235 (1927).

Free

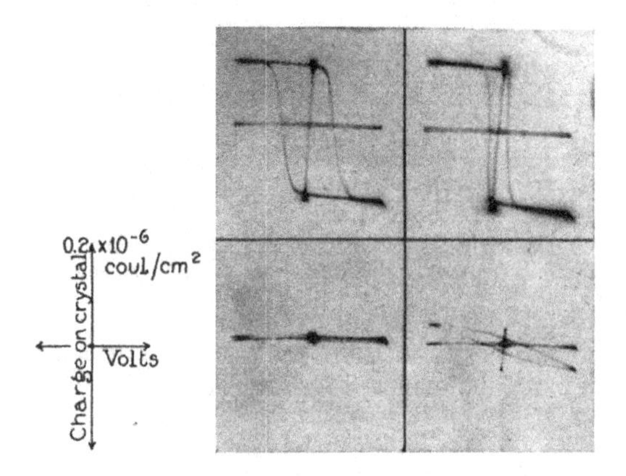

387 v. per cm   1161 v. per cm

Restrained

Fig. 2. Oscillograms of crystal plate, free and restrained. Temperature 15° C; frequency 60 cycles per sec.

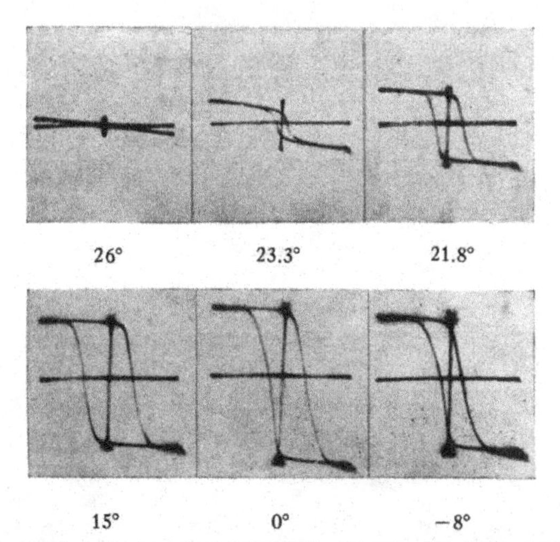

15° 0° −8°

Fig. 3. Hysteresis and saturation of Rochelle salt plate II. Potential gradient i dielectric 387 (peak) volts per cm; frequency 60 cycles per sec.

Ferroelectrics, 1987, Vol. 71, pp. 15–16
Photocopying permitted by license only

A NEW SEIGNETTE-ELECTRIC SUBSTANCE

Translated from *Naturwiss.* **23** 737 (1935) *by G. Busch*

G. BUSCH and P. SCHERRER
Physikalisches Institut der Technischen Hochschule, Zürich,

(Received August 26, 1935)

So far the anomalous dielectric and piezoelectric properties of Rochellesalt have also been observed only on mixed crystals of potassium–sodium– and potassium–ammonium–tartrate. For a better understanding of the phenomenon of seignette-electricity it would be very helpful to find other substances with analogous properties, but if possible with a simpler crystal structure than Rochellesalt. In this connection the question whether crystal water is of importance or even crucial for inducing seignette-electricity is of special interest.

Systematic experimental research, based on a theoretical model have lead us to a new substance with a temperature variation of the dielectric constant similar to Rochellesalt. This is the primary potassium phosphate (KH_2PO_4) crystallising with tetragonal-scalenoedric symmetry, but containing no crystal water. The principal dielectric constant ε_{33}, measured parallel to the crystallographic c-axes is of special interest. Figure 1 shows its temperature-variation as a result of measurements with 50 cycles a.c. at a field strength of 1000 V/cm. At room temperature ε_{33} has a value of 30 approximately, then rises steeply, reaches a maximal value of 155 at −130°C and stays approximately constant down to −190°C. Below −200°C ε_{33} drops again to 7 approximately. Like Rochellesalt KH_2PO_4 exhibits two Curie points θ_1 and θ_2 at −130°C and −195°C respectively. Although the absolute values of the dielectric constants are much lower than for Rochellesalt, the typical temperature variation is the same.

A qualitative check of the piezo-electric activity based on Giebe and Scheibe's method revealed the same temperature variation of the piezo-electric modul d_{33} down to liquid air temperature. Debye–Scherrer diagrams did not show any significant change of the crystal structure at the upper Curie point.

The crystals were grown from an aqueous solution saturated at the boiling point by slowly lowering the temperature. Plates of 1 cm^2 and a thickness of 1 mm were covered with aluminum foils. The dielectric constant was determined with an ac-capacity bridge and the temperatures were measured by means of a copper-constantan thermocouple.

Further investigations, as for instance the influence of the field strength and the exact variation of the piezo-electric constants with temperature are in preparation. Investigations of the isomorphous crystals $NH_4H_2PO_4$, KH_2AsO_4 and $NH_4H_2AsO_4$ are planned.

The measurements at temperatures below the boiling point of liquid air were

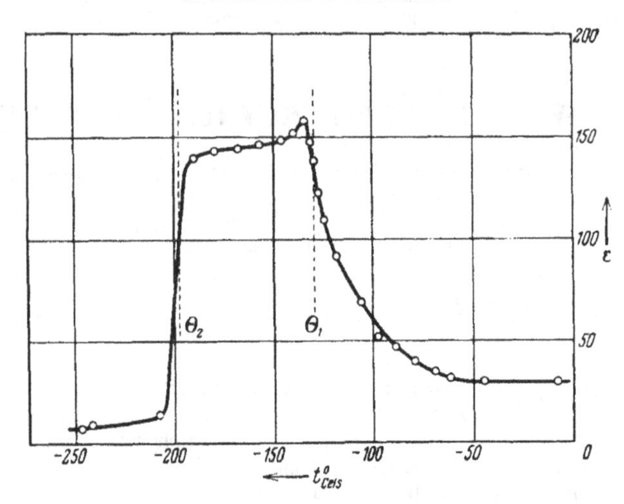

FIGURE 1 Temperature variation of the dielectric constant ε_{33} of KH_2PO_4.

carried out at Charlottenburg, because no liquid hydrogen was available in Zürich. We are indepted to Professor Stark, president of the "Physikalisch-Technische Reichsanstalt" at Berlin for providing liquid hydrogen and to Prof. Westphal and Dr. J. Engl for a place to work at the Physics Institute of the "Technische Hochschule" at Charlottenburg.

PHYSICAL REVIEW VOLUME 74, NUMBER 11 DECEMBER 1, 1948

Theoretical Model for Explaining the Ferroelectric Effect in Barium Titanate

W. P. MASON AND B. T. MATTHIAS
Bell Telephone Laboratories, Murray Hill, New Jersey

(Received August 26, 1948)

In order to explain the properties of a barium titanate single domain crystal, a previous theory of the ferroelectric effect in rochelle salt has been extended to the three-dimensional structure of barium titanate. This involves six equilibrium positions and results in significant differences from the single bond type of structure of rochelle salt. The theoretical features considered are a calculation of the spontaneous polarization as a function of temperature, the dielectric constants along the $a=y$ and $c=z$ axes as a function of temperature, the relaxation of the dielectric constant at high frequencies, and the hysteresis loops. All of these features are explained by the three-dimensional model considered here.

IN a previous paper,[1] a theoretical explanation was given for the ferroelectric effect in rochelle salt, which depended on the motion of a hydrogen nucleus between the two equilibrium positions of a hydrogen bond. It is the purpose of this paper to show that the principal features of the barium titanate single domain crystal can be explained by an extension of this model to the three-dimensional structure of barium titanate involving six equilibrium positions.

I. EXPERIMENTAL DATA

Barium titanate above the transition temperature of 120°C has the cubic cell shown by Fig. 1. The barium atoms occupy the corners of the cell, the oxygens the face-centered positions, while the titanium is usually pictured as being in the center of the cell. As a matter of fact, it probably makes a covalent bond with one of the face-centered oxygens and is displaced in the direction of that oxygen by about 0.16A[2] from the center of the cell. Above 120°C the thermal energy is sufficient to cause any one of the six positions to be equally probable and the cell appears to be cubic from x-ray measurements. Below 120°C thermal energy is no longer sufficient to cause any position to be equally probable, and most of the molecules in a given region or domain line up along one of the six directions, a dipole moment develops in that direction and the crystal becomes ferroelectric. The axis along which the titanium has been displaced becomes larger than the other two, as shown by the x-ray measurements of Miss Megaw[3] (as shown by Fig. 2) and the crystal changes from cubic to tetragonal form.

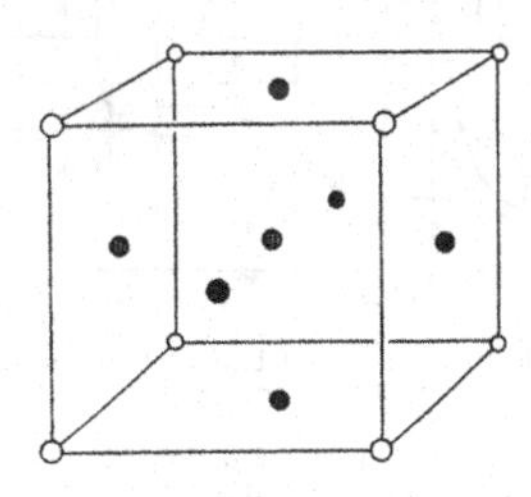

FIG. 1. Unit cell for barium titanate.

The dielectric measurements of multicrystalline ceramics, multi-domain crystals, and single domain crystals all show the presence of a ferroelectric material below 120°C. Dielectric displacement—electric field curves occur in the form of hysteresis loops. The dielectric constant at low field strengths for multicrystal ceramics,[4] as shown by Fig. 3, rises to a high value at the temperature of 120°C. Above 120 degrees, the dielectric constant follows a Curie-Weiss law approximately, and the dielectric constant decreases inversely as the difference between the tempera-

[1] W. P. Mason, Phys. Rev. 72, 854 (1947).

[2] This value for the displacement of the titanium atom from the center of the unit cell has recently been measured by x-ray methods by Gordon Danielson, Phys. Rev. 74, 986 (1948).

[3] H. D. Megaw, Proc. Roy. Soc. 189, 261–283 (1947).

[4] Von Hippel, Breckenridge, Chesley, and Tisza, Ind. Eng. Chem. 38, 1097–1109 (1946).

ture and the Curie temperature or

$$\epsilon = \epsilon_0 + C/(T - T_0), \qquad (1)$$

where ϵ_0 is the constant dielectric constant for temperatures much higher than the Curie temperature. C is a constant, T the temperature, and T_0 the Curie temperature. Below the Curie temperature the dielectric constant decreases from its high value to a value of about 350 near absolute zero. The steady decrease is interrupted at two temperatures 10°C and −70°C. At these temperatures no discontinuities occur in the axis length and hence these points cannot be associated with a change in dipole moment and hence with the position of the titanium nucleus. It has been suggested by Matthias and von Hippel[5] that these are due to a change from octahedral bonding of the titanium atom to a hybrid type of bonding which may become more probable at the lower temperature. Since this does not involve an appreciable change in the position of the titanium nucleus, this appears to be a reasonable suggestion. As the result is a small second-order change in the dielectric constant, it is neglected in the theory presented here.

The dielectric constant for multi-domain crystals is not too different from those for the multicrystalline ceramics. Figure 4 shows the measurements of Matthias and von Hippel[5] for the a and c axes. The dielectric constant along the a axis is higher than that along the c axis. The lowering of the Curie point is probably caused by the impurities introduced. By introducing larger amounts of mineralizers, single domain crystals of a relatively large size have recently been grown, and these show a very marked difference between the dielectric constants along the two axes. As shown by Fig. 5, the dielectric constant along the c axis is less than that for a ceramic material. When the dielectric constant along the a axis is measured over a frequency range, a relaxation occurs at about 15 megacycles and the dielectric constant drops to about 1200 or less, as shown by Fig. 6. A similar relaxation in the dielectric constant of the ceramic occurs at about 10^9 cycles as shown by the measurements of Nash[6] and Yager

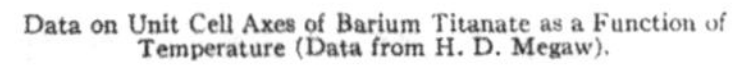
Data on Unit Cell Axes of Barium Titanate as a Function of Temperature (Data from H. D. Megaw).

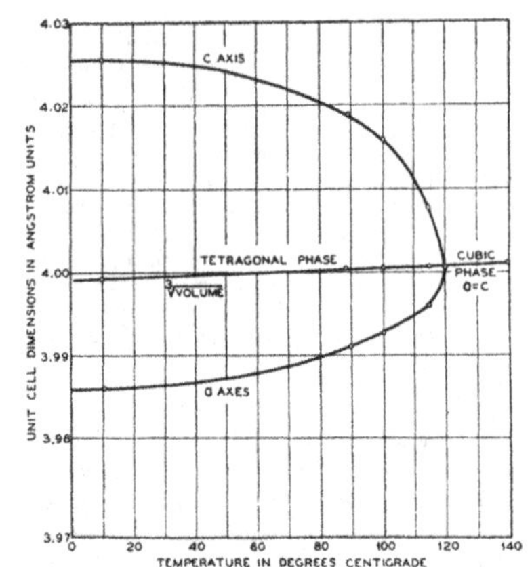

FIG. 2. Cell dimensions as a function of temperature.

(unpublished).[7] At 23.7-centimeter wave-lengths, the former found a dielectric constant and tanδ of

$$\epsilon = 1250 \text{ to } 1420, \quad \tan\delta \doteq 0.2, \qquad (2)$$

while at 1.25 centimeters Yager found a dielectric

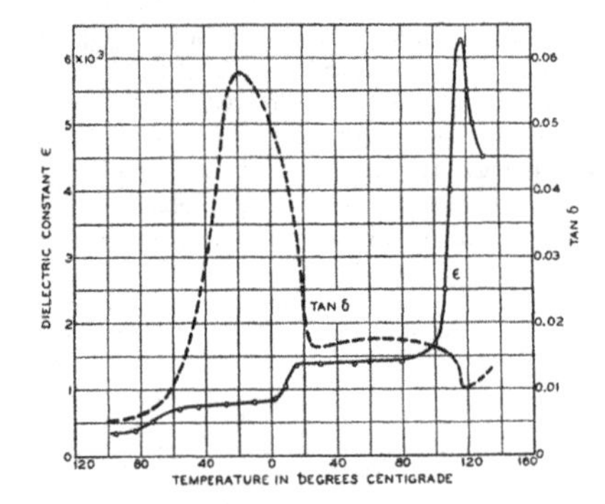

FIG. 3. Dielectric constant of barium titanate ceramic as a function of temperature.

[5] B. T. Matthias and A. von Hippel, Phys. Rev. **73**, 1378-1384 (1948).

[6] D. E. Nash, Jr., J. Exper. Theor. Phys. Acad. Sci. U.S.S.R. **17**, 537 (1947).

[7] The dielectric constants of barium titanate ceramics have recently been measured at 1.5 megacycles and 9450 megacycles over a temperature range from 20°C to 160°C by J. G. Powles of Imperial College of Science and Technology. The results are described in a note sent to Nature. From the variation of the relaxation frequency with temperature, one can calculate that the activation energy is 3.65 kilocalories per mole in fair agreement with the value found in Eq. (63).

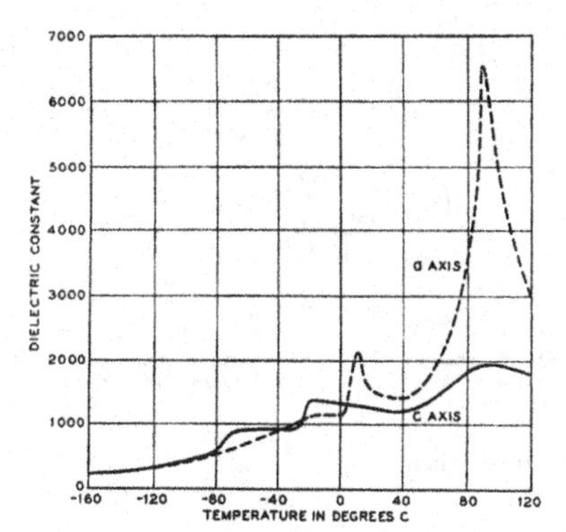

FIG. 4. Dielectric constants for the two crystallographic axes for multi-domain crystals of barium titanate.

constant of approximately

$$\epsilon = 250 \text{ to } 320, \quad \tan\delta \doteq 0.70. \qquad (3)$$

From these measurements it can be calculated that the dielectric constant has a relaxation frequency of about 6.2×10^9 cycles.

The relaxation of the dielectric constant at these frequencies shows definitely that the high dielectric constant is due to a temperature movable dipole rather than a high dielectric constant of the type due to the near vanishing of the factor $(1-\beta\gamma)$ in the dielectric equation

$$\frac{\epsilon-1}{4\pi} = \frac{\gamma}{1-\beta\gamma}, \qquad (4)$$

where γ is the polarizability and β the Lorentz factor, since the polarizability γ due to electrons, ions and atoms should not vary with frequency up to the infra-red frequencies. Hence, a temperature variable dipole of the type discussed in the next section is required to give a relaxation frequency as low as 15 megacycles.

II. SPONTANEOUS POLARIZATION AND DIELECTRIC CONSTANT UNDER EQUILIBRIUM CONDITIONS

The model considered here is the one shown by Fig. 7. Here there are six potential minima in the direction of the six oxygens which are displaced a distance δ from the center of the unit cell. If the titanium nucleus is taken from a position such as 1 to position 2 directly across the unit cell, the form of the potential barrier may be as shown by Fig. 8 in which ΔU represents the height of the potential curve at the center with respect to that at the minima. If the nucleus went directly from position 1 to position 3, it would in general have to cross a higher potential barrier than ΔU, but equilibrium between the two positions can be established by the nucleus jumping to a position slightly to one side of the center in the direction 3 and hence it is thought that the potential barrier determining the relaxation frequency for a 1 to 3 jump will not be much higher than for a 1 to 2 jump, namely ΔU.

For low frequencies, i.e., for frequencies well under the relaxation frequency, equilibrium values can be calculated by using Boltzmann's principle that the equilibrium ratios of numbers of nuclei in two potential wells are in the ratio

$$N_1/N_2 = e^{E/kT} \qquad (5)$$

where E is the potential difference between well 2 and well 1, k is Boltzmann's constant and T the absolute temperature.

Suppose now that all the minima of Fig. 7 have initially the same potential, which is set equal to zero. Then if we apply a field E_z in the z direction, a polarization P_z in this direction results. This polarization causes an internal field F of the Lorentz type given by the equation

$$F = E + \beta P \qquad (6)$$

where β is $4\pi/3$ for an isotropic material but will be much less than this when the titanium nucleus comes close to the oxygen atom. The total polarization consists of a part P_e due to electrons and atoms and a part P_d due to the dipole caused by the displacement of the titanium nucleus from

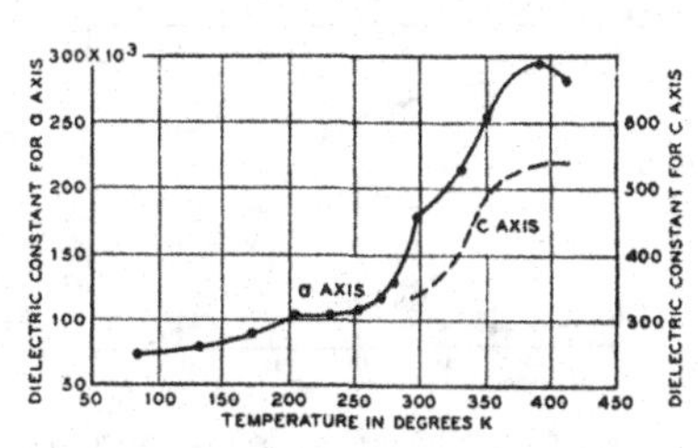

FIG. 5. Dielectric constants for a single domain crystal.

the mid-position of the unit cell. The dipole moment introduced by this change is

$$\mu = 4e\delta, \tag{7}$$

since the valence of the titanium is 4 for the structure, e is the electronic charge, and δ the distance the titanium nucleus moves in going from the center of the unit cell to the equilibrium position. An addition to the dipole may also result from a displacement of the oxygen in the direction of the titanium. The electronic and atomic polarization exerted will be proportional to the local field F, so that

$$F = E + \beta[P_e + P_d] = E + \beta[\gamma F + P_d]$$

or

$$F = \frac{E + \beta P_d}{1 - \beta\gamma} \tag{8}$$

where γ is the polarizability per unit volume due to all polarization except that of the titanium dipoles. The polarizability γ can be determined from the dielectric constant ϵ_0 measured at very low or very high temperatures, for since

$$(\epsilon_0 -)1/4\pi = P_E/E = \gamma F/E \tag{9}$$

and for P_d suppressed, $F = E/(1 - \beta\gamma)$, hence

$$\frac{\epsilon_0 - 1}{4\pi} = \frac{\gamma}{1 - \beta\gamma} \quad \text{and} \quad 4\pi\gamma = \frac{(\epsilon_0 - 1)}{1 + \dfrac{\beta}{4\pi}(\epsilon_0 - 1)}. \tag{10}$$

The dielectric constant ϵ_0 near absolute zero is

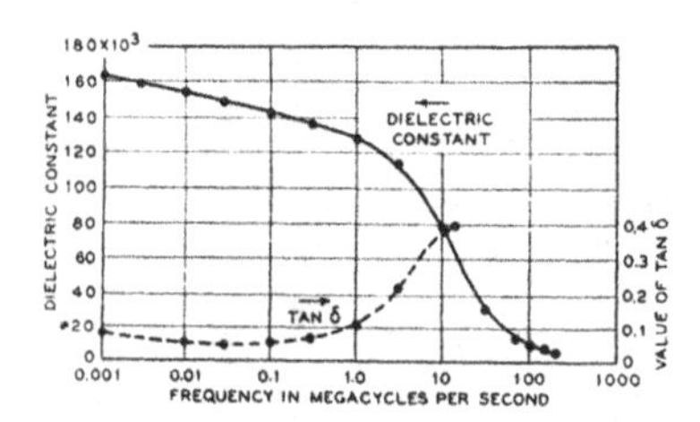

FIG. 6. Dielectric constant of an axis as a function of frequency.

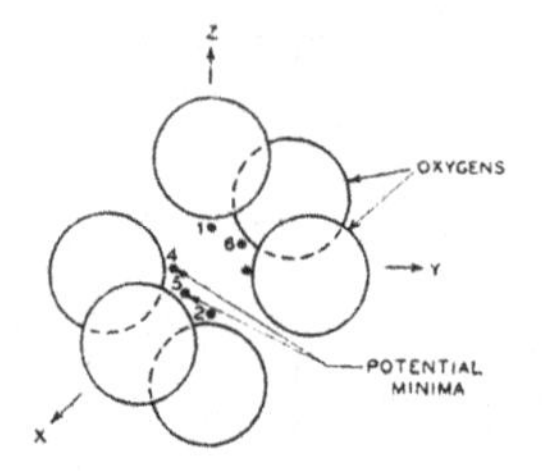

FIG. 7. Theoretical model for barium titanate, showing positions of oxygens and potential minima for the titanium nucleus.

about 350, hence

$$\frac{1}{1 - \beta\gamma} = 1 + \frac{\beta}{4\pi}(\epsilon_0 - 1) = 1 + \beta(27.8). \tag{11}$$

This internal field caused by the applied field E_z causes a decrease in the potential at the minima 1 and an increase in the potential at 2 equal, respectively, to

$$U_1 = -F\mu = -\left(\frac{E_z + \beta P_z}{1 - \beta\gamma}\right)\mu;$$
$$U_2 = +\left(\frac{E_z + \beta P_z}{1 - \beta\gamma}\right)\mu. \tag{12}$$

The potentials for the other four wells are unchanged by this field and hence,

$$U_3 = U_4 = U_5 = U_6 = 0. \tag{13}$$

By Boltzmann's principle (Eq. (5)), the relative number of nuclei in the six potential wells, all expressed relative to N_5 are

$$N_1 = N_5 \exp\left[\left(\frac{E_z + \beta P_z}{1 - \beta\gamma}\right)\frac{\mu}{kT}\right];$$
$$N_2 = N_5 \exp -\left[\left(\frac{E_z + \beta P_z}{1 - \beta\gamma}\right)\frac{\mu}{kT}\right]; \tag{14}$$
$$N_3 = N_4 = N_5 = N_6.$$

Then, since the total number of nuclei is equal to N where N is the number per cubic centimeter, we have

$$N = N_1 + N_2 + N_3 + N_4 + N_5 + N_6. \tag{15}$$

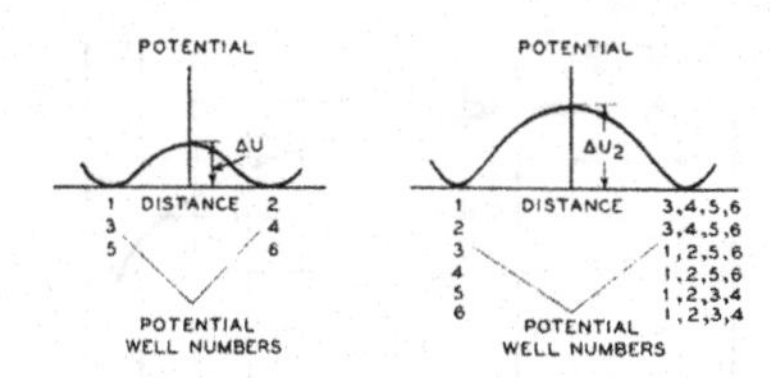

FIG. 8. Potential distribution as a function of distance from the center of the cell.

Substituting in the values from Eqs. (14) we have

$$N_1=\frac{N\exp\left[\left(\frac{E_z+\beta P_z}{1-\beta\gamma}\right)\frac{\mu}{kT}\right]}{2\left[2+\cosh\left(\frac{E_z+\beta P_z}{1-\beta\gamma}\right)\right]};$$

$$N_2=\frac{N\exp-\left[\left(\frac{E_z+\beta P_z}{1-\beta\gamma}\right)\frac{\mu}{kT}\right]}{2\left[2+\cosh\left(\frac{E_z+\beta P_z}{1-\beta\gamma}\right)\frac{\mu}{kT}\right]};\tag{16}$$

$$N_3=N_4=N_5=N_6=\frac{N}{2\left[2+\cosh\left(\frac{E_z+\beta P_z}{1-\beta\gamma}\right)\frac{\mu}{kT}\right]}.$$

The polarization of a dipole nature excited along the Z axis will be then

$$P_z=(N_1-N_2)\mu=\frac{N\mu\sinh[(E_z+\beta P_z)/(1-\beta\gamma)]\mu/kT}{2+\cosh[(E_z+\beta P_z)/(1-\beta\gamma)]\mu/kT}.\tag{17}$$

All the equilibrium values of spontaneous polarization, coercive fields, dielectric constants, etc. can be determined from this equation.

Let us first consider the condition for spontaneous polarization and the ferroelectric effect. This can be obtained by setting E_z equal to zero and determining the conditions for which the polarization P_z is different from zero. Setting E_z equal to zero and introducing the substitution

$$A=[\beta N\mu^2/(1-\beta\gamma)]1/kT.\tag{18}$$

Equation (17) becomes

$$\frac{P_z}{N\mu}=\frac{\sinh(AP_z/N\mu)}{2+\cosh(AP_z/N\mu)}.\tag{19}$$

Examining this equation, we see that $P_z/N\mu$ will have a solution different from zero only if A is equal to 3 or greater. If A is greater than 3, $P_z/N\mu$ can have a positive or negative value lying between zero and 1. This represents a spontaneous polarization along the positive or negative Z axis due to the internal field generated by charge displacements of the titanium nuclei from the central position. In general any one of the oxygen atoms can be considered as lying along the Z axis and only chance determines in which direction the spontaneous polarization occurs.

If we solve for $P_z/N\mu$ as a function of A, the relation shown by Fig. 9 results. This is a very much larger increase of $P_z/N\mu$ with increase in A than occurs for a single bond of the hydrogen bond type which is determined by an equation of the type

$$P_z/N\mu=\tanh(AP_z/N\mu).\tag{20}$$

The relative increase for this type is shown by the dashed line of Fig. 9 for the same percentage increase in A. Some confirmation for this sudden increase in polarization is obtained from the cell dimensions shown by Fig. 2. The changes in cell dimension, which are independent of the direction of polarization along the Z axis, can be regarded as due to the electrostrictive effect in barium titanate. The electrostrictive effect for the barium titanate ceramic has been investigated in a previous paper[8] and it is there shown

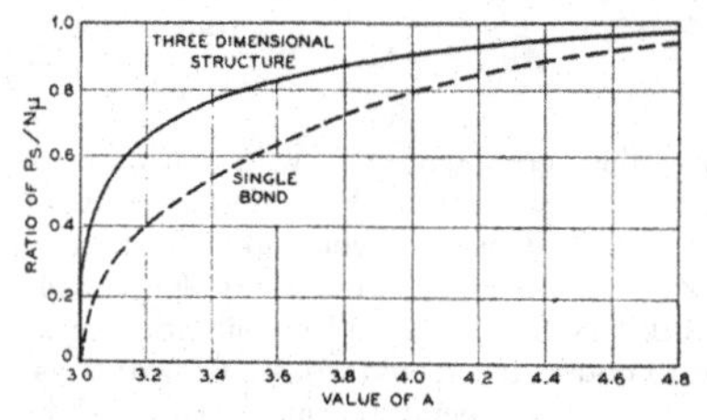

FIG. 9. Theoretical curve for ratio of spontaneous polarization P_s to the total polarization $N\mu$ as a function of the factor A.

[8] W. P. Mason, "Electrostrictive effect in barium titanate ceramics," Phys. Rev. (to be published).

that the ceramic has an increase in thickness and a decrease in radial dimension given by the strain equations

$$S_{33}=Q_{11}(P_z)^2;\quad S_{11}=S_{22}=+Q_{12}(P_z)^2 \quad (21)$$

where

$$Q_{11}=6.9\times10^{-12}\ (\text{cm}^2/\text{stat coulomb})^2;$$
$$Q_{12}=-2.15\times10^{-12}\ (\text{cm}^2/\text{stat coulomb})^2.$$

While the value of Q_{11}/Q_{12} is not exactly equal to -2 for the ceramic, a guide to the spontaneous polarization is obtained from these values. At 20°C, S_{33} the longitudinal thickness strain is equal to 6.7×10^{-3} while the radial thickness strain is equal to $S_{11}=S_{22}=-3.3\times10^{-3}$ from the measurements of Fig. 2. With these values and the electrostrictive constants of Eq. (21), the indicated spontaneous polarization for the two effects is

$$\left.\begin{aligned}P_z&=31{,}500\frac{\text{stat coulomb}}{\text{cm}^2}\\&=10.5\times10^{-6}\frac{\text{coulomb}}{\text{cm}^2}\ (\text{long.}),\\P_z&=39{,}000\frac{\text{stat coulomb}}{\text{cm}^2}\\&=12.9\times10^{-6}\frac{\text{coulomb}}{\text{cm}^2}\ (\text{radial}).\end{aligned}\right\}\quad(22)$$

Taking the average of these

$$P_z=35{,}250\frac{\text{stat coulomb}}{\text{cm}^2}=11.7\times10^{-6}\frac{\text{coulomb}}{\text{cm}^2}.\quad(23)$$

This value agrees quite well with that measured electrically by means of the hysteresis loops. For this value Matthias and von Hippel[6] find a value 12×10^{-6} coulomb/cm² while Hulm[9] finds a value 16×10^{-6} coulomb/cm². This calibration allows one to obtain the spontaneous polarization as a function of temperature, and this is shown plotted by Fig. 10. The very sudden rise in spontaneous polarization just below the Curie temperature is evident, and this agrees qualitatively with that shown by Fig. 9.

[9] F. Hulm, Nature **160**, 126 (1947).

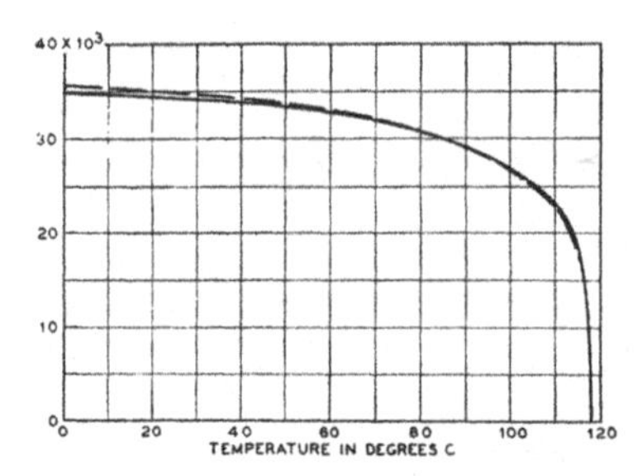

FIG. 10. Measured spontaneous polarization as a function of the temperature.

To find if the spontaneously generated polarization agrees quantitatively with that calculated from Eq. (19) we have to evaluate A and μ by other methods. One method for doing this is to measure the dielectric constants at low field strengths as a function of temperature. The calculated value can be obtained from Eq. (17) by dividing the polarization P_z into the spontaneous part P_S and a very small alternating part $P_0e^{j\omega t}$. The applied field $E_ze^{j\omega t}$ is assumed very small and hence we have

$$\begin{aligned}\sinh&\left[\frac{(E_z+\beta P_0)e^{j\omega t}+\beta P_S}{1-\beta\gamma}\right]\frac{\mu}{kT}\\&=\sinh\left[\frac{(E_z+\beta P_0)e^{j\omega t}}{1-\beta\gamma}\right]\frac{\mu}{kT}\cosh\left(\frac{\beta P_S}{1-\beta\gamma}\right)\frac{\mu}{kT}\\&\quad+\cosh\left[\frac{(E_z+\beta P_0)e^{j\omega t}}{1-\beta\gamma}\right]\frac{\mu}{kT}\sinh\left(\frac{\beta P_S}{1-\beta\gamma}\right)\frac{\mu}{kT}\\&\doteq\frac{(E_z+\beta P_0)e^{j\omega t}\mu}{(1-\beta\gamma)kT}\cosh\frac{AP_S}{N\mu}+\sinh\frac{AP_S}{N\mu}.\end{aligned}\quad(24)$$

Similarly,

$$\begin{aligned}\cosh\left[\frac{(E_z+\beta P_0)e^{j\omega t}+\beta P_S}{1-\beta\gamma}\right]\frac{\mu}{kT}&=\cosh\frac{AP_S}{N\mu}\\&\quad+\left[\frac{(E_z+\beta P_0)e^{j\omega t}}{1-\beta\gamma}\right]\frac{\mu}{kT}\sinh\frac{AP_S}{N\mu}.\end{aligned}\quad(25)$$

Inserting Eqs. (24) and (25) in (17) and solving for the constant and time variable parts, we obtain Eq. (19) for the constant part, and for the time variable part we have

$$\frac{P_0 e^{j\omega t}}{N\mu^2}=\frac{(E_z+\beta P_0)e^{j\omega t}}{(1-\beta\gamma)kT}\times\left\{\frac{2\cosh(AP_S/N\mu)+1}{[2+\cosh(AP_S/N\mu)]^2}\right\}. \quad (26)$$

Solving for P_0, multiplying by 4π and adding the dielectric constant for electrons and atoms, the dielectric constant for the z axis becomes

$$\epsilon_z=\epsilon_0+\frac{\dfrac{4\pi A}{\beta}\left[\dfrac{2\cosh\dfrac{AP_S}{N\mu}+1}{2+\cosh\dfrac{AP_S}{N\mu}}\right]}{2+\cosh\dfrac{AP_S}{N\mu}-A\left[\dfrac{2\cosh\dfrac{AP_S}{N\mu}+1}{2+\cosh\dfrac{AP_S}{N\mu}}\right]}. \quad (27)$$

Above the Curie point, the spontaneous polarization P_S disappears and this equation reduces to

$$\epsilon_z=\epsilon_0+\frac{(4\pi A/\beta)}{3-A}=\epsilon_0+\frac{C}{T-T_0} \quad (28)$$

upon introducing the value of A from Eqs. (18) and (11), where

$$C=\frac{4\pi N\mu^2[1+\beta(\epsilon_0-1)/4\pi]}{3k};$$
$$T_0=\frac{\beta N\mu^2}{3k}\{1+[\beta(\epsilon_0-1)/4\pi]\}. \quad (29)$$

The single domain crystals have so many impurities in them to prevent the breaking up of the crystal into multi-domains that they do not revert to a cubic crystal above the Curie point. This is shown by the different dielectric constant for the two directions above the Curie point. The same is true to a lesser extent for the multi-domain crystals, but the ceramic pieces show a pronounced maximum and a Curie region above 120°C, much in agreement with Eq. (28). Since above the Curie temperature the crystal becomes cubic and all directions equivalent, it is thought that the best values for C and T_0 will be obtained from a dense ceramic piece. From the dielectric constant above 120°C of Fig. 3, we obtain the values

$$C=40{,}000;\quad T_0=393°\mathrm{K} \quad (30)$$

and from low temperature measurements

$$\epsilon_0=350. \quad (31)$$

Taking the ratio of C/T_0 of Eq. (29) we find

$$\beta=(4\pi T_0/C)=0.124 \quad (32)$$

upon inserting the experimental values. Then, since the number of dipoles per cubic centimeter (as determined from the size of the unit cell) is $N=1.56\times10^{22}$; $k=1.38\times10^{-16}$, we have

$$C=40{,}000=\frac{4\pi(1.56\times10^{22})\mu^2[1+0.124(350/4\pi)]}{3\times1.38\times10^{-16}}$$

or

$$\mu=4.34\times10^{-18}. \quad (33)$$

This value of μ agrees fairly well with the value one would obtain from the recent x-ray observations that the titanium atom is displaced by 0.16A from the center of the unit cell. If the oxygen atom moves an equal distance to meet it (which could not be determined by x-ray observations), the dipole moment would be

$$(4e+2e)(0.16\times10^{-8})=6\times4.8\times10^{-10}\times0.16\times10^{-8}=4.6\times10^{-18}. \quad (34)$$

If all the dipoles pointed in one direction, the total polarization would be

$$N\mu=1.56\times10^{22}\times4.34\times10^{-18}=67{,}500\ \mathrm{e.s.u.}=22.5\times10^{-6}\ \mathrm{coulomb\ cm^2}. \quad (35)$$

The measured value of approximately 35,500 e.s.u. is 53 percent of this. If all the quantities entering Eq. (18) for A were independent of temperature except T, the absolute temperature, the value of A for 27°C = 300 K would be 3.94, and from Fig. 9 the theoretical value of the polarization $P_S/N\mu$ should be 0.90, rather than the measured value of 0.53, which corresponds to a value of $A=3.090$. This result indicates that some of the quantities in the expression for A decrease as the temperature is lowered. A similar result is also required for the variation of dielectric constant with temperature. A value of $A=3.090$, $P_S/N\mu=0.53$, and β set at 0.096 (in order to give a value of $A=3.090$), and all the other quantities unchanged, results in a dielectric constant of 1390 which agrees well with the dielectric constant for a ceramic or for a multidomain barium titanate crystal. The variation may be ascribed to β or to ϵ_0 because the measured temperature expansion coefficients indicate that N and μ should be relatively constant. From the x-ray data of Fig. 2 it is seen that from 120°C to 0°C, N should increase by 0.15 percent. Since the titanium atom is tightly bound to the oxygen, the distance between the center of oxygen and titanium should not change appreciably because of temperature contraction, and hence μ also will not change much with temperature. The value of ϵ_0, however, may be different for the a and c axes since the a axis decreases while the c axis increases. Hence, ϵ_c may be smaller and ϵ_a larger than ϵ_0. The Lorentz factor β, also, may vary considerably depending on the condition of the surrounding electrical charge configurations. For isotropic conditions, the theoretical value is $4\pi/3=4.19$. For the case of the titanium surrounded closely by the oxygens the experimental value is only 0.124. As the temperature is decreased, all the oxygen atoms come closer together and hence a decrease in β is to be expected. Assuming all the variation due to β, the values to agree with the dielectric constant measurements are shown plotted by Fig. 11. With these values of β (assuming all the other quantities in A are independent of the temperature), A can be evaluated as a function of temperature and the theoretical values of spontaneous polarization can be determined from Eq. (19). These are shown plotted by the dashed line of Fig. 10, and these agree closely with those determined from the electrostrictive effect. Hence, two independent sets of data are satisfied by the β-curve.

III. DIELECTRIC CONSTANT ALONG a AXIS

Measurements for the dielectric constants along the a axis for single domain crystals show that the dielectric constant along this axis is very much larger than that along the c axis. To determine the dielectric along the a axis, according to the model shown by Fig. 7, with a field applied along the Y axis, and a spontaneous polarization occurring along Z, the potentials for all six wells are

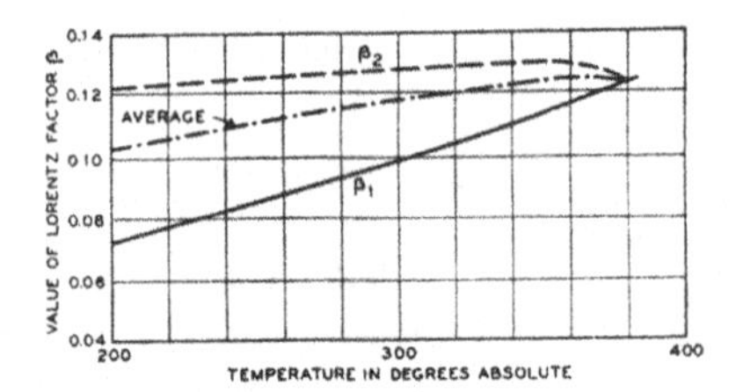

FIG. 11. Value of Lorentz factors β_1 and β_2 for $z=c$ and $y=a$ axes as a function of temperature.

$$U_1=-\left(\frac{\beta_1 P_S\mu}{1-\beta_1\nu}\right);\qquad U_2=\frac{\beta_1 P_S\mu}{1-\beta_1\gamma};$$
$$U_3=-\frac{[E_y+\beta_2 P_y]\mu}{(1-\beta_2\gamma)};\quad U_4=\left(\frac{E_y+\beta_2 P_y}{1-\beta_2\gamma}\right)\mu;\tag{36}$$
$$U_5=U_6=0.$$

We assume that β_2 along the Y axis may be different from β_1, along the Z axis. Applying the Boltzmann principle and relating N_1, N_2, N_3 and N_4 to $N_5=N_6$ we find

$$\begin{aligned}
N_1&=N_5\exp\left[\left(\frac{\beta_1 P_S}{1-\beta_1\gamma}\right)\frac{\mu}{kT}\right];\\
N_2&=N_5\exp-\left[\left(\frac{\beta_1 P_S}{1-\beta_1\gamma}\right)\frac{\mu}{kT}\right];\\
N_3&=N_5\exp\left[\frac{E_y+\beta_2 P_y}{1-\beta_2\gamma}\right]\frac{\mu}{kT};\\
N_4&=N_5\exp-\left[\frac{E_y+\beta_2 P_y}{1-\beta_2\gamma}\right]\frac{\mu}{kT};\quad N_6=N_5.
\end{aligned}\tag{37}$$

Since

$$N_1+N_2+N_3+N_4+N_5+N_6=N\tag{38}$$

we find for N_5, the value

$$N_5=\frac{N}{2[1+\cosh[(\beta_1 P_S)/(1-\beta_1\gamma)]\mu/kT+\cosh[(E_y+\beta_2 P_y)/(1-\beta_2\gamma)]\mu/kT]}.\tag{39}$$

Inserting the value of N_3, N_4, and N_5 in the expression for the polarization along the Y axis, we have

$$P_y=(N_3-N_4)\mu=\frac{N\mu\sinh[(E_y+\beta_2P_y)/(1-\beta_2\gamma)]\mu/kT}{[1+\cosh[(\beta_1P_S)/(1-\beta_1\gamma)]\mu/kT+\cosh[(E_y+\beta_2P_y)/(1-\beta_2\gamma)]\mu/kT]}. \quad (40)$$

To determine the dielectric constant along Y for small fields, we can replace

$$\sinh\left(\frac{E_y+\beta_2P_y}{1-\beta_2\gamma}\right)\frac{\mu}{kT}\doteq\left(\frac{E_y+\beta_2P_y}{1-\beta_2\gamma}\right)\frac{\mu}{kT};\quad \cosh\left(\frac{E_y+\beta_2P_y}{1-\beta_2\gamma}\right)\frac{\mu}{kT}\doteq 1. \quad (41)$$

Then

$$P_y=\frac{[N\mu^2/kT][(E_y+\beta_2P_y)/(1-\beta_2\gamma)]}{2+\cosh[(\beta_1P_S)/(1-\beta_1\gamma)]\mu/kT}=\frac{[N\mu^2/kT][(E_y+\beta_2P_y)/(1-\beta_2\gamma)]}{2+\cosh(AP_S/N\mu)} \quad (42)$$

where

$$A=\left(\frac{N\mu^2\beta_1}{1-\beta_1\gamma}\right)\frac{1}{kT}=\frac{N\mu^2\beta_1}{kT}\left[1+\beta_1\left(\frac{\epsilon_0-1}{4\pi}\right)\right]. \quad (43)$$

Solving for the ratio of P_y to E_y, multiplying by 4π, and adding ϵ_0 the dielectric constant due to other sources than the dipole moment, the dielectric constant along y becomes

$$\epsilon_y=\epsilon_0+\frac{kT[4\pi N\mu^2/(1-\beta_2\gamma)]}{2+\cosh(AP_S/N\mu)-[N\mu^2\beta_2/(1-\beta_2\gamma)kT]}. \quad (44)$$

Now, since the crystal becomes tetragonal due to the distortion caused by the electrostrictive effect, γ may increase along the a axis and cause ϵ_0 to become larger. As before, however, we assume all variation to occur in β_2 and write

$$\frac{1}{1-\beta_2\gamma}=1+\frac{\beta_2(\epsilon_0-1)}{4\pi}. \quad (45)$$

Inserting this value in Eq. (44) for the dielectric constant

$$\epsilon_y=\epsilon_0+\frac{\dfrac{4\pi A}{\beta_1}\left[\dfrac{1+(\beta_2/4\pi)(\epsilon_0-1)}{1+(\beta_1/4\pi)(\epsilon_0-1)}\right]}{2+\cosh\dfrac{AP_S}{N\mu}-A\dfrac{\beta_2}{\beta_1}\left[\dfrac{1+(\beta_2/4\pi)(\epsilon_0-1)}{1+(\beta_1/4\pi)(\epsilon_0-1)}\right]}. \quad (46)$$

At the Curie temperature where the crystal changes from tetragonal form to cubic form the value of β_2 must be equal to β_1 and hence the dielectric constant along the Y axis will have a Curie temperature at the same temperature as the one along the Z axis. For other temperatures, β_2 will not, in general, equal β_1 on account of the shift in charge due to the electrostrictive effect. One might expect, however, that the shift in charge might to a first approximation produce additive effects and that, in general

$$2\beta_2+\beta_1=3\beta, \quad (47)$$

where β is the Lorentz factor for the cubic crystal. The factor of 2 is used for β_2 since the charge along the X and Y axis is only half that along the Z axis.

According to Eq. (46) the very high dielectric constants along Y shown on Fig. 5 have to be accounted for by the near vanishing of the denominator of Eq. (46). The values of β_2 to make the denominator vanish, with the experimentally determined values of β_1 are shown by the dashed line of Fig. 11. These values would agree with the above speculation if the average value of β fell off with temperature according to the dot-dash line. Another verification of the near vanishing of the numerator is the very low value of the relaxation frequency for the dielectric constant along the Y axis, shown by Fig. 6. As shown by the next section, this can be accounted for by the same potential barrier for both Y and Z directions, provided that the denominator of Eq. (27) for the dielectric constant along the $c=Z$ axis is about 100 times as large as that of Eq. (46) for the Y axis.

IV. RELAXATION FREQUENCIES FOR THE DIELECTRIC CONSTANTS

To determine the high frequency behavior of the dielectric constants that is predicted by the model of Fig. 7, one can no longer use the Boltzmann equilibrium relation of Eq. (5) to determine the relative number of titanium nuclei in the various potential wells. Instead, one has to relate the time rate of change of the number in a given potential well to the probability of transition for a given time from one potential well to another. $\alpha_{1,2}$ the probability of a nucleus in well 1 jumping to well 2 per unit time is, according to Eyring's reaction rate theory,

$$\alpha_{1,2}=(kT/h)e^{-\Delta U/kT} \tag{48}$$

where h is Planck's constant, k Boltzmann's constant, and ΔU the difference between the maximum height of the potential barrier and the potential of well 1.

The time rate of change of the number N_1 of nuclei in wells of type 1 is obviously

$$\frac{dN_1}{dt}=-N_1(\alpha_{1,2}+\alpha_{1,3}+\alpha_{1,4}+\alpha_{1,5}+\alpha_{1,6})+N_2\alpha_{2,1}+N_3\alpha_{3,1}+N_4\alpha_{4,1}+N_5\alpha_{5,1}+N_6\alpha_{6,1}. \tag{49}$$

Similarly,

$$\frac{dN_2}{dt}=-N_2(\alpha_{2,1}+\alpha_{2,3}+\alpha_{2,4}+\alpha_{2,5}+\alpha_{2,6})+N_1\alpha_{1,2}+N_3\alpha_{3,2}+N_4\alpha_{4,2}+N_5\alpha_{5,2}+N_6\alpha_{6,2}. \tag{50}$$

Hence, the rate of change of the polarization along the Z axis is

$$\begin{aligned}\frac{dP_z}{dt}=\frac{d(N_1-N_2)\mu}{dt} &=-N_1(2\alpha_{1,2}+\alpha_{1,3}+\alpha_{1,4}+\alpha_{1,5}+\alpha_{1,6})\mu\\ &+N_2(2\alpha_{2,1}+\alpha_{2,3}+\alpha_{2,4}+\alpha_{2,5}+\alpha_{2,6})\mu\\ &+N_3\mu(\alpha_{3,1}-\alpha_{3,2})+N_4\mu(\alpha_{4,1}-\alpha_{4,2})\\ &+N_5\mu(\alpha_{5,1}-\alpha_{5,2})+N_6\mu(\alpha_{6,1}-\alpha_{6,2}).\end{aligned} \tag{51}$$

When a field is applied along the Z axis, the potential minimum U_1 is lowered, and U_2 raised by amounts shown by Eq. (12). Hence,

$$\begin{aligned}\alpha_{1,2}&=\frac{kT}{h}\exp-\left[\Delta U+\frac{(E_z+\beta_1P_z)}{1-\beta_1\gamma}\mu\right]\Big/kT;\\ \alpha_{2,1}&=\frac{kT}{h}\exp-\left[\Delta U-\frac{(E_z+\beta_1P_z)}{1-\beta_1\gamma}\mu\right]\Big/kT.\end{aligned} \tag{52}$$

By the discussion of Section II, it appears that the highest potential barrier in going from 1 to the 3, 4, 5 or 6 potential wells is also nearly ΔU. Hence,

$$\alpha_{1,2}=\alpha_{1,3}=\alpha_{1,4}=\alpha_{1,5}=\alpha_{1,6}. \tag{53}$$

Also,

$$\alpha_{2,1}=\alpha_{2,3}=\alpha_{2,4}=\alpha_{2,5}=\alpha_{2,6}. \tag{54}$$

In going from potential wells 3, 4, 5 or 6 to any of the other wells, the highest potential barrier is ΔU, since these minima are not changed by a field along Z and hence

$$\alpha_{3,n}=\alpha_{4,n}=\alpha_{5,n}=\alpha_{6,n}=\frac{kT}{h}e^{-\Delta U/kT}, \tag{55}$$

where n has all values from 1 to 6 except the one which makes the second index equal to the first.

Therefore, introducing these values in Eq. (51), the time rate of change of polarization along the Z axis becomes, for a simple harmonic field,

$$\frac{j\omega hP_z}{kT}e^{\Delta U/kT}=6\mu\left\{N_2\exp\left[\left(\frac{E_z+\beta_1P_z}{1-\beta_1\gamma}\right)\frac{\mu}{kT}\right]-N_1\exp\left[-\left(\frac{E_z+\beta_1P_z}{1-\beta_1\gamma}\right)\frac{\mu}{kT}\right]\right\}. \tag{56}$$

If ω is zero, this reduces to the Boltzmann condition for determining the ratio of N_2/N_1.

Since we are dealing only with infinitesimal fields, the sum of N_2 and N_1 can be taken equal to their equilibrium values given by Eq. (16). Since Eq. (56) can be written in the form

$$\frac{j\omega hP_z}{kT}e^{\Delta U/kT}=6\mu\left[(N_1+N_2)\sinh\left(\frac{E_z+\beta_1P_z}{1-\beta_1\gamma}\right)\frac{\mu}{kT}-(N_1-N_2)\cosh\left(\frac{E_z+\beta_1P_z}{1-\beta_1\gamma}\right)\frac{\mu}{kT}\right], \quad (57)$$

this becomes

$$\frac{j\omega hP_ze^{\Delta U/kT}}{6kT\cosh[(E_z+\beta_1P_z)/(1-\beta_1\gamma)]\mu/kT}=\left[\frac{N\mu\sinh[(E_z+\beta_1P_z)/(1-\beta_1\gamma)]\mu/kT}{2+\cosh[(E_z+\beta_1P_z)/(1-\beta_1\gamma)]\mu/kT}-P_z\right]. \quad (58)$$

Introducing the relations of Eqs. (24) and (25) and solving for the time variable parts of the polarization, P_0, noting that

$$\frac{\partial P_z}{\partial t}=j\omega P_0e^{j\omega t},$$

we find for the dielectric constant as a function of frequency, the equation

$$\epsilon_c=\epsilon_0+\frac{\frac{4\pi A}{\beta_1}\left(\frac{2\cosh(AP_S/N\mu)+1}{2+\cosh(AP_S/N\mu)}\right)}{\left(2+\cosh\frac{AP_S}{N\mu}\right)-A\left(\frac{2\cosh(AP_S/N\mu)+1}{2+\cosh(AP_S/N\mu)}\right)+\left(\frac{2+\cosh(AP_S/N\mu)}{\cosh(AP_S/N\mu)}\right)\frac{j\omega h}{6kT}e^{\Delta U/kT}}. \quad (59)$$

When the last term in the denominator equals the sum of the other two, the dipole dielectric constant has equal resistance and reactance values and the corresponding frequency is the relaxation frequency. This frequency f_0 is given by

$$f_0=\frac{6kTe^{-\Delta U/kT}}{2\pi h}\left[\cosh\frac{AP_S}{N\mu}\left(1-\frac{A(2\cosh(AP_S/N\mu)+1)}{(2+\cosh(AP_S/N\mu))^2}\right)\right]. \quad (60)$$

For 27°C = 300°K, we found $A=3.090$; $P_S/N\mu=0.53$. Introducing these values and the values

$$k=1.38\times10^{-16};\quad T=300;\quad h=6.56,\ 10^{-27}, \quad (61)$$

we find for f_0, the value

$$f_0=1.6\times10^{12}e^{-\Delta U/kT}. \quad (62)$$

From the data of Eqs. (2) and (3), the relaxation frequency of a ceramic (which probably coincides with that for the c axis direction) is 6.2×10^9 cycles. From this one obtains a value for the potential maximum[7] of

$$e^{\Delta U/kT}=260;\quad \Delta U=3.35 \text{ kilocalories per mole}. \quad (63)$$

This value represents the amount of energy to remove the titanium nucleus from its equilibrium position to a position in the center of the barium titanate unit cell.

The data of Fig. 6 show that the dielectric constant along the a axis is relaxed at a frequency of about 15 megacycles at room temperature. Applying the same process to calculating the dielectric constant along the a axis, one finds

$$\epsilon_y=\epsilon_0+\frac{\frac{4\pi A}{\beta_1}\left[\frac{1+(\beta_2/4\pi)(\epsilon_0-1)}{1+(\beta_1/4\pi)(\epsilon_0-1)}\right]}{2+\cosh\frac{AP_S}{N\mu}-\frac{A\beta_2}{\beta_1}\left[\frac{1+(\beta_2/4\pi)(\epsilon_0-1)}{1+(\beta_1/4\pi)(\epsilon_0-1)}\right]+\left(\frac{2+\cosh(AP_S/N\mu)}{\cosh(AP_S/N\mu)}\right)\frac{j\omega h}{6kT}e^{\Delta U/kT}}. \quad (64)$$

To obtain a dielectric constant of 150,000 at 27°C = 300°K, the real part of the denominator has to be 0.0028. Hence, the indicated relaxation frequency for this temperature is

$$f_0=\frac{6kT}{2\pi h}e^{-\Delta U/kT}\frac{\left[2+\cosh\frac{AP_S}{N\mu}-\frac{A\beta_2}{\beta_1}\left(\frac{1+(\beta_2/4\pi)(\epsilon_0-1)}{1+(\beta_1/4\pi)(\epsilon_0-1)}\right)\right]\cosh\frac{AP_S}{N\mu}}{[2+\cosh(AP_S/N\mu)]}.$$

Introducing the numerical values,

$$e^{\Delta U/kT}=645 \text{ or } \Delta U=3.9 \text{ kilocalories.}$$

Thus the indicated activation energy for going from the 1, 2 wells to the 3, 4, 5 or 6 wells is only slightly higher than that between opposite wells such as 1 and 2. This calculation also checks the facts that it is the near vanishing of the denominator of Eq. (64) that causes the very high dielectric constant along the a or $X=Y$ axes.

V. COERCIVE FIELDS ALONG a AND c AXES

The coercive fields along the a and c crystallographic axes and the interaction between a field along c and a polarization generated along a can be calculated from Eqs. (36) and (40), giving the polarizations along the $c=Z$ direction and the $a=Y$ direction. In terms of complete fields and polarizations along the two directions these equations become

$$P_z=\frac{N\mu\sinh[(E_z+\beta_1P_z)/(1-\beta_1\gamma)]\mu/kT}{1+\cosh[(E_z+\beta_1P_z)/(1-\beta_1\gamma)]\mu/kT+\cosh[(E_y+\beta_2P_y)/(1-\beta_2\gamma)]\mu/kT}, \quad (66)$$

$$P_y=\frac{N\mu\sinh[(E_y+\beta_2P_y)/(1-\beta_2\gamma)]\mu/kT}{1+\cosh[(E_z+\beta_1P_z)/(1-\beta_1\gamma)]\mu/kT+\cosh[(E_y+\beta_2P_y)/(1-\beta_2\gamma)]\mu/kT}. \quad (67)$$

From these two equations and the constants evaluated previously, the coercive fields for the two directions can be approximately calculated.

The calculations show that it takes considerably more of a negative field along Z to reverse the sign of a domain along Z than it does to change

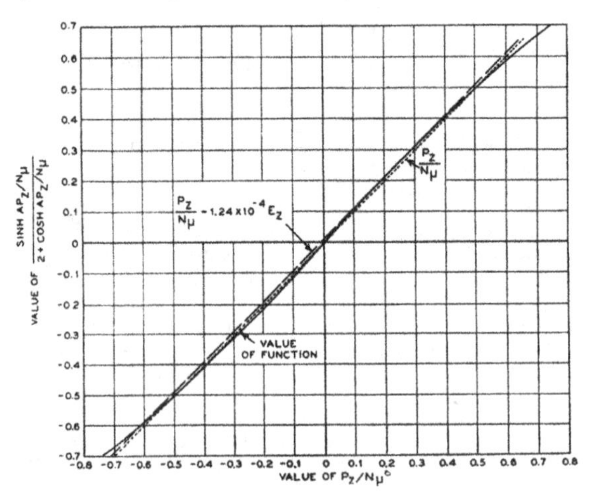

FIG. 12. Method for obtaining spontaneous polarization and coercive field.

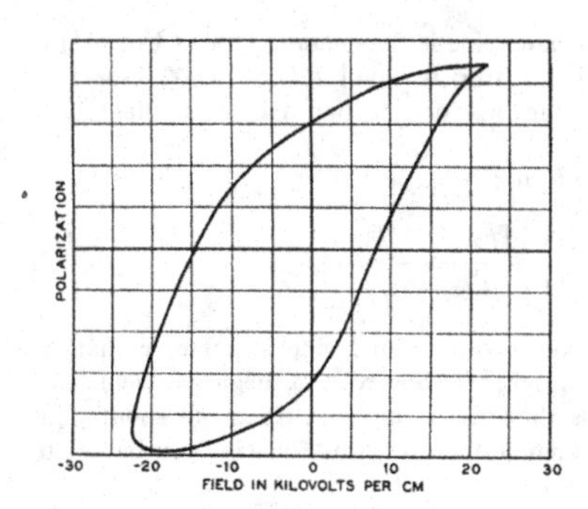

FIG. 13. Hysteresis loop showing relation between polarization along Z and field along Z for a single domain barium titanate crystal.

the direction from Z to Y. To show this, let us assume that no field or polarization exist along Y. Then Eq. (66) can be written in the form

$$\frac{P_z}{N\mu}=\frac{\sinh[(AE_Z/\beta N\mu)+(AP_Z/N\mu)]}{2+\cosh[(AE_z/\beta N\mu)+(AP_z/N\mu)]}. \quad (68)$$

Now, since $AE_Z/\beta N\mu$ is going to be a very small quantity for any field that can be applied, this can be written as

$$\frac{\sinh(AP_z/N\mu)}{2+\cosh(AP_z/N\mu)}=\frac{P_z}{N\mu}-\frac{AE_z}{\beta N\mu}\left[\frac{2\cosh(AP_z/N\mu)+1}{(2+\cosh(AP_z/N\mu))^2}\right]. \quad (69)$$

If the applied field E_z is zero, this equation reduces to that for the spontaneous polarization.

If we plot the left hand side of Eq. (69) as a function of $P_z/N\mu$ (assuming $A=3.090$ for room temperature) the curve of Fig. 12 results. The left hand side is larger than the right, up to a value of $P_z/N\mu=0.534$ when the two are equal, and this represents the theoretical value of spontaneous polarization for no applied field. If the applied field is positive, a larger ratio of $P_z/N\mu$ is required to satisfy Eq. (69). Since at room temperature, $A=3.090$; $N\mu=67{,}100$ e.s.u.; $\beta=0.096$; $\cosh(AP_z/N\mu)=2.68$, the coefficient multiplying E_z is 1.24×10^{-4}. It takes, then, a very high field to increase sensibly $P_z/N\mu$. For example, a field of 30,000 volts per cm = 100 e.s.u., will cause the polarization to increase from 35,600 stat coulombs/cm^2 to 41,500 stat coulombs, an increase of 16 percent. This agrees quite well with the increase measured by Hulm[9] who found an increase of about 13 percent for this case.

If we put on a negative voltage along the axis the ratio $P_z/N\mu$ will decrease steadily until the difference between the left hand side of Eq. (69) and $P_z/N\mu$ reaches a maximum. This occurs for $P_z/N\mu=0.405$, and it requires a negative field of

$$E_z=74 \text{ e.s.u./cm}=22{,}200 \text{ volts/cm}. \quad (70)$$

This is the theoretical field strength to switch the direction of a domain along one direction of Z to that along the other. Single domain crystals have been observed to switch at around this value of field strength.

A true single domain crystal, however, will have a hysteresis loop for a considerably smaller field strength than this. For such a crystal a typical field strength polarization curve is as shown by Fig. 13. When the voltage is in the direction of the spontaneous polarization, the curve has a tail toward the right hand side that is considerably different from the rounded relation on the left hand side. This dissymmetrical type of curve occurs down to field strengths of the order of 1000 volts per centimeter and appears to result from the fact that on the application of a negative field along Z, parts of the domain can be spontaneously polarized along Y. To see that this is possible one can examine the conditions for spontaneous polarization along Y given by Eq. (67). Here we set E_y equal to zero and solve for the conditions that will give a finite value of P_y in the presence of a field E_z, and a spontaneous polarization P_z. The onset of P_y will be determined when P_y approaches zero, and thus we can replace the hyperbolic sinh by the argument, and the hyperbolic cosh by unity. Then the equations to solve are

$$\frac{P_y}{N\mu}=\frac{(\beta_2)/(1-\beta_2\gamma)(\mu/kT)P_y}{2+\cosh[(E_z+\beta_1P_z)/(1-\beta_1\gamma)]\mu/kT}. \quad (71)$$

If $E_z=0$, this reduces to the case

$$P_y\left(2+\cosh\frac{AP_z}{N\mu}\right)=\frac{A\beta_2}{\beta_1}\left\{\frac{1+[\beta_2(\epsilon_0-1)/4\pi]}{1+[\beta_1(\epsilon_0-1)/4\pi]}\right\}P_y.$$

The difference between the left hand side and the right hand side is the denominator of Eq. (44) for the dielectric constant along the Y axis. This denominator is small (about 0.0028 for room temperature) but is always positive, hence no spontaneous polarization can exist along Y as long as there is no static field $-E_z$.

For the addition of a static field, Eq. (71) takes the form

$$P_y\left[2+\cosh\left(\frac{AE_z}{\beta_1 N\mu}+\frac{AP_z}{N\mu}\right)\right]=\frac{A\beta_2}{\beta_1}\left\{\frac{1+[\beta_2(\epsilon_0-1)/4\pi]}{1+[\beta_1(\epsilon_0-1)/4\pi]}\right\}P_y. \tag{72}$$

A positive field E_z in the same direction as P_z makes the left hand side still larger than the right, and no possibility exists for polarization along Y. If, however, a negative field E_z is applied, the left hand side can be made equal or less than the right hand side, and spontaneous polarization can exist along Y. Since $AE_z/\beta_1 N\mu$ is a small quantity, this equation can be written in the form

$$\frac{AE_z}{\beta_1 N\mu}=-\frac{\left[2+\cosh\frac{AP_z}{N\mu}-\frac{A\beta_2}{\beta_1}\left\{\frac{1+(\beta_2/4\pi)(\epsilon_0-1)}{1+(\beta_1/4\pi)(\epsilon_0-1)}\right\}\right]}{\sinh(AP_z/N\mu)}. \tag{73}$$

Since for room temperature the numerator is equal to 0.0028, the denominator to 2.53, the field E_z to cause a domain to switch to the Y direction is

$$E_z=\frac{0.0028}{2.53}\times\frac{0.108\times 67{,}100}{3.090}=2.6\text{ e.s.u./cm}=780\text{ volts/cm}, \tag{74}$$

which is considerably less than the voltage required to shift a domain along Z.

The question arises as to why the whole domain does not go over in the Y direction. This appears to be owing to the fact that when parts of the large domains change direction, they exert an E_y field on the remainder of the domain that is still directed along Z. Then the term $\cosh[(E_y+\beta_2 P_y)/(1-\beta_2\gamma)]\times(\mu/kT)$ can no longer be replaced by unity, and the equation for the field to produce a spontaneous polarization along Y becomes

$$\frac{AE_z}{\beta_1 N\mu}=-\frac{\left[1+\cosh\frac{AP_z}{N\mu}+\cosh\left(\frac{E_y+\beta_2 P_y}{1-\beta_2\gamma}\right)\frac{\mu}{kT}-A\frac{\beta_2}{\beta_1}\left\{\frac{1+(\beta_2/4\pi)(\epsilon_0-1)}{1+(\beta_1/4\pi)(\epsilon_0-1)}\right\}\right]}{\sinh(AP_z/N\mu)} \tag{75}$$

and the field E_z becomes larger. There is no definite saturation for the effect which accounts for the rounded shape of the left side of the hysteresis loop of Fig. 13. When a positive E_z voltage is applied, all the Y domains revert back to the Z direction, which accounts for the tail-like shape of the right hand side of the curve of Fig. 13.

When a field is applied along Y, the relation between P_y and E_y is very linear and shows no hysteresis effects up to a field strength of 300 volts per centimeter, at which field the crystal usually breaks down because of the high conductivity along the a axis. Up to that voltage, no domain shift in the Y direction has occurred. To obtain the field for the shift requires that both Eqs. (66) and (67) shall be solved simultaneously for the P_y and P_z polarizations and this is not attempted here.

VI. SPECIFIC HEAT ANOMALY OF BARIUM TITANATE

The specific heat anomaly of barium titanate ceramics for the 120°C transition has been measured by Harwood, Popper and Rushman,[10] and

[10] Nature **160**, 58 (1948).

Blattner and Merz.[11] The former obtain a value of 0.14 cal./gram whereas the latter obtain 0.2 cal./gram. It has been shown by Mueller[12] that the specific heat anomaly is related to the spontaneous polarization by the equation

$$Q=\beta/2P_s^2 \tag{76}$$

where Q is the specific heat anomaly in ergs/cc, β is the Lorentz factor and P_s the spontaneous polarization. Since the specific heat anomaly was the integrated increase from about 100°C to a temperature above the Curie temperature we have from Fig. 10, that $P_s=27{,}000$ c.g.s. units of charge per square cm. Q, the specific heat anomaly, is 0.2 cal./gram $=1.2$ cal./cc $=5\times10^7$ ergs/cc. This gives a value of β determined by the specific heat anomaly of

$$\beta=0.138 \tag{77}$$

which agrees reasonably well with the value given in Eq. (32), obtained from dielectric measurements.

[11] Helv. Phys. Acta, Vol. XXI, *Fasciculus Tertius et Quartus* (1948).

[12] Annals New York Academy of Sciences, Vol. XL (Art. 5), page 353.

JOURNAL OF THE PHYSICAL SOCIETY OF JAPAN Vol. 7, No. 1, JAN.-FEB. 1952

Phase Transitions in Solid Solutions of $PbZrO_3$ and $PbTiO_3$ (II) X-ray Study

By Gen SHIRANE, Kazuo SUZUKI and Akitsu TAKEDA
Tokyo Institute of Technology, Oh-okayama, Tokyo

(Received March 23, 1951)

The phase diagram of the whole range of the $PbZrO_3$-$PbTiO_3$ system was determined by the dielectric and dilatometric measurements. According to the reason shown in part I, this phase diagram can be divided into three regions; paraelectric, ferroelectric and antiferroelectric phases. The crystal structure of this system was determined by the Debye photographs. In the antiferroelectric region solid solutions have a tetragonal modification of perovskite structure with $c/a<1$ and have some superstructure which seems to have intimate relation to the antiferroelectricity. In the ferroelectric region, except a region near antiferroelectric phase in which pseudo-cubic structure is observed, they have ordinary tetragonal structure of $c/a>1$ without any superstructure.

§1. Introduction.

In part I[1] of this paper, an investigation was made on solid solutions of $PbZrO_3$ and $PbTiO_3$ ranging from pure $PbZrO_3$ to Pb(Zr90−Ti10)O_3. Now, a more thorough study has been made of the whole range of Pb(Zr−Ti)O_3 system by using dielectric, dilatometric and X-ray methods.

According to Megaw[2], crystal structure of pure $PbZrO_3$ and $PbTiO_3$ belong to tetragonal modification of perovskite structure with the following lattice constants:

	a(A)	c/a
$PbZrO_3$	4.1503 (±0.001)	0.988 (±0.001)
$PbTiO_3$	3.8966 (±0.001)	1.0635 (±0.001)

It is a noticeable fact that the tetragonality of $PbZrO_3$ is less than unity in marked contrast with the case of lead titanate which has a tetragonality considerably larger than unity, so that it seems to be very interesting to study the variation of tetragonality of Pb(Zr−Ti)O_3 series as a function of concentration and its bearing upon the dielectric properties.

§2. Phase Diagram.

For the first place, we have determined the phase diagram of the whole range of this Pb(Zr−Ti)O_3 system by permittivity measurements. Fig. 1 shows permittivity versus rising temperature curves for a number of solid solutions. Dielectric constant of these specimens was measured at 1 MC/sec by a resonance method. Measurements were rather difficult for solid solutions containing $PbTiO_3$ more than 50% because of their high Curie temperature and of their increasing porosity. Above the Curie point, it was found that the Curie-Weiss law $\varepsilon=C/(T-T_p)$ holds sufficiently well for all specimens with the Curie constant $C=8\sim 10\times10^4$°C. The phase diagram of this system thus determined is shown in Fig. 2.

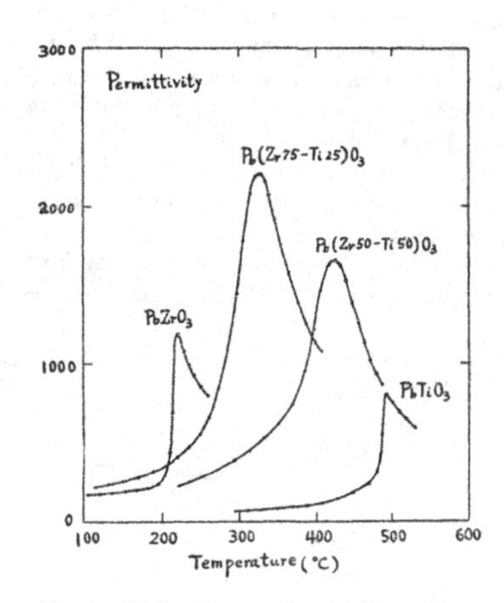

Fig. 1. Dielectric constant of Pb(Zr−Ti)O_3 compositions at rising temperature,

As for the reason why we have classified the left hand lowest corner of this phase diagram as an antiferroelectric region, the reader may refer to the part I, in which it was shown that all anomalies at this transition point can be well explained only if the above classification were accepted.

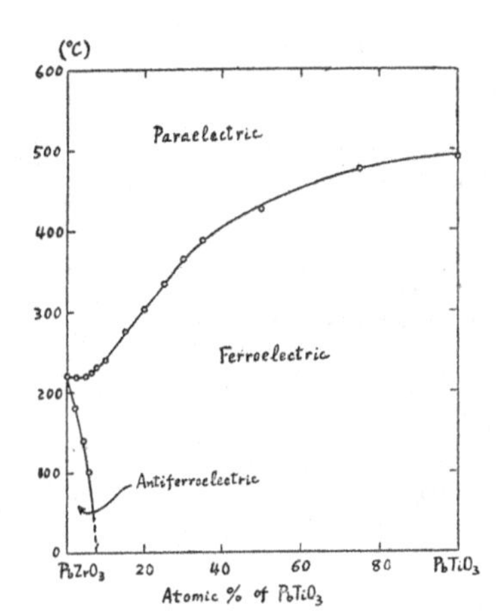

Fig. 2. Phase diagram of the $PbZrO_3$–$PbTiO_3$ system.

§ 3. Crystal Structure at Room Temperature

A series of Debye photographs of powdered specimens were taken over whole range of Pb(Zr–Ti)O_3 compositions by using Cu Kα radiation. The camera radius was about 35 mm. All the photographs taken at room temperature were well explained by assuming tetragonal structures of perovskite type, except some weak lines found in solid solutions near the pure $PbZrO_3$ side. The lattice constants and the axial ratio c/a were calculated mainly from (*510*), (*143*) and (*134*) lines on photographs

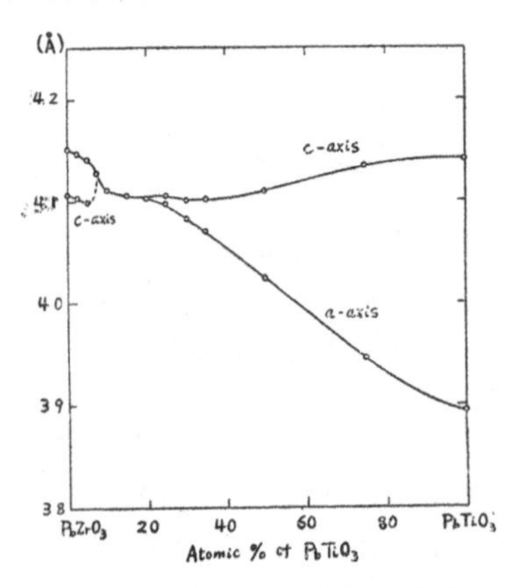

Fig. 3. Lattice spacing of $PbZrO_3$–$PbTiO_3$.

taken by back reflection method. The results are shown in Figs. 3 and 4. Figs 5 shows a unit cell volume a^2c versus composition curve. All the numerical data are presented in Table I.

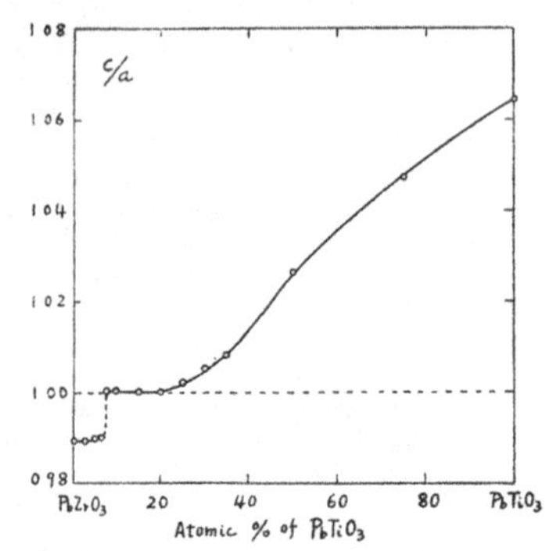

Fig. 4. Tetragonality c/a as a function of composition.

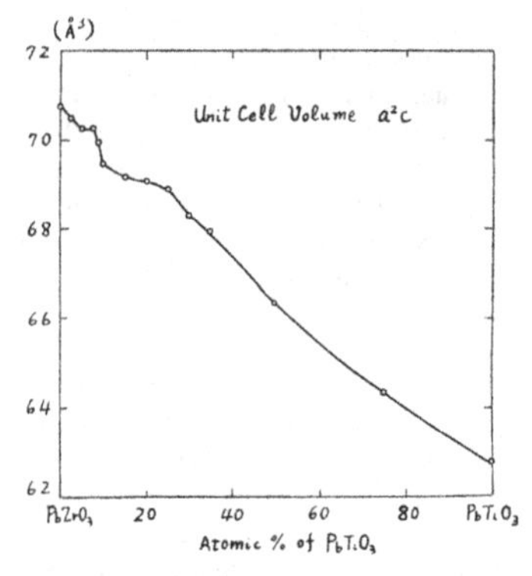

Fig. 5. Unit cell volume as a function of composition.

a) Pure $PbZrO_3$

Our results on pure $PbZrO_3$, $a=4.151$A and $c/a=0.989$, are in good agreement with the values due to Megaw. In the Debye photograph of this crystal, there are some extra lines besides the ordinary tetragonal lines. The existence of these extra lines was already reported by Hoffmann[3] and by Megaw[4]. To illustrate them, a Debye photograph of pure $PbZrO_3$ and its roughly estimated intensity diagram are shown in Figs. 6 and 8. In the latter figure, too weak lines are omitted. The extra lines are indicated by open circles. For

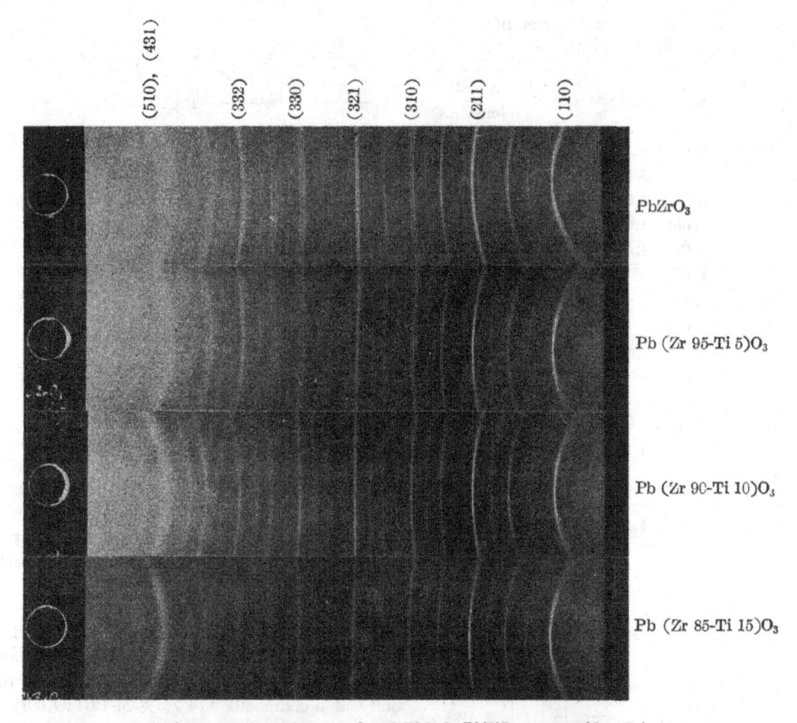

Fig. 6. Debye photographs of $PbZrO_3$-$PbTiO_3$ system (Cu $K\alpha$)

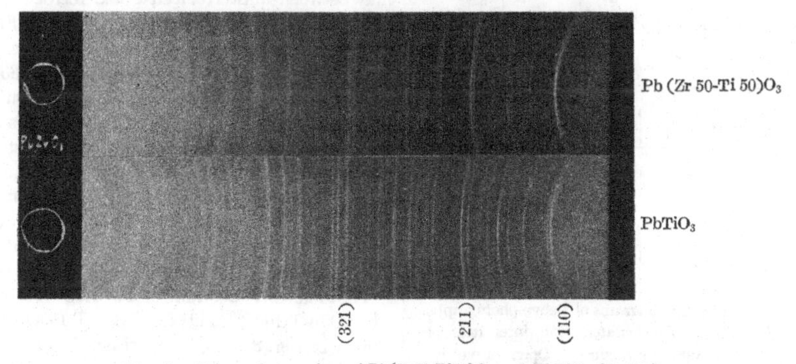

Fig. 7. Debye photograhps of Pb(Zr50-Ti50)O_3 and $PbTiO_3$ (Cu $K\alpha$).

comparison, intensities of Debye lines of perovskite structure with $a=4.150$ and $c/a=0.99$ are calculated by the following equation:

$$I \propto F^2 \, p \cdot \frac{1+\cos^2 2\theta}{\sin^2\theta \cdot \cos^2\theta} \, A$$

where F denotes the structure factor, p the number of multiplicity of the Debye line, θ the Bragg angle, A the absorption factor. The result is also shown in Fig. 8.

A comparison of observed lines with calculated ones revealed the fact that, besides the existence of the extra lines, observed intensities of

Table I. Lattice parameter of $Pb(Zr-Ti)O_3$

Atomic % of $PbTiO_3$	a(Å)	c(Å)	c/a	unit cell volume(Å3)
$PbZrO_3$	4.151	4.105	0.989	70.74
2.5	4.146	4.101	0.989	70.49
3.8	4.143	4.098	0.989	70.35
5	4.140	4.097	0.9895	70.23
6.3	4.139	4.098	0.990	70.21
7.5	4.126			70.23
8.8	4.120			69.92
10	4.111			69.49
15	4.105			69.15
20	4.102			69.03
25	4.097	4.105	1.002	68.90
30	4.081	4.100	1.005	68.28
35	4.070	4.101	1.008	67.94
50	4.014	4.119	1.026	66.37
75	3.946	4.133	1.047	64.36
$PbTiO_3$	3.894	4.140	1.063	62.77

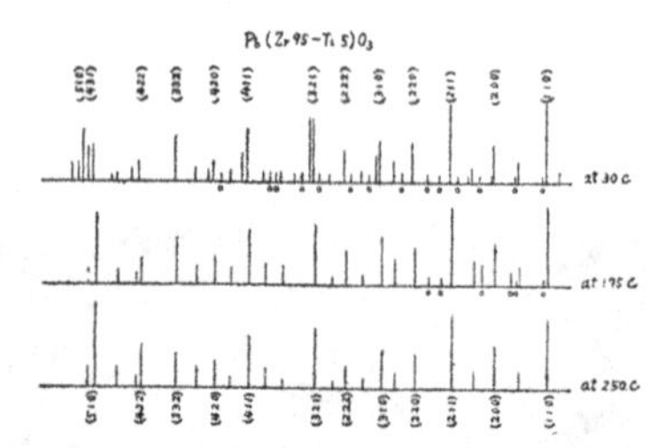

Fig 9 Intensity diagrams of Debye photographs of Pb (Zr95-Ti5) O_3 as a function of temperature Open circle indicates extra line

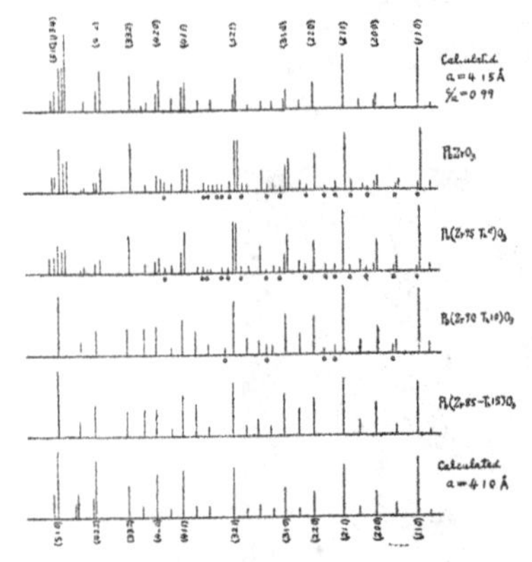

Fig. 8. Intensity diagrams of Debye photographs and the calculated line intensities of perovskite structure. Open circle indicates extra line.

some tetragonal lines, say (510) (143) and (321) etc , are not in good agreement with calculated values, though their spacings coincide completely. A study of $PbZrO_3$ single crystal, recently performed by E. Sawaguchi, H. Maniwa and S Hoshino in our laboratory using the method of oscillation photographs (unpublished yet), suggests that this crystal has superstructure. The discrepancy in the line intensities, accordingly, may be accounted for by the existence of the superstructure lines superposed on the tetragonal lines

b) Solid Solution.

With increasing $PbTiO_3$ concentration, the tetragonality remains at first almost constant ($c/a<1$) and then shows sudden increase to unity at a composition of 7.5% of $PbTiO_3$ (Fig. 4), which is just corresponding to the border line between antiferroelectric and ferroelectric regions at room temperature as shown in Fig. 2. The lattice parameter of solid solutions in a range from Pb(Zr92.5−Ti7 5)O_3 to Pb(Zr80−Ti20)O_3 was calculated on assuming their cubic structure, but the Debye lines of these substances are in fact diffuse (Fig. 6) suggesting that the crystal lattice of these substances might be not exactly cubic This broadening may partly be due to the strain inhomogeneity. For the sake of brevity, we will hereafter simply denote this structure as ***pseudo-cubic***.

The intensity diagram of Pb(Zr95−Ti5)O_3, Pb(Zr90−Ti10)O_3 and Pb(Zr85−Ti15)O_3 are shown in Fig. 8. For comparison, the calculated line intensities of cubic perovskite structure of $PbZrO_3$ with $a=4.10$A is also shown in the same figure. The higher the Ti concentration, the better becomes the agreement between the observed and calculated intensities. Enlarged figure of lattice spacings of solid solutions on the $PbZrO_3$ side is shown in Fig. 10.

Superstructure lines as found in pure $PbZrO_3$

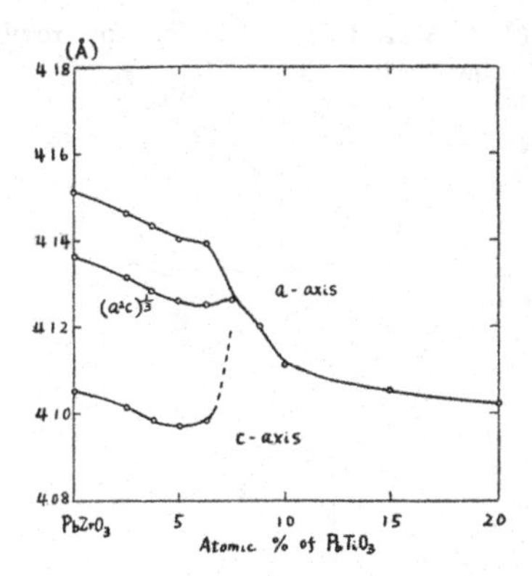

Fig 10. Lattice parameters of solid solutions on the $PbZrO_3$ side.

appear also in solid solutions as far as they have tetragonality $c/a<1$, though accompanied by only slight change of their intensities They become gradually weak and then some of them suddenly disappear when tetragonality attains to unity, and at last in $Pb(Zr85-Ti15)O_3$ all of them completely disappear (see Figs. 6 and 8) It seems reasonable to assume that this superstructure has intimate relation with the antiferroelectric properties When the solid solutions contain $PbTiO_3$ more than 20%, the tetragonality becomes $c/a>1$ thus similar to the case of barium titanate, and its axial ratio increases with Ti concentration, attaining to 1 063 for pure $PbTiO_3$ (Fig 7) The relation of the dielectric property and the crystal structure of this system is considered as follows:

Antiferroelectric	Ferroelectric
$c/a<1$ with superstructure	pseudo-cubic—$c/a>1$

As shown in Fig. 5, a unit cell volume of this system decreases with increasing $PbTiO_3$ concentration, and this result is reasonable consequence of the difference of ionic radii between Ti^{+4}ion (0 68A) and Zr^{+4}ion (0 80A). Small anomalies of volume change were found at the compositions containing 7.5% and 25% $PbTiO_3$. These compositions coincide with the two ends of pseudo-cubic region. The anomaly at 7 5% $PbTiO_3$ is clearly corresponding to the boundary of the antiferroelectric and ferroelectric regions and is of the nature similar to the anomalous large expansion observed at 140°C in the thermal expansion measurement for $Pb(Zr95-Ti5)O_3$, as shown in part I and also in Fig 11 of the present paper We have at present, on the other hand, no satisfactory explanation for anomaly at composition of 25% of $PbTiO_3$

§ 4 Thermal Expansion

Linear thermal expansion of ceramic specimens was measured on the several solid solutions The specimen has a cylindrical form, 3cm in length and 0 5cm in diameter Differential dilatation of the specimen to a silica rod was magnified by an optical lever by about 1000 times as large Fig 11 shows $l-l_0/l_0$ versus rising temperature for several $Pb(Zr-Ti)O_3$ compositions, where l_0 is the length at room temperature.

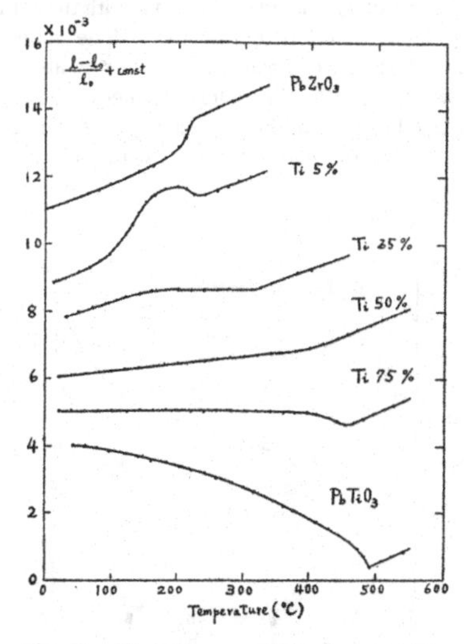

Fig 11. Thermal expansion curves of Pb $(Zr-Ti)O_3$ compositions

As shown in part I, $Pb(Zr95-Ti5)O_3$ shows an anomalous volume expansion at 140°C corresponding to the transition from antiferroelectric to ferroelectric region At first we had expected that the ferroelectric region may increase with increasing $PbTiO_3$ concentration accompanying more striking volume changes at the Curie point as well as at the lower transition point But the true situation was found to be more complicated: though the lower transition temperature, which is 140°C in Pb $(Zr95-Ti5)O_3$, is indeed lowered as the $PbTiO_3$ concentration increases, yet the transition becomes less sharp and the volume change smaller Similarly, though the Curie temperature, which is 215°C in $Pb(Zr95-Ti5)O_3$ increases with the concentration of $PbTiO_3$ as already shown by the permittivity measurements, yet no marked volume contraction at the Curie point can be detected with solid solutions of $Pb(Zr75-Ti25)O_3$ and $Pb(Zr50-Ti50)O_3$

Though the thermal expansion coefficient in the ferroelectric region shows complicated behaviour with composition, the linear thermal expansion coefficient in the cubic region is nearly the same for all the compositions, that is, $6\,5\sim8\,5\times10^{-6}$/°C We have estimated the unstrained volume at room temperature by extrapolation of the linear part of dilatation curve in the cubic region and compared it with the actual volume at room temperature. The volume strain $\Delta v/v$ thus estimated at room temperature is shown in Fig 12 The volume change at the border line between antiferroelectric and ferroelectric states can clearly be seen in this figure

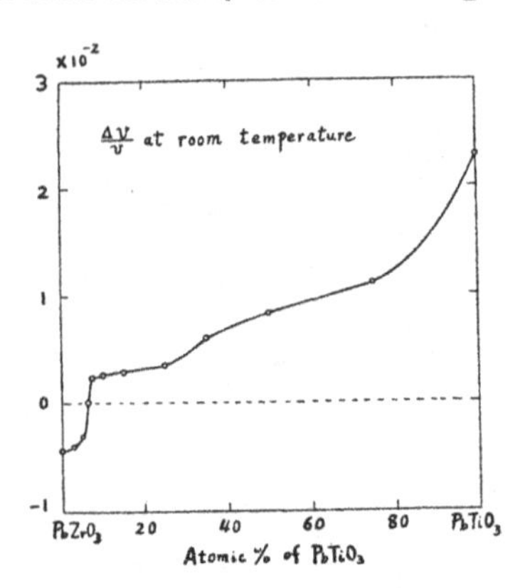

Fig 12 Volume strain at room temperature estimated by extrapolation from the cubic region

§ 5 X-ray Study near the Transition Points

The two phase transitions in $Pb(Zr95-Ti5)O_3$ were studied in detail in part I Now, a temperature dependence of lattice spacing of the same substance has been investigated at temperatures from 10° to 280°C A series of Debye photographs were taken on powdered specimens by back reflection method, using Cu $K\alpha$ radiation A camera length was about 40mm Lattice constants were calculated from (510), (143) and (134) lines and the results are shown in Fig 13

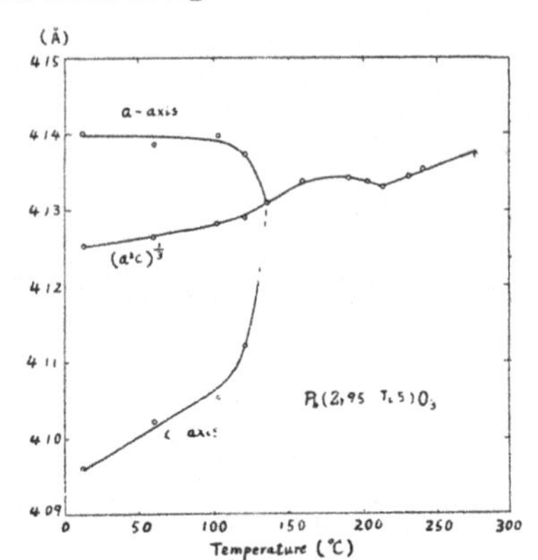

Fig. 13 Lattice spacing of $Pb(Zr95-Ti5)O_3$ as a function of temperature

At rising temperature, the tetragonality becomes unity at the lower transition point of 140°C, but the Debye lines are very diffuse even at temperatures above this transition point. This broadening of lines may be of the same nature as that found for pseudo-cubic solid solutions ranging, at room temperature, from $Pb(Zr92\,5-Ti7\,5)O_3$ to $Pb(Zr80-Ti20)O_3$ Superstructure lines are found in the antiferroelectric region below 140°C and they become faint as the structure becomes pseudo-cubic

If the temperature is raised further, the diffuseness of the powder reflection reduces, and the crystal structure of this substance has an exact cubic symmetry without any superstructure at temperatures above the Curie point. The behaviour of the Debye lines with temperature is shown schematically in Fig 9 The temperature change of unit cell volume is just in accordance with that obtained by dilatometric measurement (Fig 11) It must be noticed that a volume contraction occurs at the Curie point in similar way as in the case of pure barium titanate[4], though the crystal structure in the ferroelectric region is not tetragonal but pseudo-cubic. Above results should be compared with the case of pure lead zirconate[5].

Phase transition of $Pb(Zr85-Ti15)O_3$ was also roughly studied by the Debye photographs. In this substance, the powder reflections are very diffuse below the Curie point (pseudo-cubic) in marked contrast with the sharp lines above it. A volume contraction at the Curie point is also observed. No extra lines however found in this solid solution at any temperature

§ 6. Discussions

From the investigation already described, the crystal structure of the $PbZrO_3-PbTiO_3$ system can be divided into regions as shown in Fig 14 Some of the superstructure lines appear even in the pseudo-cubic region near the border line to the antiferroelectric region, but they disappear gradually as the temperature is raised far above the line

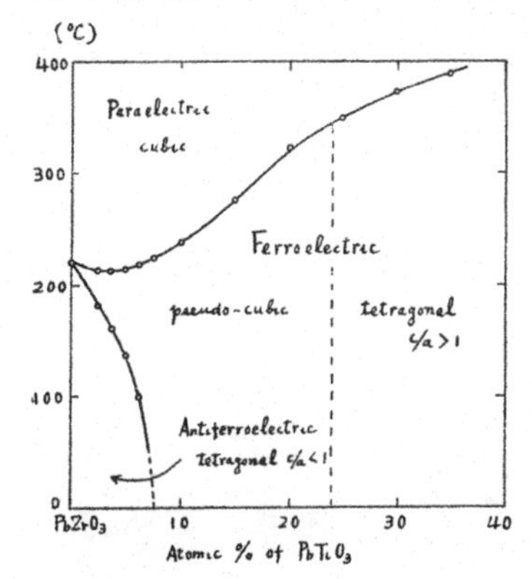

Fig 14, Phase diagram and crystal structure of the $PbZrO_3$-$PbTiO_3$ system.

As shown in part I, solid solutions in pseudo-cubic region, for example Pb(Zr90−Ti10)O_3, show clearly ferroelectric characteristics and distinguish themselves from antiferroelectric ones. Of considerable interest may be a mechanism of ferroelectricity in this pseudo-cubic crystal, especially if it were compared with that of barium titanate type ferroelectrics with a net tetragonality, $c/a>1$. At present, we are unable to detect any essential difference between the properties of pseudo-cubic region and those of tetragonal region, except small anomaly of unit cell volume found at their border line (Fig 5)

In connection with the above phenomenon, it should be emphasized that in the Pb(Zr−Ti)O_3 system, we are replacing B atoms in ABO_3 double oxide, in contrast with the cases of (Ba−Sr)TiO_3[6], (Ba−Pb)TiO_3[7] or (Pb−Sr)TiO_3[8] where A atoms are being replaced

We have assumed the lowest phase in Figs 2 and 14 to be in an antiferroelectric state Of course, there may be alternative interpretation that the lowest phase is another ferroelectric state which has very large coercive field We believe, however, our assumption of antiferroelectric state to be more reasonable and natural than the interpretation of another ferroelectric state. Of course, in order to decide the two alternatives, however, some crucial experiments, such as determination of the exact atomic positions by the X-ray analysis of $PbZrO_3$ single crystal, are necessary.

In conclusion, we wish to express our sincere thanks to Professors Y Takagi and S Miyake for their kind guidance and continued encouragement in the course of this research We are also grateful to Mr. S Hoshino and Mr. E Sawaguchi for their helpful discussions. This research was helped by the research grant of the Ministry of Education.

References

1) G Shirane and A. Takeda, J Phys Soc Japan, **7** (1952) 5.

2) H D Megaw, Proc Phys London, **58** (1946) 133

3) A Hoffmann, Z Phys Chem **B28** (1935) 65

4) S Sawada and G Shirane, J Phys Soc Japan, **4** (1949) 52

5) R Ueda and G. Shirane, J Phys Soc Japan, **6** (1951) 209.

6) D. F Rushman and M A Strivens, Trans. Faraday Soc., **42A** (1946) 231.

7) S Nomura and S Sawada, J Phys Soc Japan, **6** (1951) 36 G. Shirane and K Suzuki, ibid **6** (1951) 274.

8) S Nomura and S Sawada, ibid **5** (1950) 270.

Ferroelectricity of Glycine Sulfate

B. T. MATTHIAS, C. E. MILLER, AND J. P. REMEIKA
Bell Telephone Laboratories, Murray Hill, New Jersey
(Received August 24, 1956)

WE have discovered that glycine sulfate, $(CH_2NH_2\text{-}COOH)_3H_2SO_4$,[1] and its isomorphous selenate are ferroelectric. The Curie point of the sulfate is 47°C and that of the selenate is 22°C. As in all ferroelectrics which are not cubic above their Curie points, there is only one ferroelectric axis. For glycine sulfate at room temperature the spontaneous polarization is 2.2×10^{-6} coul/cm^2 and the coercive field is 220 v/cm.

It is tempting to regard this result as confirmation of a general mechanism operative in such ferroelectric materials as ammonium sulfate,[2] and with it the same for the guanidine aluminum sulfate and its isomorphs (GASH)[3] and some of the alums.[4]

In the infrared absorption of ammonium sulfate[5] there are two characteristic features below its Curie point. The 3.3μ band due to the NH_4^+ shows a change similar to that in the transition region of ammonium chloride. At the same time, the absorption spectrum due to the SO_4 tetrahedra shows a similar rearrangement. One could therefore assume that an ordinary ammonium transition acts as a trigger to induce a moment in the SO_4^{--} groups. The mechanism would be thus strongly reminiscent of the one causing ferroelectricity in KH_2PO_4 as described by Slater.[6] This point of view suggests a formal similarity among the ferroelectric sulfates thus far encountered regardless of crystal structure or water of crystallization. The role of ammonium is played by guanidinium in GASH, by methylamine or urea in some alums, and by glycine in the material here reported.

We want to thank A. N. Holden for his critical reading of this paper.

[1] Nicklès, Compt. rend. trav. chim. (1849).
[2] B. T. Matthias and J. P. Remeika, Phys. Rev. **103**, 262 (1956).
[3] Holden, Matthias, Merz, and Remeika, Phys. Rev. **98**, 546 (1955).
[4] Pepinsky, Jona, and Shirane, Phys. Rev. **102**, 1181 (1956).
[5] R. Pohlman, Z. Physik **79**, 394 (1932).
[6] J. Slater, J. Chem. Phys. **9**, 16 (1941).

CRYSTAL STABILITY AND THE THEORY OF FERROELECTRICITY

W. Cochran*
Atomic Energy of Canada, Limited, Chalk River, Ontario, Canada
(Received August 7, 1959; revised manuscript received October 1, 1959)

The condition that a crystal should be stable for all small deformations is that all the normal modes should have real frequencies.[1] The limit of stability against a particular mode of vibration is approached as the corresponding frequency approaches zero. In what follows we show that there is reason to believe that a ferroelectric transition, at least in certain crystals, is associated with such an instability or near-instability. The condition for a ferroelectric transition is therefore a problem in lattice dynamics, and when it is treated as such new insight into the problem is obtained. We illustrate this by means of a simple example, in which it is shown that it is possible for a diatomic cubic crystal to exhibit properties which are remarkably similar to those of barium titanate.

A shell model for the ions in an alkali halide crystal, proposed by Dick and Overhauser,[2] has been found to give a satisfactory explanation of the dielectric properties of alkali halides[3,4] and of the lattice dynamics of sodium iodide.[5] A similar model accounts quite well for the observed relation between frequency ω $(=2\pi\nu)$ and wave vector $\vec{q}$ $(q=2\pi/\lambda)$ of certain of the normal modes of germanium.[6] The theory of Woods, Cochran, and Brockhouse,[5] based on this model, gives the following expressions for the frequencies of the transverse optic (T.O.) and longitudinal optic (L.O.) modes of wave vector zero in a diatomic cubic crystal:

$$\mu\omega_T^{\ 2}=R_0'-4\pi(\epsilon+2)(Z'e)^2/9v, \tag{1}$$

$$\mu\omega_L^{\ 2}=R_0'+8\pi(\epsilon+2)(Z'e)^2/9v\epsilon. \tag{2}$$

Here μ is the reduced mass of the ions, ϵ the high-frequency dielectric constant (the square of the optical refractive index), and v the volume of the unit cell. $Z'e$ is the effective ionic charge, while $R_0'\vec{u}$ is the restoring "short-range" force on any one atom when the two lattices carrying the nuclei of different type are displaced a small relative distance $\vec{u}$. Primed quantities depend explicitly on the parameters of the shell model. In a crystal such as sodium iodide the two quantities on the right of Eq. (1) are of the same order of magnitude, but the term R_0' which arises from the short-range interaction is about twice as great as the other, which arises from the Coulomb interaction. Let us consider the situation that would arise if their difference approached zero.

It is found from the theory[5] that ω_T may approach zero without the crystal necessarily becoming unstable against other vibration modes; indeed all transverse optic modes for which $\vec{q}$ is not close to zero may retain quite usual frequencies. The quantities which appear on the right of Eq. (1) may be temperature dependent, as the lattice vibrations are in practice not completely harmonic, so that near $T = T_c$ we may postulate

$$\frac{\mu\omega_T^2}{R_0'} = 1 - \frac{4\pi(\epsilon+2)(Z'e)^2}{9vR_0'} = \gamma(T - T_c), \qquad (3)$$

where γ is a temperature coefficient and T_c is the temperature at which the crystal would become unstable. It has been shown by Lyddane, Sachs, and Teller[7] that

$$\omega_L^2/\omega_T^2 = \epsilon_0/\epsilon, \qquad (4)$$

where ϵ_0 is the static dielectric constant. Combining these equations, one finds

$$\frac{\epsilon_0 - 1}{4\pi} \simeq \frac{\epsilon_0 - \epsilon}{4\pi} = \frac{(\epsilon+2)^2(Z'e)^2}{9vR_0'\gamma(T - T_c)}, \qquad (5)$$

so that a Curie-Weiss law is followed with a Curie constant

$$C = (\epsilon+2)^2(Z'e)^2/9vR_0'\gamma \simeq (\epsilon+2)/4\pi\gamma. \qquad (6)$$

The condition $\epsilon_0 \to \infty$ is thus $\omega_T^2 \to 0$. The ionic polarizability of one unit cell is found to be given by

$$\alpha_i = (Z'e)^2/R_0', \qquad (7)$$

while the electronic polarizability is as usual given by

$$4\pi\alpha_e/3v = (\epsilon - 1)/(\epsilon + 2). \qquad (8)$$

The condition $\omega_T^2 = 0$ is therefore the same as

$$(4\pi/3v)(\alpha_i + \alpha_e) = 1, \qquad (9)$$

so that the terms "instability" and "polarizability catastrophe" are synonymous in this instance.

If one now postulates that the short-range potential between the two Bravais lattices is neither precisely harmonic nor precisely isotropic for comparatively large displacements, it can be shown that the crystal may become spontaneously polarized without becoming completely unstable. Instead it makes a transition to another phase. When the short-range potential is given explicitly by

$$V_R = \tfrac{1}{2}R_0'(u_x^2 + u_y^2 + u_z^2) + \tfrac{1}{4}B(u_x^4 + u_y^4 + u_z^4) + \tfrac{1}{6}B'(u_x^6 + u_y^6 + u_z^6) + \tfrac{1}{2}B''(u_x^2u_y^2 + u_y^2u_z^2 + u_z^2u_x^2), \qquad (10)$$

with $B < 0$, B' and $B'' > 0$, the crystal makes a first-order transition to a tetragonal phase at a temperature exceeding T_c. Equation (10) refers to the "free" crystal; the potential V_R cannot therefore be used as it stands to give ω_T in a noncubic phase, since, for the lattice vibrations, the crystal is "clamped" by its inertia. At the cubic-tetragonal transition the atoms become displaced a relative amount

$$u_0 = (3|B|/4B')^{1/2},$$

along a crystallographic axis, say [001]. The minimum T.O. frequency is reached just before the transition, and is given by

$$\mu(\omega_T^2)_{\min} = 3B^2/16B'.$$

Expressions for the spontaneous polarization as a function of temperature, etc., are also obtained in terms of the above atomic parameters. As the temperature decreases still further the crystal makes a second transition in which the atoms become relatively displaced along [011], followed by a third which leaves them displaced along [111]. These results will be understood when it is pointed out that Eqs. (3) and (10) lead eventually to an expression for the free energy G_1 of the unstressed crystal which is almost identical with that postulated by Devonshire.[8]

The theory has been extended to apply to antiferroelectric transitions in diatomic crystals, and to ferroelectric transitions in other cubic crystals, including barium titanate. The equations which apply to the latter are greatly simplified by assuming that in the T.O. mode of lowest frequency (there are three T.O. modes in all), or in a static field, the framework of oxygen atoms is not distorted. The crystal structure analyses[9,10] of tetragonal barium titanate and lead titanate support this assumption. The dielectric properties of barium titanate, and the movements of barium and titanium atoms relative to the oxygen octahedron, may be accounted for by assuming a temperature dependence of certain atomic parameters analogous to that given by Eq. (3), and a short-range po-

VOLUME 3, NUMBER 9 PHYSICAL REVIEW LETTERS NOVEMBER 1, 1959

tential for relative movement of titanium and oxygen atoms given by Eq. (10). A very small departure from a harmonic potential is found to be sufficient to account for the dielectric properties, and the numerical values required for other atomic parameters are physically reasonable.

The equation corresponding to Eq. (4), applicable to a perovskite-type crystal, is found to be

$$(\omega_2\omega_3\omega_4)_L^2/(\omega_2\omega_3\omega_4)_T^2 = \epsilon_0/\epsilon. \qquad (11)$$

(This result does not depend on the assumption of an undistorted oxygen framework.) The infrared absorption frequencies $(\omega_3)_T$ and $(\omega_4)_T$ have been measured by Last,[11] and are in no way unusual. It follows from the theory that $(\omega_2)_T^2$ should be proportional to $(T - T_c)$ in the cubic phase, and should reach an abnormally low value estimated as $\nu = 2$ or 3×10^{11} cps just before the first transition. This frequency should split appreciably in the tetragonal phase, and each frequency in this phase should vary inversely as the square root of the corresponding "clamped" dielectric constant. Although relaxation of the dielectric constant of barium titanate has been reported for frequencies of the order 10^{10} cps, Benedict and Durand[12] found that when a single crystal is used, the dielectric constant of the cubic phase is the same as the static value up to $\nu = 2.4\times10^{10}$ cps, the limit of their experiment. The resonance frequency predicted here lies in the millimeter wavelength range. A study of the properties of barium titanate in this difficult frequency range should yield interesting results.

The statement that the problem of the onset of ferroelectric properties in any crystal is a problem in lattice dynamics is probably correct, but at present there is little prospect of detailed application of the theory of lattice dynamics to low-symmetry or disordered crystals.

I am indebted to a number of colleagues, especially Dr. B. N. Brockhouse, for helpful comments on this work.

---

*On leave from Crystallographic Laboratory, Cavendish Laboratory, Cambridge, England.

[1] M. Born and K. Huang, Dynamical Theory of Crystal Lattices (Oxford University Press, Oxford, 1954).
[2] B. G. Dick and A. W. Overhauser, Phys. Rev. 112, 90 (1958).
[3] J. E. Hanlon and A. W. Lawson, Phys. Rev. 113, 472 (1959).
[4] W. Cochran, Phil. Mag. (to be published).
[5] Woods, Cochran, and Brockhouse, Bull. Am. Phys. Soc. 4, 246 (1959).
[6] W. Cochran, Phys. Rev. Letters 2, 495 (1959).
[7] Lyddane, Sachs, and Teller, Phys. Rev. 59, 673 (1941).
[8] A. F. Devonshire, Advances in Physics, edited by N. F. Mott (Taylor and Francis, Ltd., London, 1954), Vol. 3, p. 85.
[9] Shirane, Danner, and Pepinsky, Phys. Rev. 105, 856 (1957).
[10] Shirane, Pepinsky, and Frazer, Phys. Rev. 97, 1179 (1955).
[11] J. T. Last, Phys. Rev. 105, 1740 (1957).
[12] T. S. Benedict and J. L. Durand, Phys. Rev. 109, 1091 (1958).

2

Ferroelectrics 1961–2001

Acta Cryst. (1972). B28, 3384

The Classification of Tilted Octahedra in Perovskites

BY A. M. GLAZER
Crystallographic Laboratory, Cavendish Laboratory, Cambridge, England

(*Received* 4 *May* 1972)

A simple method for describing and classifying octahedral tilting in perovskites is given and it is shown how the tilts are related to the unit-cell geometries. Several examples from the literature are listed and predictions about hitherto unknown structures of some materials are made.

Introduction

The perovskite structure is very commonly found in compounds of general formula ABX_3 and many of these materials have interesting and important properties, such as ferroelectricity, piezoelectricity, non-linear optical behaviour and so on.

Fig. 1 shows the basic unit, which consists of corner-linked octahedra of X anions (usually oxygen or fluorine) with B cations at their centres and A cations between them; the cations have been left out of the diagram since, in this paper, only the octahedra will be considered. The ideal structure thus depicted is found in some materials, for example $SrTiO_3$ at room temperature; more usually the structure is modified by cation displacements as in $BaTiO_3$, or by the tilting of octahedra as in $CaTiO_3$, or by a combination of both as in $NaNbO_3(P)$. The cation displacements, which are directly linked with ferroelectricity and antiferroelectricity, are relatively simple to deal with and in any case do not directly affect the lattice parameters except by a relatively small distortion of the octahedra. The tilting of the octahedra has usually a far greater effect on lattice parameters but is more difficult to describe. However, attempts have been made to discuss these phenomena (Megaw, 1966, 1969) and it is as a result of these studies that the present work has evolved.* It is the aim of this paper to show how the various tilt systems may be classified and how they affect the crystal symmetries. Displacements of cations are not discussed here at any length. The derived results are based on the assumption that the octahedra are regular throughout. Very commonly the overall symmetry follows that of the tilts in spite of displacements and distortions; and even when it does not, the symmetry due to the tilts can be considered separately.

This classification has already proved useful in a recent study of the $T_2 \rightarrow$ cubic transition in $NaNbO_3$ (Glazer & Megaw, 1972), in which it readily suggested a likely model for the T_2 structure. More recently, it has also been successfully used in studying the $T_1 \rightarrow T_2$ transition (Ahtee, Glazer & Megaw, 1972). In connexion with phase transitions, dynamic (as opposed to static) tilting of octahedra can be interpreted in terms of lattice modes. In fact, it seems probable that the most obvious tilt configurations generally correspond to some of the most important modes associated with phase transitions in perovskites. The present classification should therefore find some use in the lattice dynamical studies of these transitions.

Effect of octahedral tilts

When an octahedron in the perovskite structure is tilted in some particular way it causes tilting of the neighbouring octahedra. However, it is, in practice, extremely difficult to visualize the total effect, and in

* A scheme very similar to the one described here has recently been derived independently by J. K. Brandon (private communication) in connexion with the structure of $Ca_2Nb_2O_7$.

any case, there are several possibilities for the final structure.

Physically it is useful to consider the tilting of an octahedron about any one of its symmetry axes. For the purposes of a general classification, however, it is preferable to consider all tilts as combinations of component tilts about the three tetrad axes.* For small angles of tilt, the component tilts can be taken to be about the pseudocubic axes prior to tilting, the magnitudes of the tilts being indicated symbolically by a set of three letters which refer to the axes in the order [100], [010], [001], and which in the general case of unequal tilts are denoted *abc*. Equality of tilts is denoted by repeating the appropriate letter, *e.g.* *aac* means equal tilts about [100] and [010] with a different tilt about [001]. We consider only those tilts that are 'freely variable' and independent.

* A note of caution must be sounded here. The three separate tilt operations about the tetrad axes do not belong to an Abelian group. In other words, the final tilt arrangement depends on the order in which the tilt operations are carried out. This does not greatly affect the arguments of the present paper, since the scheme outlined here is intended for the description of structures and not their derivation. In any case, for small tilt angles ($<15°$) the dependence of the result on the sequence of operations is only a second-order effect.

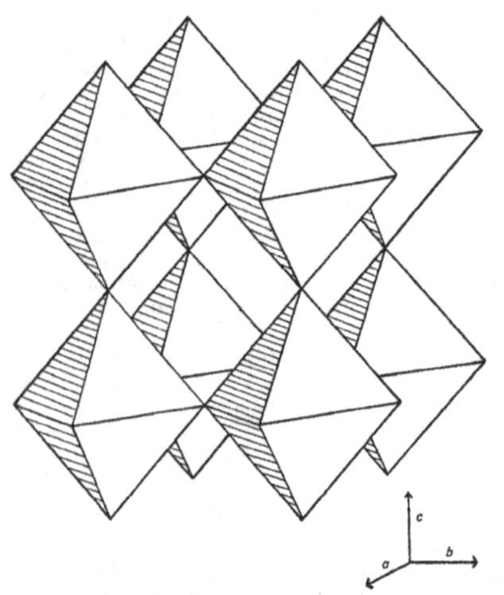

Fig. 1. Octahedral framework of cubic perovskite. The *A* cation is in the interstice formed by the anion octahedra; the *B* cation is at the centre of each octahedron.

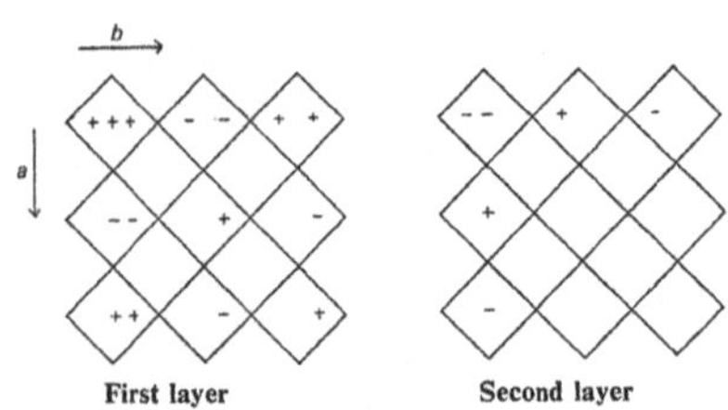

Fig. 2. Schematic diagram of two adjacent layers of octahedra. The + and − signs indicate the relative senses of tilt in the octahedra (see text).

In Fig. 2 two adjacent layers of octahedra are shown schematically. Consider the octahedron at the top-left position in the first layer and let it be tilted in an arbitrarily defined positive sense about the pseudocubic [100], [010] and [001] axes. This is indicated on the diagram by $+++$.

We must now see how this affects the surrounding octahedra. If we choose any particular axis, then in directions *perpendicular* to it, successive octahedra are constrained to have opposite tilts about that axis. Thus, the positive tilt about [100] of magnitude a in the first octahedron makes a negative tilt about [100] of equal magnitude in the nearest-neighbour octahedra along [010] and [001]. This operation can be carried through to all the octahedra and the final result is shown in this figure. It is obvious that there are some missing signs, indicating some choice in how to complete the structure. This arises because successive octahedra *along* an axis can have either the same or opposite sense of tilt. We can then indicate the particular choice by the superscripts $+$, $-$ or 0, to show whether successive octahedra along an axis have the same tilt, opposite tilt or no tilt about that axis. Assuming no repeat period consists of more than two octahedra, there are 10 distinct possibilities:

$a^+b^+c^+$	$a^+b^+c^-$	$a^+b^-c^-$	$a^-b^-c^-$	3 tilts
$a^0b^+c^+$	$a^0b^+c^-$	$a^0b^-c^-$		2 tilts
$a^0a^0c^+$	$a^0a^0c^-$			1 tilt
$a^0a^0a^0$				no tilts

where the repetition of the symbol a^0 is used when there is zero tilt about more than one axis.

In the above scheme, combinations such as $a^+a^+c^+$, $a^+a^-a^+$ *etc.* have been omitted, since, for the moment, we shall deal only with the senses of tilt. In Fig. 3 these arrangements are drawn out in full. It can be seen that each case gives a self-consistent structure, *i.e.* there are now no missing signs and each structure has a regular repeat. This simple diagram permits the lattice-centring conditions to be obtained merely by inspection. For example, $a^+b^+c^+$ is body-centred and $a^-b^-c^-$ is all-face-centred.

In order to demonstrate that all these structures are physically possible and that the predicted centring does occur, a flexible model was constructed. In Fig. 4 a series of stereo-photographs of the eight representative octahedra of each arrangement is shown. The correctness of the prediction can easily be verified. With some experience, it is also a simple matter from Fig. 4 to find other symmetry elements, such as mirror planes and rotation axes.

We must now go on to discuss the relation between the tilts and the crystal systems, and here the magnitude of the tilts becomes important. Any tilt, as defined here, results in a decrease in the distances between octahedron centres (pseudocubic subcell edges) *perpendicular* to the tilt axis. The corresponding distance *along* the tilt axis is unchanged by the tilt operation. Denoting the angles of tilt about [100], [010] and [001] by α, β and γ respectively (less than 15°, say) the pseudocubic axial lengths are given by

$$a_p=\xi \cos\beta \cos\gamma$$
$$b_p=\xi \cos\alpha \cos\gamma$$
$$c_p=\xi \cos\alpha \cos\beta$$

where ξ is the anion–anion distance through the centre of the octahedron.

These equations show that three unequal tilts produce three unequal pseudocubic spacings, two equal tilts produce two equal spacings and one different and three equal tilts produce three equal spacings. Thus

$a^ib^jc^k$ has 3 unequal spacings
$a^ib^jb^k$ has 2 equal spacings
$a^ia^ja^k$ has 3 equal spacings

where superscripts i, j, k refer to the senses of tilt about [100], [010] and [001] respectively, and may stand for +, −, or 0. It should be noted that three equal tilts correspond to tilting about one of the triad axes of the octahedron and two equal tilts correspond to tilting about one of its diad axes, provided that the tilt angles are small; or, with larger tilt angles, that the operations are carried out in an alternating sequence of small steps.

In order to deal with the interaxial angles it is useful to note that whenever a superscript is 0 or + this implies the existence of a mirror plane perpendicular to the relevant axis.

For the one-tilt systems $a^0a^0c^+$ and $a^0a^0c^-$, therefore, there are respectively three and two mutually perpendicular mirror planes and hence both tilt systems give rise to orthogonal axes; it is also obvious that a single tilt about [001] must leave a 4-fold axis and therefore both cases have tetragonal symmetry.

For the two-tilt systems it is clear that whenever the superscripts are 0++ or 0+− (in any order) the unit cell must have orthogonal axes, since the +'s and 0's necessarily imply at least two mirror planes mutually perpendicular. When two superscripts are − the problem becomes a little more difficult. Fig. 5(*a*) shows part of the unit cell of $a^0b^-c^-$. *A*, *B* and *C* are three anion atoms of one octahedron. Here *A* lies above the (001) plane through $z=0$ and *B* lies on it; *C* is at a height $z=\frac{1}{4}$. The next octahedron along [001] is denoted by *A'*, *B'* and *C'*. *A'* lies below the (001) plane through $z=\frac{1}{2}$ and *B'* lies on it. *C'* is the atom at the bottom vertex of this octahedron and must therefore be the same atom as *C*. Since the diagram has been drawn as if the c_p axis were perpendicular to (001) we see that there is a displacement vector along [010] between *C'* and *C*. Therefore in order that *C'* and *C* be the same atom the c_p axis must be inclined to b_p but must remain perpendicular to a_p. A special case arises in $a^0b^-b^-$. Here the equality of tilts about [010] and [001] implies equality between b_p and c_p, whence it is possible to construct an orthogonal unit cell by transforming from the axes $a_o=2a_p$, $b_o=2b_p$, $c_o=2c_p$ to a new set a_n, b_n, c_n defined by

$$\begin{array}{c|ccc} & a_o & b_o & c_o \\ \hline a_n & 1 & 0 & 0 \\ b_n & 0 & \frac{1}{2} & -\frac{1}{2} \\ c_n & 0 & \frac{1}{2} & \frac{1}{2} \end{array}.$$

First Layer			Second Layer			Symbol	Lattice
+++	−+−	+++	−−+	+−−	−−+		
+−−	−−+	+−−	−+−	+++	−+−	$a^-b^-c^-$	*I*
+++	−+−	+++	−−+	+−−	−−+		
+++	−+−	+++	−−−	+−+	−−−		
+−−	−−+	+−−	−++	++−	−++	a^-b^-c	*P*
+++	−+−	+++	−−−	+−+	−−−		
+++	−−−	+++	−−−	+++	−−−		
+−−	−++	+−−	−++	+−−	−++	$a^-b\,c^-$	*A*
+++	−−−	+++	−−−	+++	−−−		
+++	−−−	+++	−−−	+++	−−−		
−−−	+++	−−−	+++	−−−	+++	$a\,b\,c$	*F*
+++	−−−	+++	−−−	+++	−−−		
0++	0+−	0++	0−+	0−−	0−+		
0−−	0−+	0−−	0+−	0++	0+−	$a^0b^-c^-$	*I*
0++	0+−	0++	0−+	0−−	0−+		
0++	0+−	0++	0−−	0−+	0−−		
0−−	0−+	0−−	0++	0+−	0++	a^0b^-c	*B*
0++	0+−	0++	0−−	0−+	0−−		
0++	0−−	0++	0−−	0++	0−−		
0−−	0++	0−−	0++	0−−	0++	$a^0b\,c$	*F*
0++	0−−	0++	0−−	0++	0−−		
00+	00−	00+	00+	00−	00+		
00−	00+	00−	00−	00+	00−	$a^0a^0c^-$	*C*
00+	00−	00+	00+	00−	00+		
00+	00−	00+	00−	00+	00−		
00−	00+	00−	00+	00−	00+	$a^0a^0c^-$	*F*
00+	00−	00+	00−	00+	00−		
000	000	000	000	000	000		
000	000	000	000	000	000	$a^0a^0a^0$	*P*
000	000	000	000	000	000		

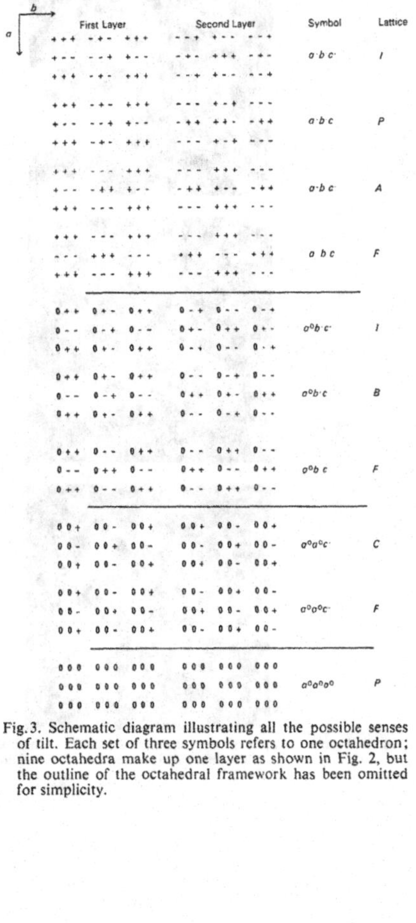

Fig. 3. Schematic diagram illustrating all the possible senses of tilt. Each set of three symbols refers to one octahedron; nine octahedra make up one layer as shown in Fig. 2, but the outline of the octahedral framework has been omitted for simplicity.

ACTA CRYSTALLOGRAPHICA, Vol. B28, 1972—Glazer PLATE 8

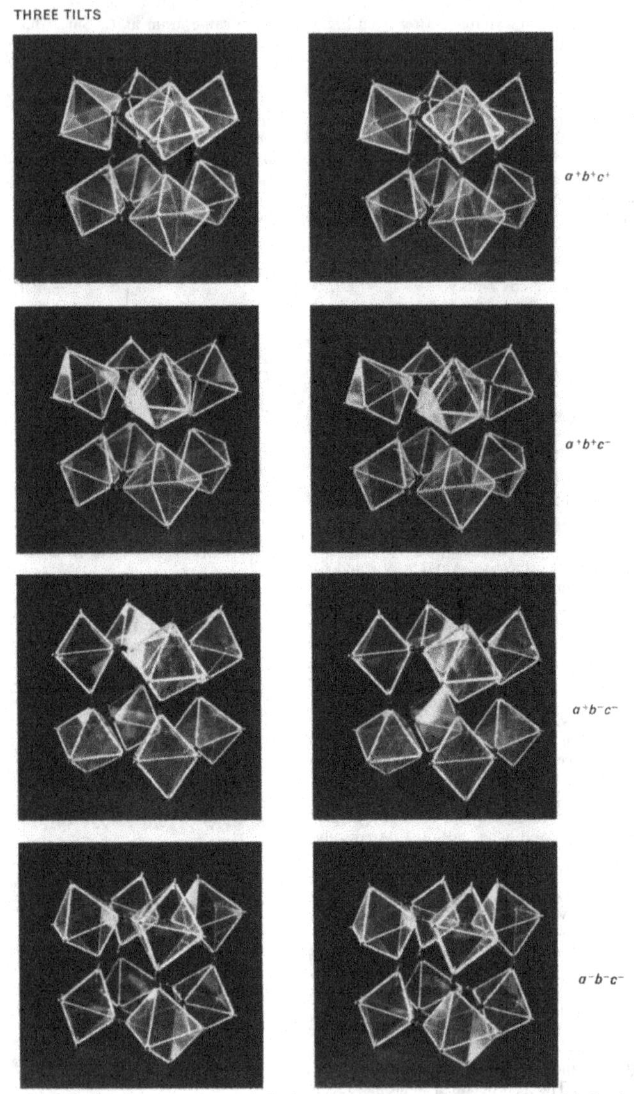

Fig. 4. Stereo-photographs of the eight representative octahedra in each arrangement. The axes used are right-handed with [001] vertical and [010] to the right. The origin is taken at the centre of any octahedron.

[*To face p.* 3386

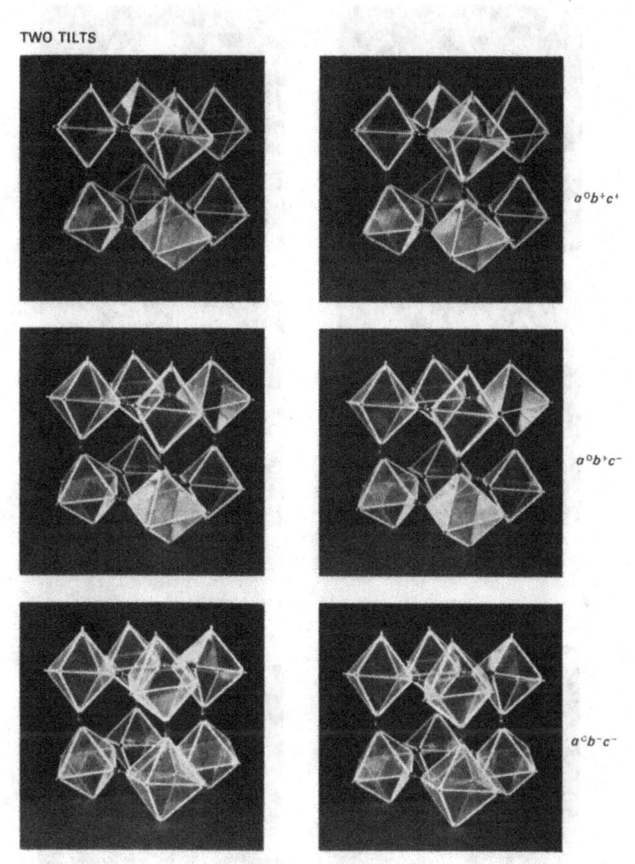

Fig. 4 (*cont.*)

ACTA CRYSTALLOGRAPHICA, VOL. B28, 1972—GLAZER PLATE 10

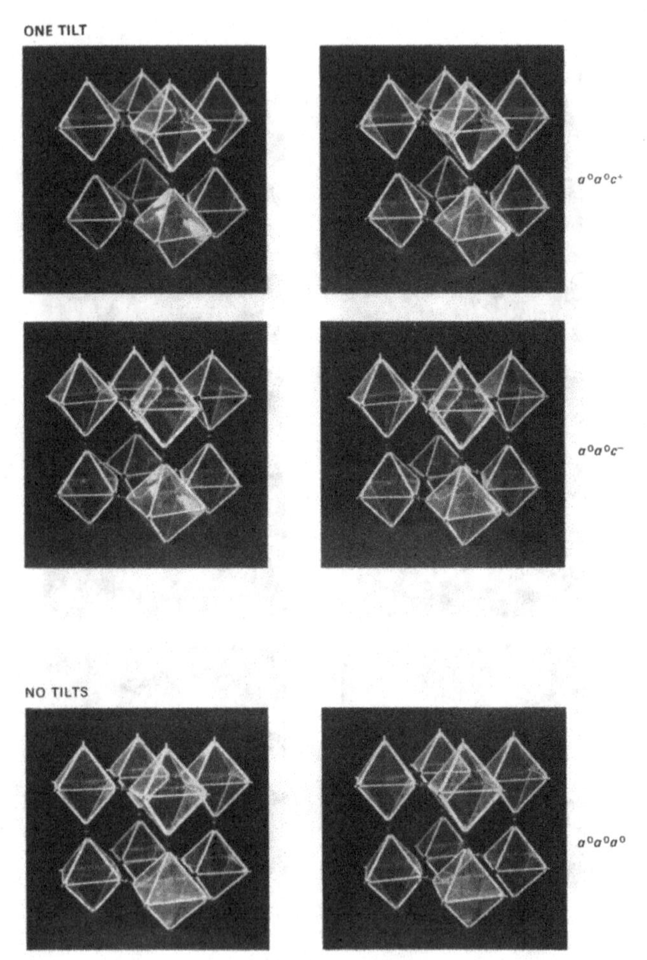

Fig. 4 (*cont.*)

For the three-tilt systems, similar arguments can be applied. Thus $a^+b^+c^+$ and $a^+b^+c^-$ are orthogonal systems, since they have at least two mutually-perpendicular mirror planes, while $a^+b^-c^-$ must have two axes b_p and c_p inclined and one (a_p) perpendicular, as in $a^0b^-c^-$. When all three superscripts are – it is therefore reasonable to expect all three axes to be inclined to one another. Fig. 5(*b*) shows the interesting case of $a^-b^-b^-$. Since the tilts about [010] and [001] are of the – type, the b_p and c_p axes are shown inclined to one another. In this case we see that the displacement vector $\overrightarrow{A'A}$ lies along [0$\bar{1}$1] and as expected all three axes are inclined to one another. In this particular example, transformation by the above matrix gives a unit cell with c_n perpendicular to a_n and b_n and with a_n and b_n inclined to one another. The special case of $a^-a^-a^-$ has three axes equal in length and equally inclined to one another. Since three equal tilts correspond to tilting of the octahedra about their triad axes, this system must be rhombohedral.

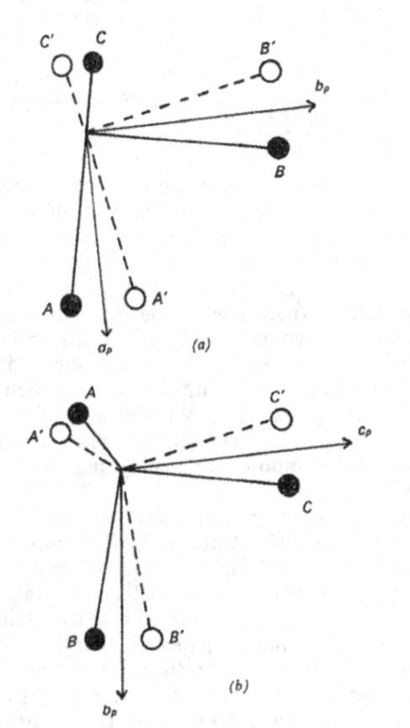

Fig. 5. (*a*) Projection on (001) of part of the unit cell of $a^0b^-c^-$, showing the anions in two successive octahedra along [001]. The relative coordinates of the anions are indicated in the key to the diagram. The vector $\overrightarrow{C'C}$ implies that c_p is inclined to (001) (see text). *z* coordinates: *A*, $+z$; *A'*, $\frac{1}{2}-z$; *B*, 0; *B'*, $\frac{1}{2}$; *C*, *C'*, $\frac{1}{4}$. (*b*) Projection on (100) of part of the unit cell of $a^-b^-b^-$, showing the anions in two successive octahedra along [100]. The vector $\overrightarrow{A'A}$ implies that a_p is inclined to (100). *x* coordinates; *A*, *A'*, $\frac{1}{4}$; *B*, $+x$; *B'*, $\frac{1}{2}-x$; *C*, $+x$; *C'*, $\frac{1}{2}-x$.

In this way it is possible to formulate some general rules to help in correlating the tilt systems with the relative lattice parameters.

(1) Equality of tilts about two or more pseudocubic axes leads to equality in the respective pseudocubic axial lengths, *e.g.* $a^+a^+a^+$ has $a_p=b_p=c_p$.

(2) If, in the symbolic notation used here, two or more superscripts are + or 0, the pseudocubic axes are orthogonal.

(3) If two, and only two, superscripts are –, then the two respective pseudocubic axes are inclined to one another whilst the third is perpendicular to them both.

(4) If all three superscripts are –, then all three pseudocubic axes are inclined to one another.

It finally remains necessary to determine the space-group symmetry of the tilt systems. The simplest way to do this is to draw a plan of the structure and then fill in the symmetry elements. As an example, Fig. 6 shows two (001) layers of the system $a^-b^+c^+$ and the presence of mirror and *n*-glide planes shows that its space group is *Pnmm* (No. 63).

In Table 1, the resulting tilt systems have been summarized together with the relevant symmetry information. The different tilt systems are numbered serially from 1 to 23. The lattice-centring refers to a unit cell $a_o=ma_p$, $b_o=nb_p$, $c_o=qc_p$ as given in the fourth column. The space-group symbol in each case refers to the axes a_o, b_o, c_o except where this would give lower symmetry than the true crystal symmetry. In these cases, indicated by an asterisk, the true space group has been given, the axes a_n, b_n, c_n being defined by the matrix given above.

This Table enables many of the structures to be determined uniquely from X-ray evidence. For example, the systems $a^+b^-b^-$, $a^+a^-a^-$ and $a^0b^-b^-$, which are orthorhombic, are all distinguishable from one another either because of their pseudocubic axial lengths or because of their lattice-centring when referred to the pseudocubic multiple-cell axes a_o, b_o, c_o. Similarly, the three tetragonal systems $a^0b^+b^+$, $a^0a^0c^+$ and $a^0a^0c^-$ are distinguishable from one another; use of this was made in the derivation of the $NaNbO_3(T_2)$ structure (Glazer & Megaw, 1972).

Experimental determination of the lattice-centring can be difficult, since the X-ray reflexions that characterize it are very weak in intensity. Because the tilts produce doubling of the pseudocubic axial lengths, these reflexions are found on half-integral reciprocal lattice planes, if the original subcell axes are used. Unless specifically looked for, they can easily be missed, particularly in studies on powdered materials. Nevertheless, from the observation of very few half-integral reflexions it is possible to derive a trial model for the structure.

The scheme of simple tilt systems outlined so far has relied on the condition that no repeat period consists of more than two octahedra. It is obvious that more complicated compound tilt systems can be built up by stacking the simple systems together in an infinite variety of ways. In practice only two different compound tilt systems have been observed up till now, the stacking being found to occur along one crystallographic direction, [010]. For this reason we consider only this kind of compound tilt, the others being beyond the scope of the present work. The structures may best be described by considering them in terms of successive pairs of octahedron layers, each sharing a layer in common, or single layers when the tilt about [010] is zero.

In one of the two cases the repeat period along [010] is $4b_p$. Numbering the layers 1 to 4, the structure is described by $a^-b^+a^-$ for layers 1 and 2, $a^-b^-a^-$ for layers 2 and 3, $a^-b^+a^-$ for layers 3 and 4. This can be written thus

$$(a^-b^+a^-)^2_1 \quad (a^-b^-a^-)^3_2 \quad (a^-b^+a^-)^4_3 .$$

In the other case the repeat period along [010] is $6b_p$. Here the structure consists of $a^-b^+c^+$ for layers 1 and 2, $a^-b^0c^+$ for layer 3, $a^-b^+c^+$ for layers 4 and 5, $a^-b^0c^+$ for layer 6, thus:

$$(a^-b^+c^+)^2_1 \quad (a^-b^0c^+)^3_3 \quad (a^-b^+c^+)^5_4 \quad (a^-b^0c^+)^6_6 .$$

The actual multiplicities of $4b_p$ and $6b_p$ in these cases are consequences of the tilts about [100] and [001].

Symmetry-determination for compound tilt systems is more difficult than for the simple systems and will not be attempted here. In any case, the known examples of such tilt systems ($NaNbO_3$, phases *P* and *R*) also have cation displacements to further complicate matters. In fact, it has only been possible to determine their structures by carrying out a complete refinement using many reflexions (Sakowski-Cowley, Łukaszewicz & Megaw, 1969; Sakowski-Cowley, 1967).

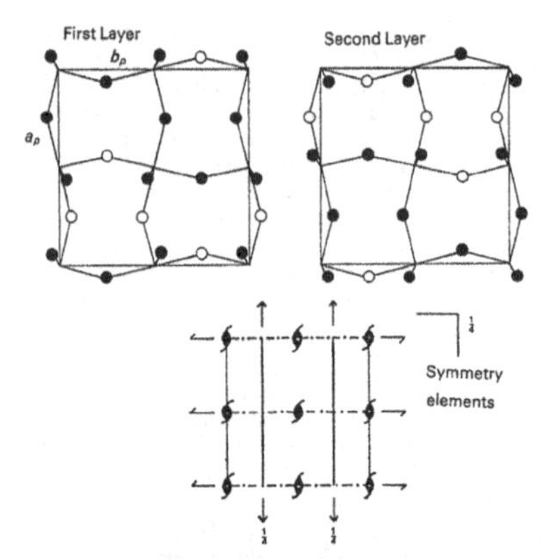

Fig. 6. Plan of two (001) layers of octahedra for the tilt system (4) $a^-b^+c^+$. Closed and open circles indicate whether the anions are above or below the planes through the centres of the octahedra; the centres of the octahedra of the first layer are at height 0 and those of the second layer at height $\frac{1}{2}$. The symmetry elements, shown below the plan, indicate that the space group is *Pnmm*.

Examples of tilt systems

A search of the literature has revealed many perovskite structures with octahedral tilts and these are summarized in Table 2. The assignment of the tilt systems was made usually by reference to the published atomic coordinates supplied by the various authors. Sometimes, as in $AgNbO_3$ for example, the structures have not actually been determined except by analogy with already known structures; as there is no reason to doubt their correctness, they too have been included in the Table, but are marked with an asterisk.

In Table 2 the materials are arranged according to tilt system, together with an indication of which cations, if any, are displaced. The observed space groups are also given for comparison with the space groups that would be expected from the anion framework alone.

It is immediately apparent that the majority of tilted structures belong to the systems (10) $a^-b^+a^-$ and (14) $a^-a^-a^-$. It is not clear why this should be; in fact, it is surprising that out of the 23 possible tilt systems only 9 have been found to occur. It would be extremely interesting to find some of the other structures, in particular the cubic (3) $a^+a^+a^+$ and the tetragonal (16) $a^0b^+b^+$ with axial ratio less than unity.

The differences between the observed space groups and those due to the tilted framework alone arise out of the cation displacements. For example, we find that there are two space groups for $a^-a^-a^-$ materials; when there are no *B* cation displacements the space group is $R\bar{3}c$, whereas when the *B* cation is displaced the centre of symmetry is lost and the space group becomes *R3c*. Again, in $PrAlO_3$ below 135°K, the observed space group (with reference to pseudocubic axial directions) is $F\bar{1}$ as opposed to $F12/m1$ for the framework alone. Here the Pr displacements are responsible for imposing triclinic symmetry on a monoclinic framework.

One example of an untilted structure has been given in Table 2, namely $KCuF_3$, since it sounds a note of caution. In this material the observed space group is *F4/mmc*, as in $a^0a^0c^-$, and yet the axial ratio is less than 1. It is therefore very difficult to ascertain by the arguments of the present paper whether the structure is one with no tilts plus *large* distortion of the octahedra or whether it is one with tilts as in $a^0a^0c^-$ plus *small* distortions. In fact it is known to be the former; this is hardly surprising in view of the large Jahn–Teller effects that are normally shown by copper compounds.

Distortions of the octahedra, then, can lead to ambiguities, although it appears that in the majority of cases they do not. Because of them the actual magnitudes of the lattice parameters may not lead to an accurate measure of the tilts, although there is evidence that they do, at least, provide a reasonable estimate of them (Megaw, 1971, with reference to the rare-earth orthoferrites). In $LaAlO_3$, which is $a^-a^-a^-$, the flattening of the oxygen octahedra (Megaw, 1971) actually leads to a rhombohedral angle greater than 90°. Nevertheless, this still does not affect the nature of the tilt system: the lattice-centring conditions and relative pseudo-cubic lattice parameters would still lead to the correct assignment of $a^-a^-a^-$.

$SrTiO_3$ and $KMnF_3$ possess the $a^0a^0c^-$ structure and this has been shown to be associated with the condensation of a Γ_{25} soft-phonon mode at $q=\frac{1}{2},\frac{1}{2},\frac{1}{2}$ on transforming from the cubic phase (Shirane & Yamada, 1969; Shirane, Minkiewicz & Linz, 1970). This mode has components consisting of rigid oscillations of the anion octahedra about [001] with successive octahedra along this direction oscillating in antiphase. It is for this reason that, when the mode condenses at the transition temperature, $a^0a^0c^-$ results.

On the other hand, $NaNbO_3(T_2)$ possesses the $a^0a^0c^+$ structure (determined with the present technique, using three reflexions), and condensation of an M_3 soft-phonon mode at $q=\frac{1}{2},\frac{1}{2},0$ has been postulated to explain the transition from the cubic phase (Glazer & Megaw, 1972). This mode consists of successive octahedra along [001] oscillating in phase about this direction, thus giving rise to the $a^0a^0c^+$ structure on condensation.

More recent work on the T_1 phase of $NaNbO_3$ (Ahtee, Glazer & Megaw, 1972) showed that its structure could be described by $a^-b^0c^+$, and the transition from T_2 to T_1 has been explained in terms of condensation of a Γ_{25} mode. The two transitions can then be summarized thus:

$$a^0a^0a^0\ (\text{cubic}) \xrightarrow{M_3} a^0a^0c^+\ (T_2) \xrightarrow{\Gamma_{25}} a^-b^0c^+\ (T_1)$$

and we see that condensation of soft-modes produces a sequence of transitions to give the scheme: zero-tilt system → one-tilt system → two-tilt system.

The simple tilt systems derived in this paper are therefore of importance in the discussion of soft-phonon modes in connexion with phase transitions in perovskites, since rigid oscillations of the octahedra constitute important modes near the transitions.

As mentioned earlier, there are not many known examples of compound tilts. Actually, only $NaNbO_3$ is definitely known to possess such complicated tilt systems, although it is very likely that $AgNbO_3$ is isomorphous with it. The evidence for this (Francombe & Lewis, 1958) is based on the presence of superlattice lines corresponding to a unit cell of $2a_p \times 4b_p \times 2c_p$, with $a_p=c_p$ and $\beta>90°$, as is found in $NaNbO_3(P)$.

Table 1. *Complete list of possible simple tilt systems*

Serial number	Symbol	Lattice centring	Multiple cell	Relative pseudocubic subcell parameters	Space group
3-tilt systems					
(1)	$a^+b^+c^+$	*I*	$2a_p \times 2b_p \times 2c_p$	$a_p \neq b_p \neq c_p$	*Immm* (No. 71)
(2)	$a^+b^+b^+$	*I*		$a_p \neq b_p = c_p$	*Immm* (No. 71)
(3)	$a^+a^+a^+$	*I*		$a_p = b_p = c_p$	*Im*3 (No. 204)
(4)	$a^+b^+c^-$	*P*		$a_p \neq b_p \neq c_p$	*Pmmn* (No. 59)
(5)	$a^+a^+c^-$	*P*		$a_p = b_p \neq c_p$	*Pmmn* (No. 59)
(6)	$a^+b^+b^-$	*P*		$a_p \neq b_p = c_p$	*Pmmn* (No. 59)
(7)	$a^+a^+a^-$	*P*		$a_p = b_p = c_p$	*Pmmn* (No. 59)
(8)	$a^+b^-c^-$	*A*		$a_p \neq b_p \neq c_p \alpha \neq 90°$	$A2_1/m11$ (No. 11)
(9)	$a^+a^-c^-$	*A*		$a_p = b_p \neq c_p \alpha \neq 90°$	$A2_1/m11$ (No. 11)
(10)	$a^+b^-b^-$	*A*		$a_p \neq b_p = c_p \alpha \neq 90°$	*Pnma* (No. 62)*
(11)	$a^+a^-a^-$	*A*		$a_p = b_p = c_p \alpha \neq 90°$	*Pnma* (No. 62)*
(12)	$a^-b^-c^-$	*F*		$a_p \neq b_p \neq c_p \alpha \neq \beta \neq \gamma \neq 90°$	$F\bar{1}$ (No. 2)
(13)	$a^-b^-b^-$	*F*		$a_p \neq b_p = c_p \alpha \neq \beta \neq \gamma \neq 90°$	*I*2/*a* (No. 15)*
(14)	$a^-a^-a^-$	*F*		$a_p = b_p = c_p \alpha = \beta = \gamma \neq 90°$	$R\bar{3}c$ (No. 167)
2-tilt systems					
(15)	$a^0b^+c^+$	*I*	$2a_p \times 2b_p \times 2c_p$	$a_p < b_p \neq c_p$	*Immm* (No. 71)
(16)	$a^0b^+b^+$	*I*		$a_p < b_p = c_p$	*I*4/*m* (No. 78)
(17)	$a^0b^+c^-$	*B*		$a_p < b_p \neq c_p$	*Bmmb* (No. 63)
(18)	$a^0b^+b^-$	*B*		$a_p < b_p = c_p$	*Bmmb* (No. 63)
(19)	$a^0b^-c^-$	*F*		$a_p < b_p \neq c_p \alpha \neq 90°$	*F*2/*m*11 (No. 12)
(20)	$a^0b^-b^-$	*F*		$a_p < b_p = c_p \alpha \neq 90°$	*Imcm* (No. 74)*
1-tilt systems					
(21)	$a^0a^0c^+$	*C*	$2a_p \times 2b_p \times c_p$	$a_p = b_p < c_p$	*C*4/*mmb* (No. 127)
(22)	$a^0a^0c^-$	*F*	$2a_p \times 2b_p \times 2c_p$	$a_p = b_p < c_p$	*F*4/*mmc* (No. 140)
Zero-tilt system					
(23)	$a^0a^0a^0$	*P*	$a_p \times b_p \times c_p$	$a_p = b_p = c_p$	*Pm*3*m* (No. 221)

* These space group symbols refer to axes chosen according to the matrix transformation

$$\begin{pmatrix} 1 & 0 & 0 \\ 0 & \frac{1}{2} & -\frac{1}{2} \\ 0 & \frac{1}{2} & \frac{1}{2} \end{pmatrix}.$$

Table 2. *Examples of known tilt systems*

Tilt system	Substance	Cation displacements	Observed space group	Space group of tilted framework alone	Reference
(4) $a^-b^+c^+$	$NaNbO_3(S)$	Na?	*Pnmm*	*Pnmm*	Ahtee, Glazer & Megaw (1972)
(10) $a^-b^+a^-$	$YAlO_3$	Y	*Pbnm*	*Pbnm*	Geller & Wood (1956)
	$SmAlO_3$	Sm	*Pbnm*	*Pbnm*	Marezio, Dernier & Remeika (1972)
	$EuAlO_3$*, $GdAlO_3$*	Eu?, Gd?	*Pbnm*	*Pbnm*	Geller & Bala (1956)
	$DyAlO_3$	Dy	*Pbnm*	*Pbnm*	Bidaux & Mériel (1968)
	$BaCeO_3$	Ba	*Pbnm*	*Pbnm*	Jacobson, Tofield & Fender (1972)
	$YCrO_3$, $YFeO_3$, $LaFeO_3$, $PrFeO_3$, $NdFeO_3$, $SmFeO_3$, $EuFeO_3$, $GdFeO_3$	Y, Y, La, Pr, Nd, Sm, Eu, Gd	*Pbnm*	*Pbnm*	Geller & Wood (1956)
	$TbFeO_3$, $DyFeO_3$, $HoFeO_3$, $ErFeO_3$, $TmFeO_3$, $YbFeO_3$, $LuFeO_3$	Tb, Dy, Ho, Er, Tm, Yb, Lu	*Pbnm*	*Pbnm*	Marezio, Remeika & Dernier (1970)
	$NaMgF_3$(<760°C)*		*Pbnm*	*Pbnm*	Chao, Evans, Skinner & Milton (1961)
	$YNiO_3$, $SmNiO_3$, $EuNiO_3$, $GdNiO_3$, $DyNiO_3$, $HoNiO_3$, $ErNiO_3$, $TmNiO_3$, $YbNiO_3$, $LuNiO_3$	Y, Sm, Eu, Gd, Dy, Ho, Er, Tm, Yb, Lu	*Pbnm*	*Pbnm*	Demazeau, Marbeuf, Pouchard & Hagenmuller (1971)
	$BaPrO_3$	Ba	*Pbnm*	*Pbnm*	Jacobson, Tofield & Fender (1972)
	$CaTiO_3$		*Pbnm*	*Pbnm*	Kay & Bailey (1957)
	$CdTiO_3$	Cd, Ti	$Pbn2_1$	*Pbnm*	Kay & Miles (1957)
(14) $a^-a^-a^-$	$LaAlO_3$	La	$R\bar{3}c$	$R\bar{3}c$	Derighetti, Drumheller, Laves, Müller & Waldner (1965); de Rango, Tsoucaris & Zelwer (1966)
	$PrAlO_3$(172–293°K)		$R\bar{3}c$	$R\bar{3}c$	Burbank (1970)
	$NdAlO_3$		$R\bar{3}c$	$R\bar{3}c$	Derighetti, Drumheller, Laves, Müller & Waldner (1965); Marezio, Dernier & Remeika (1972)
	$LaCoO_3$		$R\bar{3}c$	$R\bar{3}c$	Menyuk, Dwight & Raccah (1967)
	FeF_3, CoF_3, RuF_3, RhF_3, PdF_3, IrF_3		$R\bar{3}c$	$R\bar{3}c$	Hepworth, Jack, Peacock & Westland (1957)
	VF_3		$R\bar{3}c$	$R\bar{3}c$	Jack & Gutmann (1951)
	$BiFeO_3$	Bi, Fe	$R3c$	$R\bar{3}c$	Michel, Moreau, Achenbach, Gerson & James (1969*a*)
	$LiNbO_3$	Li, Nb	$R3c$	$R\bar{3}c$	Abrahams, Reddy & Bernstein (1966); Megaw (1968)
	$NaNbO_3(N)$	Na, Nb	$R3c$	$R\bar{3}c$	Darlington (1971)
	$LiTaO_3$	Li, Ta	$R3c$	$R\bar{3}c$	Abrahams & Bernstein (1967)
	$BaTbO_3$		$R\bar{3}c$	$R\bar{3}c$	Jacobson, Tofield & Fender (1972)
	$PbZr_{0.9}Ti_{0.1}O_3$	Pb, (Zr, Ti)	$R3c$	$R\bar{3}c$	Michel, Moreau, Achenbach, Gerson & James (1969*b*)

Table 2 (*cont.*)

(17)	$a^-b^0c^+$	$NaNbO_3(T_1)$	Na?	*Ccmm*	*Ccmm*	Ahtee, Glazer & Megaw (1972)
(19)	$a^-b^0c^-$	$PrAlO_3(<135°K)$	Pr	$F\bar{1}$	*F*12/*m*1	Burbank (1970)
(20)	$a^-b^0a^-$	$PrAlO_3(135–172°K)$	Pr	*I*12/*m*1	*Icmm*	Burbank (1970)
(21)	$a^0a^0c^+$	$NaNbO_3(T_2)$		*C*4/*mmb*	*C*4/*mmb*	Glazer & Megaw (1972)
(22)	$a^0a^0c^-$	$KMnF_3(88–184°K)$		*F*4/*mmc*	*F*4/*mmc*	Minkiewicz, Fujii & Yamada (1970)
		$SrTiO_3(<110°K)$		*F*4/*mmc*	*F*4/*mmc*	Unoki & Sakudo (1967)
(23)	$a^0a^0a^0$	$KCuF_3$		*F*4/*mmc*	*Pm*3*m*	Okazaki & Suemune (1961)
Compound tilts						
(10) (13)	$(a^-b^+a^-)_1^2(a^-b^-a^-)_2^3(a^-b^+a^-)_3^4$	$AgNbO_3(<325°C)$*	Ag, Nb	Not given		Francombe & Lewis (1958)
		$NaNbO_3(P)$	Na, Nb	*Pbma*		Sakowski-Cowley, Łukaszewicz & Megaw (1969)
		$AgTaO_3(<370°C)$*	Ag? Ta?	Not given		Francombe & Lewis (1958)
(4) (17)	$(a^-b^+c^+)_1^2(a^-b^0c^+)_3^3(a^-b^+c^+)_4^5(a^-b^0c^+)_6$	$NaNbO_3(R)$	Na, Nb	*Pmnm*		Sakowski-Cowley (1967)

* Structures that have been determined by analogy with alreadyknown structures and for which no atomic coordinates have been given.

In the case of $AgTaO_3$, the evidence for the compound system is even more tenous since there has been no observation of the necessary superlattice lines.

It is tempting to make some predictions about the nature of the tilts in hitherto undetermined structures by making use of the known lattice parameters and symmetries; examples are shown in Table 3.

$SrZrO_3$ below 730°C has pseudocubic subcell parameters $a_p=c_p>b_p, \beta\neq 90°$ and is consistent, therefore, with $a^-b^+a^-$ or $a^-b^0a^-$. If the lattice type were known it would be possible to decide between these two possibilities, since they are *B*-face-centred and all-face-centred respectively; the former structure is the most likely in view of the high frequency with which it is found in practice. Between 730 and 860°C the unit cell is tetragonal with axial ratio less than 1, indicating that the structure may be $a^+a^+c^0$. Similar considerations apply to $NaMgF_3$ between 760 and 900°C, to $KCoF_3$ at 78°K and to $RbCoF_3$ below 101°K.* Between 860 and 1170°C, $SrZrO_3$ is tetragonal with axial ratio greater than 1 and therefore may be $a^0a^0c^+$ or $a^0a^0c^-$. The presence or absence of difference reflexions on half-integral reciprocal lattice planes has not been explicitly reported in these compounds. The predictions, therefore, must be viewed with caution, since in making them it has been assumed that there are no distortions of the octahedra.

It is clear that, when studying the structures of perovskites, a special effort must be made to record the very weak difference reflexions found on half-integral reciprocal lattice planes. Very often in the past, the perovskite structures have been 'determined' by observing that the X-ray powder patterns are very similar to those of already-known structures. The present study has shown that this is unreliable and some of the structures quoted in Table 3, therefore, may be subject to reconsideration. It is to be hoped that the interpretation of future studies of these materials will be made more reliable and simpler by the use of the scheme reported here.

I wish to thank Dr Helen D. Megaw for introducing me to the problem of octahedral tilts and for her constant inspiration. Also, I gratefully acknowledge a grant from the Science Research Council which made this work possible.

* One must bear in mind that the original workers with these materials have assigned tetragonal symmetry to these phases from the measured lattice parameters and not from direct observations of symmetry. This means that the structures could, in fact, be orthorhombic with $a_p=b_p\neq c_p$ and would be then consistent with $a^+a^+c^+$, $a^+a^+c^-$, $a^+a^-c^+$ or $a^+a^-c^0$.

References

Abrahams, S. C. & Bernstein, J. L. (1967). *J. Phys. Chem. Solids*, **28**, 1685.

Abrahams, S. C., Reddy, J. M. & Bernstein, J. L. (1966). *J. Phys. Chem. Solids*, **27**, 997.

Table 3. *Tilt systems suggested by known lattice parameters (and symmetry, when known)*

Tilt system	Substance	Observed space group	Reference
(10) $a^-b^+a^-$	$BaPbO_3$, $SrPbO_3$	?	Shuvaeva & Fesenko (1970)
	$PrRhO_3$, $NdRhO_3$, $SmRhO_3$, $EuRhO_3$, $GdRhO_3$, $TbRhO_3$, $DyRhO_3$, $HoRhO_3$, $ErRhO_3$, $TmRhO_3$, $LuRhO_3$	*Pbnm*	Shannon (1970)
	$SrZrO_3$(< 730°C)	?	Carlsson (1967)
(14) $a^-a^-a^-$	$KFeF_3$(78°K)	?	Okazaki, Suemune & Fuchikami (1959)
	$LaNiO_3$	$R\bar{3}c$	Demazeau, Marbeuf, Pouchard & Hagenmuller (1971)
(16) $a^+a^+c^0$	$KCoF_3$(78°K)	?	Okazaki, Suemune & Fuchikami (1959)
	$RbCoF_3$(< 101°K)	?	Nouet, Kleinberger & de Kouchkovsky (1969)
	$NaMgF_3$(760–900°C)	?	Chao, Evans, Skinner & Milton (1961)
	$SrZrO_3$(730–860°C)	?	Carlsson (1967)
(21) $a^0a^0c^+$ or (22) $a^0a^0c^-$	$SrZrO_3$(860–1170°C)	?	Carlsson (1967)

AHTEE, M., GLAZER, A. M. & MEGAW, H. D. (1972). Abstracts of the IXth International Congress of Crystallography, Kyoto, Japan; *Phil. Mag.* **26**, 995.

BIDAUX, R. & MÉRIEL, P. (1968). *J. Phys. Radium*, **29**, 220.

BURBANK, R. D. (1970). *J. Appl. Cryst.* **3**, 112.

CARLSSON, L. (1967). *Acta Cryst.* **23**, 901.

CHAO, E. C. T., EVANS, H. T., SKINNER, B. J. & MILTON, C. (1961). *Amer. Min.* **46**, 379.

DARLINGTON, C. N. W. (1971). Thesis, Univ. of Cambridge.

DEMAZEAU, G., MARBEUF, A., POUCHARD, M. & HAGENMULLER, P. (1971). *J. Solid. State Chem.* **3**, 582.

DERIGHETTI, B., DRUMHELLER, J. E., LAVES, F., MÜLLER, K. A. & WALDNER, F. (1965). *Acta Cryst.* **18**, 557.

FRANCOMBE, M. H. & LEWIS, B. (1958). *Acta Cryst.* **11**, 175.

GELLER, S. & BALA, V. B. (1956). *Acta Cryst.* **9**, 1019.

GELLER, S. & WOOD, E. A. (1956). *Acta Cryst.* **9**, 563.

GLAZER, A. M. & MEGAW, H. D. (1972). *Phil. Mag.* **25**, 1119.

HEPWORTH, M. A., JACK, K. H., PEACOCK, R. D. & WESTLAND, G. J. (1957). *Acta Cryst.* **10**, 63.

JACK, K. H. & GUTMANN, V. (1951). *Acta Cryst.* **4**, 246.

JACOBSON, A. J., TOFIELD, B. C. & FENDER, B. E. F. (1972). *Acta Cryst.* **B28**, 956.

KAY, H. F. & BAILEY, P. C. (1957). *Acta Cryst.* **10**, 219.

KAY, H. F. & MILES, J. L. (1957). *Acta Cryst.* **10**, 213.

MAREZIO, M., DERNIER, P. D. & REMEIKA, J. P. (1972). *J. Solid State Chem.* **4**, 11.

MAREZIO, M., REMEIKA, J. P. & DERNIER, P. D. (1970). *Acta Cryst.* **B26**, 2008.

MEGAW, H. D. (1966). *Proceedings of the International Meeting on Ferroelectricity*, Prague, Vol. 1.

MEGAW, H. D. (1968). *Acta Cryst.* **A24**, 583.

MEGAW, H. D. (1969). *Proceedings of the European Meeting on Ferroelectricity*, Saarbrücken.

MEGAW, H. D. (1971). Abstracts on the Second European Meeting on Ferroelectricity, Dijon; (1972). *J. Phys. Radium*, **33**, C2·1–C2·5.

MENYUK, N., DWIGHT, K. & RACCAH, P. M. (1967). *J. Phys. Chem. Solids*, **28**, 549.

MICHEL, C., MOREAU, J.-M., ACHENBACH, G., GERSON, R. & JAMES, W. J. (1969*a*). *Solid State Commun.* **7**, 701.

MICHEL, C., MOREAU, J.-M., ACHENBACH, G., GERSON, R. & JAMES, W. J. (1969*b*). *Solid. State Commun.* **7**, 865.

MINKIEWICZ, V. J., FUJII, Y. & YAMADA, Y. (1970). *J. Phys. Soc. Japan*, **28**, 443.

NOUET, J., KLEINBERGER, R. & DE KOUCHKOVSKY, R. (1969). *C. R. Acad. Sci. Paris*, **B269**, 986.

OKAZAKI, A. & SUEMUNE, Y. (1961). *J. Phys. Soc. Japan*, **16**, 176.

OKAZAKI, A., SUEMUNE, Y. & FUCHIKAMI, TS. (1959). *J. Phys. Soc. Japan*, **14**, 1823.

RANGO, C. DE, TSOUCARIS, G. & ZELWER, C. (1966). *Acta Cryst.* **20**, 590.

SAKOWSKI-COWLEY, A. C. (1967). Thesis, Univ. of Cambridge.

SAKOWSKI-COWLEY, A. C., ŁUKASZEWICZ, K. & MEGAW, H. D. (1969). *Acta Cryst.* **B25**, 851.

SHANNON, R. D. (1970). *Acta Cryst.* **B26**, 447.

SHUVAEVA, E. T. & FESENKO, E. G. (1970). *Kristallografiya*, **15**, 379.

SHIRANE, G., MINKIEWICZ, V. J. & LINZ, A. (1970). *Solid State Commun.* **8**, 1941.

SHIRANE, G. & YAMADA, Y. (1969). *Phys. Rev.* **177**, 858.

UNOKI, H. & SAKUDO, T. (1967). *J. Phys. Soc. Japan*, **23**, 546.

Simple Ways of Determining Perovskite Structures

By A. M. Glazer

Wolfson Unit for the Study of Phase Transitions in Dielectrics, Cavendish Laboratory, Cambridge, England

(*Received* 2 *May* 1975; *accepted* 9 *May* 1975)

A simple technique is described for ascertaining trial models for the structures of perovskites. The method relies on an understanding of the fundamental components of the structure. Rules are given for determining trial models rapidly.

Introduction

The class of materials known as perovskites is of considerable technological importance, particularly with regard to physical properties such as pyro- and piezo-electricity, dielectric susceptibility, linear and non-linear electrooptic effects. Many of these properties are gross effects, varying enormously from one perovskite to another, and yet the differences in the crystal structures are hardly apparent. The changes in physical properties are particularly large when the external conditions, such as temperature or pressure, are altered. The huge rise in dielectric constant, by as much as a factor of 10^4, found on heating barium titanate is a well known example of this. Generally speaking, such effects occur in connexion with the simultaneous presence of phase transitions in the system, where the atomic structure of the perovskite changes either discontinuously or continuously into another form. In order to be able to understand the origin of the behaviour of the physical properties near phase transitions it is necessary to have as complete a description as possible of what is happening to the atoms in the structure. In the case of the perovskites, because the structural differences between one phase and another are so slight, it can be extremely difficult to carry out a precise structure determination. To the casual worker all perovskites look the same; it is only by a careful study of the splittings of certain X-ray diffraction spots or lines and of the presence of any weak 'extra' (often called difference or superlattice) reflexions, that one can derive a model for the crystal structure. This can be particularly difficult when the splittings of the originally cubic reflexions are small and the difference reflexions arise from weak scatterers. For example, polycrystalline specimens of $PbZr_{0.9}Ti_{0.1}O_3$ show a very small splitting of the powder lines consistent with a rhombohedral distortion of a cube but no difference reflexions (except with neutron diffraction); a single crystal oscillating through 2° for 20 h has been used in our laboratory in order to show their presence. Clearly, such small effects can easily be missed and this accounts for the many incorrect structure determinations of perovskites carried out in the past.

The purpose of this paper is to set out in one place a general method for obtaining a suitable trial model of a particular perovskite, a 'recipe' in effect. The method outlined here will not be suitable for *all* perovskites (which have a habit of doing the unexpected) but it is certainly applicable to a great many of them. The technique has been evolved through many years of experience and owes a great deal to the work of Megaw (1966, 1969, 1971, 1973), who first showed how the structures could be broken up into their various components. These components are:

(i) tilting of the anion octahedra

(ii) displacements of the cations; these can be parallel (ferroelectric structure) or antiparallel (antiferroelectric structure).

(iii) distortions of the octahedra.

These components are also the important parameters involved in the soft modes often found near the phase transitions. Component (iii) is usually associated with component (ii) and so will not be discussed separately here; in any case, distortions are normally of a second-

Fig. 1. The ideal cubic (aristotype) perovskite of formula ABX_3 (A, B = cation, X = anion). The anions are at the vertices of the octahedra. Black circles B cations, hatched circle A cation (taken from Megaw, 1973).

order nature. As has been shown earlier (Glazer, 1972), component (i), when it is present, is the most important in establishing the overall space-group symmetry of the particular perovskite.

If the actual kind of tilting can be established then one is well on the way to understanding the diffraction pattern, and it is on this aspect that we shall concentrate. Throughout, unless specifically stated otherwise, all unit cells will be referred to the pseudocubic axes.

Tilting of anion octahedra

In Fig. 1 is shown a diagram of the ideal perovskite (usually the highest-temperature phase), the *aristotype* in Megaw's terminology. When the cations are displaced or the octahedra are tilted (or rotated), different types of structures are produced, *hettotypes*, which are always of lower symmetry than the aristotype. One may ask what happens when a particular octahedron is tilted about some direction; it turns out that there is no unique answer to this. In fact, there are 23 possible simple tilt systems (Glazer, 1972). In this earlier work is was possible to derive a convenient nomenclature for the tilt systems based on breaking the tilting into component tilts about the three pseudocubic axes.

The tilting of the octahedra has several effects which we shall deal with separately.

(*a*) *Unit-cell lengths*

The most important consequence of the tilting is to double certain cell axes. This can be seen in Fig. 2 where the tilt axis is vertically out of the plane of the diagram. The tilting causes the B cation–anion bond in one octahedron to rotate in an opposite direction to that in a neighbouring octahedron, thus giving rise to a doubling of the repeat distances perpendicular to the tilt axis. At the same time if we attempt to maintain the B cation–anion bond distance we must expect the B cation–B cation distance to become shorter and hence reduce the axial lengths. If we denote the angles of tilt about the pseudocubic [100], [010] and [001] directions by α, β and γ respectively (not to be confused with the unit-cell angles), then the new axial lengths are given by

$$a_p = a_o \cos\beta \cos\gamma \qquad (1a)$$
$$b_p = a_o \cos\alpha \cos\gamma \qquad (1b)$$
$$c_p = a_o \cos\alpha \cos\beta \qquad (1c)$$

where a_p, b_p and c_p are the pseudocubic subcell lengths and a_o the cell edge length of the aristotype. The actual

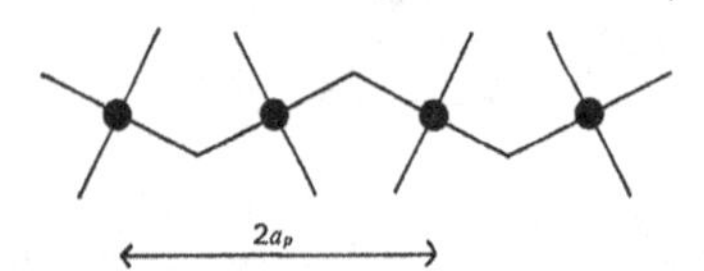

Fig. 2. Schematic diagram of the tilting of octahedra about an axis normal to the plane of the paper. Black circles B cations.

repeat distances, of course, are twice these subcell lengths. If we rearrange these equations thus:

$$\frac{a_p}{b_p} = \frac{\cos\beta}{\cos\alpha} \quad \frac{b_p}{c_p} = \frac{\cos\gamma}{\cos\beta} \quad \frac{c_p}{a_p} = \frac{\cos\alpha}{\cos\gamma}, \qquad (2)$$

it can be seen immediately that equality of any two tilt angles means equality of the cell axes *coincident* with the tilt axes. For example, if the tilts about [100] and [010], *i.e.* α and β, are equal in magnitude then a_p must of necessity be equal to b_p. We have, of course, ignored the effect of distortions of the octahedra in this process, but since it is the tilts that govern the symmetry so much, we can expect that our argument will be reasonable as far as equality or non-equality of tilts/lattice parameters is concerned. It does mean that we cannot expect to use formulae such as (2) to calculate absolute values of the lattice parameters unless we are sure that the distortions are negligible. We therefore see that if the lattice parameters are known we may infer the tilt angles or, at least, whether any two are equal. This only holds true, of course, when there are tilts present; when they are not, the lattice parameters are correlated directly with octahedral distortions normally produced by cation displacements.

In the tilt-system nomenclature, equality of tilt magnitudes is denoted by repetition of the letters appropriate to the particular tilt axes. Thus **aac** means equal tilts about [100] and [010], **aaa** means equal tilts about all three axes, and **abc** means three unequal tilts. These symbols therefore tell us immediately which unit cell axes are equal. In these examples, we have

$$a_p = b_p \neq c_p \text{ for } \mathbf{aac}$$
$$a_p = b_p = c_p \text{ for } \mathbf{aaa}$$

and

$$a_p \neq b_p \neq c_p \text{ for } \mathbf{abc}\,.$$

(*b*) *Unit-cell angles*

As was shown in the earlier work, there are basically two types of tilt: tilts where the octahedra *along* a tilt axis are tilted *in-phase* about the axis, denoted with the superscript +, and tilts where the octahedra are tilted in *antiphase*, denoted by the superscript −. When there are no tilts about a particular axis, 0 is used. The signs of the tilts are important for determining the lattice type and the unit-cell angles. The latter are correlated with the signs in the following way. Any two + tilts, or one + and one − tilt, mean that the relevant axes are normal to each other; any two − tilts mean that the relevant cell axes are inclined to each other. Thus $\mathbf{a}^-\mathbf{b}^+\mathbf{c}^+$ has three axes of unequal length normal to one another. On the other hand $\mathbf{a}^-\mathbf{a}^-\mathbf{c}^+$ has two equal axes, $a_p = b_p$, which are inclined to each other but which are both normal to c_p. Similarly $\mathbf{a}^-\mathbf{a}^-\mathbf{a}^-$ has three equal axes inclined to one another (obviously with equal angles). In the 1972 paper the space groups of all the 23 possible tilt systems were worked out and for convenience they are repeated here (Table 1).

(c) Difference reflexions

Since tilting of the octahedra causes a doubling of the unit-cell axes, extra reflexions are produced which lie on half-integral reciprocal-lattice planes. With reference now to the 'doubled' unit cell these reflexions can be indexed with some indices odd, whilst the ordinary reflexions (the main reflexions) have all *hkl* even. It turns out, fortunately, that the two types of tilts, in-phase (+) and antiphase (−), result in two distinct classes of difference reflexion. It is a simple matter to show that + tilts give rise to reflexions of the type odd-odd-even, whilst − tilts produce odd-odd-odd reflexions. More specifically we can write the following rules:

$\mathbf{a}^+$ produce reflexions even-odd-odd with $k \neq l$ *e.g.* 013, 031 (3*a*)

$\mathbf{b}^+$ produce reflexions odd-even-odd with $h \neq l$ *e.g.* 103, 301 (3*b*)

$\mathbf{c}^+$ produce reflexions odd-odd-even with $h \neq k$ *e.g.* 130, 310 (3*c*)

$\mathbf{a}^-$ produce reflexions odd-odd-odd with $k \neq l$ *e.g.* 131, 113 (3*d*)

$\mathbf{b}^-$ produce reflexions odd-odd-odd with $h \neq l$ *e.g.* 113, 311 (3*e*)

$\mathbf{c}^-$ produce reflexions odd-odd-odd with $h \neq k$ *e.g.* 131, 311 . (3*f*)

Actually it is possible to go still further and derive the relationships between the intensities of the difference reflexions and the tilt angles, α, β, and γ. These are

$$I(\mathbf{a}^+) \propto (ki^l - li^k)^2\alpha^2 \quad (4a)$$

$$I(\mathbf{b}^+) \propto (-li^h + hi^l)^2\beta^2 \quad (4b)$$

$$I(\mathbf{c}^+) \propto (hi^k - ki^h)^2\gamma^2 \quad (4c)$$

$$I(\mathbf{a}^-, \mathbf{b}^-, \mathbf{c}^-) \propto [(ki^l - li^k)\alpha \pm (-li^h + hi^l)\beta \pm (hi^k - ki^h)\gamma]^2 \quad (4d)$$

where $i = \sqrt{-1}$ and the $\pm$ signs depend on the particular choice of origin for the tilt system (there are two such choices in general, but for the present purposes we shall not need to consider them). The reader will immediately see that these relationships are consistent with the rules given before. Another point worth noting is that + tilts produce Bragg reflexions at reciprocal-lattice points corresponding to the one-face-centred positions, whilst for − tilts they occur at the all-face-centred points. When these tilts arise, as they often do, through the 'freezing-in' of a soft mode at a

Table 1. *Complete list of possible simple tilt systems*

Serial number	Symbol	Lattice centring	Multiple cell	Relative pseudocubic subcell parameters	Space group
Three-tilt systems					
(1)	$a^+b^+c^+$	*I*	$2a_p \times 2b_p \times 2c_p$	$a_p \neq b_p \neq c_p$	*Immm* (No. 71)
(2)	$a^+b^+b^+$	*I*		$a_p \neq b_p = c_p$	*Immm* (No. 71)
(3)	$a^+a^+a^+$	*I*		$a_p = b_p = c_p$	*Im*3 (No. 204)
(4)	$a^+b^+c^-$	*P*		$a_p \neq b_p \neq c_p$	*Pmmn* (No. 59)
(5)	$a^+a^+c^-$	*P*		$a_p = b_p \neq c_p$	*Pmmn* (No. 59)
(6)	$a^+b^+b^-$	*P*		$a_p \neq b_p = c_p$	*Pmmn* (No. 59)
(7)	$a^+a^+a^-$	*P*		$a_p = b_p = c_p$	*Pmmn* (No. 59)
(8)	$a^+b^-c^-$	*A*		$a_p \neq b_p \neq c_p \quad \alpha \neq 90°$	$A2_1/m11$ (No. 11)
(9)	$a^+a^-c^-$	*A*		$a_p = b_p \neq c_p \quad \alpha \neq 90°$	$A2_1/m11$ (No. 11)
(10)	$a^+b^-b^-$	*A*		$a_p \neq b_p = c_p \quad \alpha \neq 90°$	*Pmnb* (No. 62)*†
(11)	$a^+a^-a^-$	*A*		$a_p = b_p = c_p \quad \alpha \neq 90°$	*Pmnb* (No. 62)*†
(12)	$a^-b^-c^-$	*F*		$a_p \neq b_p \neq c_p \quad \alpha \neq \beta \neq \gamma \neq 90°$	$F\bar{1}$ (No. 2)
(13)	$a^-b^-b^-$	*F*		$a_p \neq b_p = c_p \quad \alpha \neq \beta \neq \gamma \neq 90°$	*I*2/*a* (No. 15)*
(14)	$a^-a^-a^-$	*F*		$a_p = b_p = c_p \quad \alpha = \beta = \gamma \neq 90°$	$R\bar{3}c$ (No. 167)
Two-tilt systems					
(15)	$a^0b^+c^+$	*I*	$2a_p \times 2b_p \times 2c_p$	$a_p < b_p \neq c_p$	*Immm* (No. 71)
(16)	$a^0b^+b^+$	*I*		$a_p < b_p = c_p$	*I*4/*mmm* (No. 139)†
(17)	$a^0b^+c^-$	*B*		$a_p < b_p \neq c_p$	*Bmmb* (No. 63)
(18)	$a^0b^+b^-$	*B*		$a_p < b_p = c_p$	*Bmmb* (No. 63)
(19)	$a^0b^-c^-$	*F*		$a_p < b_p \neq c_p \quad \alpha \neq 90°$	*F*2/*m*11 (No. 12)
(20)	$a^0b^-b^-$	*F*		$a_p < b_p = c_p \quad \alpha \neq 90°$	*Imcm* (No. 74)*
One-tilt systems					
(21)	$a^0a^0c^+$	*C*	$2a_p \times 2b_p \times c_p$	$a_p = b_p < c_p$	*C*4/*mmb* (No. 127)
(22)	$a^0a^0c^-$	*F*	$2a_p \times 2b_p \times 2c_p$	$a_p = b_p < c_p$	*F*4/*mmc* (No. 140)
Zero-tilt system					
(23)	$a^0a^0a^0$	*P*	$a_p \times b_p \times c_p$	$a_p = b_p = c_p$	*Pm*3*m* (No. 221)

* These space-group symbols refer to axes chosen according to the matrix transformation

$$\begin{pmatrix} 1 & 0 & 0 \\ 0 & \frac{1}{2} & -\frac{1}{2} \\ 0 & \frac{1}{2} & \frac{1}{2} \end{pmatrix}.$$

† In the 1972 paper tilt systems (10) and (11) were incorrectly given the space-group symbol *Pnma* and tilt system (16) the symbol *I*4/*m*.

phase transition, the relevant mode is one with wave vector $\mathbf{q}=(\frac{1}{2}\frac{1}{2}0)$ for + tilts and $\mathbf{q}=(\frac{1}{2}\frac{1}{2}\frac{1}{2})$ for − tilts. This is often described as the condensation of a phonon at the M and R points respectively of the Brillouin zone.* We see then that this system of tilt components is naturally related to the normal modes of vibration which are so important near a phase transition.

Cation displacements

It is considerably more difficult to give any hard-and-fast rules for the cation displacements. Sometimes half-integral difference reflexions (or odd indices on the doubled pseudocubic cell) occur which do not conform to the tilt reflexions, in which case it is relatively easy to infer the type of displacement. Clearly this can only occur when they are antiparallel; when they are parallel, however, no difference reflexions are produced. The most reliable indication of the displacements is often given by the space group due to the tilt system, particularly if it is known whether the particular substance is centrosymmetric or not. If the displacements are antiparallel the structure is centrosymmetric and we normally find that the space group is the same as that for the tilt system alone. When they are parallel, the structure is non-centrosymmetric and the space group is a subgroup of that for the tilt system. It is therefore relatively easy to suggest the most likely space group for the structure and since the A and B cations usually lie at special positions we often know the displacement directions.

When there are no tilts present, as evidenced perhaps by the lack of the necessary difference reflexions, the displacements can often be ascertained from the unit-cell geometry. Thus if the unit cell is a rhombohedral distortion of the aristotype cell, and no difference reflexions are found, it is reasonable to suppose that the structure has parallel cation displacements along [111] (the threefold axis) with space group $R3m$. The lowest-temperature phase of $BaTiO_3$ is an example of this (Rhodes, 1949; Kay & Vousden, 1949).

One must be careful here, however. It is possible to have distortions of the octahedra even when there are no displacements or tilts. $KCuF_3$ is an example of this (Okazaki & Suemune, 1961), where Jahn–Teller effects distort the octahedra to produce a tetragonal structure. Fortunately this type of structure is very rare and in any case might well be expected since it is well known that Cu(d^9) shows strong Jahn–Teller distortion.

Examples

Let us consider some examples of known perovskite structures in order to see how the technique described can be applied in practice.

* With the origin chosen at the centre of the octahedra the irreducible representations for these modes are conventionally labelled M_3 and R_{25}.

(*a*) *The high-temperature phases of* $NaNbO_3$

Above room temperature, $NaNbO_3$ possesses six phases. For the present purposes we shall consider the four highest phases, labelled S, T_1, T_2 and cubic. Lattice-parameter measurements (Glazer & Megaw, 1973) gave the following results:

$$\left.\begin{array}{lll} 480\text{–}520\,^{\circ}\text{C} & S & a_p\neq b_p\neq c_p \\ 520\text{–}575\,^{\circ}\text{C} & T_1 & a_p\neq b_p\neq c_p \\ 575\text{–}641\,^{\circ}\text{C} & T_2 & a_p=b_p<c_p \\ >641\,^{\circ}\text{C} & \text{Cubic} & a_p=b_p=c_p \end{array}\right\}\ \text{all axes orthogonal}.$$

Weissenberg photographs of the first half-integral reciprocal-lattice layers of phase T_2 (Glazer & Megaw, 1972), showed the presence of only three, very weak difference reflexions, $\frac{1}{2}\frac{3}{2}0$, $\frac{1}{2}\frac{3}{2}1$ and $\frac{1}{2}\frac{5}{2}0$ (130, 131 and 150 on a pseudocubic cell $2a_p\times 2b_p\times c_p$). Since these reflexions occur at low angles it is likely that they arise from the O atoms and as they are of the odd-odd-even type the tilt system (No. 21) $\mathbf{a}^0\mathbf{a}^0\mathbf{c}^+$ is immediately suggested.† The space group from Table 1 is $C4/mmb$ and, being tetragonal, is consistent with the measured unit-cell geometry. Trivially, two equal (zero-magnitude) tilts mean two equal axes. The initial determination of the trial structure in this case took approximately five minutes allowing for indexing. Subsequent calculations of structure factors confirmed this structure.

Similar photographs of phase T_1 (Ahtee, Glazer & Megaw, 1969) showed the following reflexions 310, 130, 131, 151, 312, and 113 (on the doubled cell $2a_p\times 2b_p\times 2c_p$). It was possible to infer from this that these were tilt reflexions which could be broken down into the following:

310, 130, 312	$\mathbf{c}^+$ tilt	(odd-odd-even)
131, 151	$\mathbf{a}^-$ tilt	(odd-odd-odd).

Reference to (3) shows that there is neither a $\mathbf{b}^+$ nor a $\mathbf{b}^-$ tilt. This suggests structure (No 17) $\mathbf{a}^-\mathbf{b}^0\mathbf{c}^+$ with space group $Ccmm$; this orthorhombic space group is again consistent with the lattice parameter measurements. We have three unequal tilts and three unequal lattice parameters.

Finally in phase S, the reflexion 510 was observed (Glazer & Ishida, 1974) in addition to the above reflexions, suggesting a $\mathbf{b}^+$ tilt. This gives the structure of phase S as (No. 4) $\mathbf{a}^-\mathbf{b}^+\mathbf{c}^+$ with space group $Pnmm$, with unit-cell geometry again consistent with the measured lattice parameters. The four phases are summarized thus

$$\mathbf{a}^-\mathbf{b}^+\mathbf{c}^+(S) \overset{520\,^{\circ}\text{C}}{\rightleftharpoons} \mathbf{a}^-\mathbf{b}^0\mathbf{c}^+(T_1) \overset{575\,^{\circ}\text{C}}{\rightleftharpoons}$$
$$\mathbf{a}^0\mathbf{a}^0\mathbf{c}^+(T_2) \overset{641\,^{\circ}\text{C}}{\rightleftharpoons} \mathbf{a}^0\mathbf{a}^0\mathbf{a}^0\ \text{(cubic)},$$

showing the progressive loss of tilts one by one as the temperature is raised.

† Note that in this tilt system one of the unit-cell axes is not doubled.

(*b*) *The room-temperature structure of* $CaTiO_3$

The structure of $CaTiO_3$ was determined by Kay & Bailey (1957) from single-crystal data. Here we shall consider the powder diagram and see how much can be inferred from it. From the splittings of the main reflexions it can easily be determined that

$$a_p = b_p \neq c_p \quad \gamma \neq 90° .$$

The fact that the angle between the a_p and b_p axes is different from 90° suggests two antiphase tilts (from the rules given earlier) and the equality of a_p and b_p further suggests that the tilt angles are equal. Two possibilities arise, (No. 10) $\mathbf{a}^-\mathbf{a}^-\mathbf{c}^+$ or (No. 20) $\mathbf{a}^-\mathbf{a}^-\mathbf{c}^0$ (space groups *Pbnm* and *Imam* respectively), as given in Table 1. Examination of the difference reflexions in the powder diffraction pattern allows one to classify the reflexions into classes, and Table 2 can be constructed. We see that the presence of 00*l*-type reflexions supports the $\mathbf{a}^-\mathbf{a}^-\mathbf{c}^+$ system.

In Table 2 no attempt has been made to consider the signs of the *hkl*, except where absolutely necessary. Note that in the last column, tilt system $\mathbf{a}^-\mathbf{a}^-\mathbf{c}^+$ has been used to index certain difference reflexions, the remaining ones, $N=13$, 17 and 29 not being consistent with any tilt type. It is natural, therefore, to suggest that these reflexions arise from antiparallel cation displacements.

We can get some idea of the directions of the cation displacements in the following way. Consider Fig. 3 which shows schematically the tilt arrangement $\mathbf{a}^-\mathbf{a}^-\mathbf{c}^+$. The displacement of the Ca atom will be linked to a great extent to its environment by simple steric principles. Let us examine the environment of the Ca atom marked A. Firstly, of the O atoms numbered 5, 6, 7 and 8 the closest are 5 and 8 since they are displaced towards the Ca atom, thus tending to push it along [010]. Secondly, of the O atoms 9, 10, 11 and 12 number 12 is closest to the Ca atom and thus tends to push it along $[\bar{1}\bar{1}0]$. O atoms 1, 2, 3 and 4, however, are not sufficiently close to the Ca atom to affect its position much. We see, then, that two O atoms push the Ca along [010] and one along $[\bar{1}\bar{1}0]$. To a gross approximation we find that the resultant vector displacement is

$$2[010]+[\bar{1}\bar{1}0]=[\bar{1}10] .$$

Proceeding in a similar manner with the remaining Ca atoms, we arrive at the arrangement of displacements shown in Fig. 3. If we now assume that the tilt-system space group is the same as that of the crystal (*Pbnm*), with the origin half-way between Ca atoms along [001], it is easily seen that the Ti atoms are not displaced at all, as they lie on centres of symmetry. It is worth noting that there is an alternative choice of origin for this structure. We could place the Ca atoms on centres of symmetry and then the Ti atoms would have freedom to be displaced. This is a less likely structure since it ignores the steric effects of the O atoms on the Ca atom. We conclude, therefore, that the cation displacements must lie approximately along the $\langle 1\bar{1}0 \rangle$ pseudocubic directions (the $\langle 010 \rangle$ directions of the orthorhombic cell) in antiparallel sheets perpendicular to [001]. The extra reflexions, therefore must have *l* odd and hence they are

N	*hkl*
13	023, 203
17	401, 041
29	025, 205 .

Reference to Kay & Bailey (1957) shows that our structure is correct and, in fact, their refinement resulted in Ca atoms displaced *exactly* along the $\langle 1\bar{1}0 \rangle$ pseudocubic directions (in some other perovskites with the same type of structure the A cations are not displaced exactly along these directions, although they are always fairly close to them).

(*c*) *The structure of* $PbZr_{0.9}Ti_{0.1}O_3$

In Fig. 4 is shown the neutron diffraction pattern from a polycrystalline sample of $PbZr_{0.9}Ti_{0.1}O_3$. Apart from a few extra reflexions the pattern looks very much like that from a primitive cubic structure. However the higher resolution found in X-ray photographs shows that reflexions such as *hhh* are split whilst *h*00 are not. From this we infer that the symmetry is probably rhombohedral. It is known that the material is a ferroelectric and so the cations are displaced in a parallel fashion, obviously along the threefold axis. The extra reflexions shown in the neutron profile do not appear in the X-ray photographs since they arise from the O

Table 2. *Indexing of the powder pattern for* $CaTiO_3$

$N=h^2+k^2+l^2$ (Pseudocubic subcell)	$N=h^2+k^2+l^2$ (Doubled cell)	Type	Tilt	*hkl* for $\mathbf{a}^-\mathbf{a}^-\mathbf{c}^+$
$2\frac{1}{2}$	10	*ooe, oeo* or *eoo*	+	310, 130
$2\frac{3}{4}$	11	*ooo*	−	311, 131, 113
$3\frac{1}{4}$	13	*oee, eoe* or *eeo*		
$3\frac{1}{2}$	14	*ooe, oeo* or *eoo*	+	312, 132
$4\frac{1}{4}$	17	*oee, eoe* or *eeo*		
$4\frac{3}{4}$	19	*ooe, oeo* or *eoo*	+	$\bar{3}30$
$5\frac{1}{2}$	22	*ooe, oeo* or *eoo*	+	$\bar{3}32$
$7\frac{1}{4}$	29	*oee, eoe* or *eeo*		
$8\frac{3}{4}$	35	*ooo*	−	531, 513, 135 ...

o = odd *e* = even.

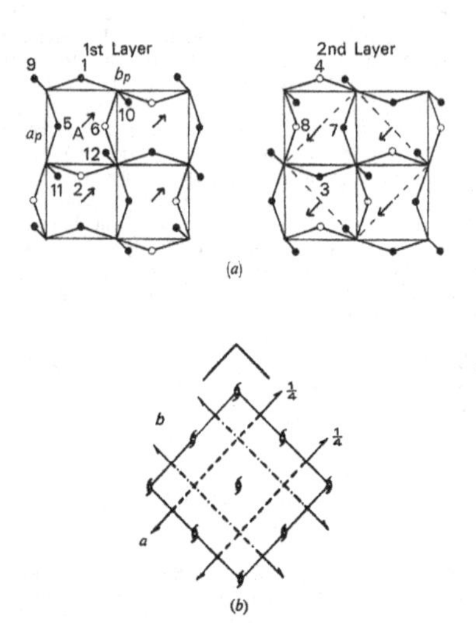

Fig. 3. (*a*) The structure of $CaTiO_3$. Full circles denote O atoms above the plane of the diagram whilst open circles are those below. Arrows indicate the directions of Ca displacements (at height $\frac{1}{4}c$); Ti atoms are at the centres of the octahedra. The dashed cell is the true crystallographic cell with space group *Pbnm*. (*b*) Symmetry elements of space group *Pbnm*.

atoms which do not scatter X-rays strongly enough. It is a very simple matter to index them as 311, 511 *etc.* (on a doubled cell $2a_p \times 2b_p \times 2c_p$) and taking the probable rhombohedral symmetry into account we expect the tilt system to be (No. 14) $\mathbf{a^-a^-a^-}$ with symmetry $R\bar{3}c$. The parallel cation (Pb and Zr/Ti) displacements mean that the space-group symmetry will be $R3c$, and this agrees with the earlier derivation by Michel, Moreau, Achenbach, Gerson & James (1969). Fig. 4, incidentally, shows the observed and calculated profiles computed with the profile refinement technique (Rietveld, 1969). On heating the powder, the extra reflexions are lost but the symmetry remains rhombohedral and the material is still ferroelectric. This means that the tilts are lost but the displacements remain to give space group $R3m$. Finally at a higher temperature the symmetry becomes cubic and the displacements disappear to give the paraelectric phase. The whole series can be described by

$$\underset{[111]\text{ displacements}}{\mathbf{a^-a^-a^-}(R3c)} \rightarrow \underset{[111]\text{ displacements}}{\mathbf{a^0a^0a^0}(R3m)} \rightarrow \mathbf{a^0a^0a^0}(Pm3m)\,.$$

Conclusion

It can be seen that by breaking the perovskite structure into simple components it is often possible to simplify greatly the task of ascertaining a trial model for the structure of these pseudosymmetric materials. The technique described works well for the vast majority

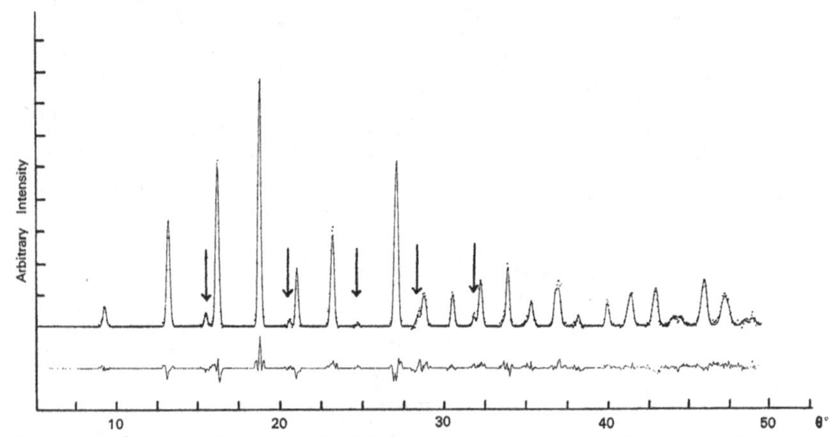

Fig. 4. Neutron diffraction profile of polycrystalline $PbZr_{0.9}Ti_{0.1}O_3$ at room temperature ($\lambda = 1.33$ Å). Arrows indicate difference reflexions arising from tilted octahedra. Points are the observed intensities, the full line is the calculated profile. The lower trace shows the difference between observed and calculated intensities (Glazer & Clarke, unpublished).

of perovskites. In fact the only perovskites known to the author in which difficulty would occur are $NaNbO_3$ (phases *P* and *R*), $PbZrO_3$, possibly $AgNbO_3$ and $AgTaO_3$, and $KCuF_3$. In the last case the difficulty, as mentioned earlier, is the large Jahn–Teller distortion of the octahedra which dominates the symmetry. In the other cases, the difficulty lies in the fact that the tilt systems are not simple ones, but consist of combinations of the simple systems. However, even here it is possible to go some way towards deriving the correct structure, but great care is needed when attempting this. Fortunately such structures seem to be quite rare.

At present this general method is being used tc analyse the complex sequence of phases in the system $Na/KNbO_3$ with considerable success. A preliminary note of this work has already been published (Ahtee & Glazer, 1974) in which tentative suggestions for the various structures have been made. Since then, many of these have been verified (manuscript in preparation), and this has shown that reliable trial models can be obtained very quickly even when there are many possible phases within a single solid-solution series.

I thank Dr H. D. Megaw for introducing me to the fascinating complexities of the perovskite structure, and the Wolfson Foundation for funds enabling this work to be carried out.

References

AHTEE, M. & GLAZER, A. M. (1974). *Ferroelectrics*, **7**, 93–95.
AHTEE, M., GLAZER, A. M. & MEGAW, H. D. (1969). *Phil. Mag.* **26**, 995–1014.
GLAZER, A. M. (1972). *Acta Cryst.* **B28**, 3384–3392.
GLAZER, A. M. & ISHIDA, K. (1974). *Ferroelectrics*, **6**, 219–224.
GLAZER, A. M. & MEGAW, H. D. (1972). *Phil. Mag.* **25**, 1119–1135.
GLAZER, A. M. & MEGAW, H. D. (1973). *Acta Cryst.* **A29**, 489–495.
KAY, H. F. & BAILEY, P. C. (1957). *Acta Cryst.* **10**, 219–226.
KAY, H. F. & VOUSDEN, P. (1949). *Phil. Mag. Ser.* 7, **40**, 1019–1040.
MEGAW, H. D. (1966). *Proceedings of the International Meeting on Ferroelectricity, Prague*, Vol. 1, pp. 314–321.
MEGAW, H. D. (1969). *Proceedings of the European Meeting on Ferroelectricity, Saarbrücken*, pp. 27–35.
MEGAW, H. D. (1971). *Abstracts of the Second European Meeting on Ferroelectricity, Dijon*: *J. de Phys.* (1972). **33**, C2. 1–C2. 5.
MEGAW, H. D. (1973). *Crystal Structures: A Working Approach*. Philadelphia: Saunders.
MICHEL, C., MOREAU, J.-M., ACHENBACH, G., GERSON, R. & JAMES, W. J. (1969). *Solid State Commun.* **7**, 865–868.
OKAZAKI, A. & SUEMUNE, Y. (1961). *J. Phys. Soc. Japan*, **16**, 176–183.
RHODES, R. G. (1949). *Acta Cryst.* **2**, 417–419.
RIETVELD, H. M. (1969). *J. Appl. Cryst.* **2**, 65–71.

PHYSICAL REVIEW B VOLUME 1, NUMBER 7 1 APRIL 1970

Equation of State for the Cooperative Transition of Triglycine Sulfate near T_c

JULIO A. GONZALO
Puerto Rico Nuclear Center, Mayaguez, Puerto Rico 00708*
(Received 13 June 1969)

Measurements of polarization versus field in the vicinity of the Curie temperature from triglycine sulfate, both *below* and *above* T_c, allow first, the determination of a number of critical exponents and second, the characterization of the ferroelectric equation of state. The relationship, below and above T_c, between the "scaled" variables $\hat{P}/[1-(T/T_c)]^\beta$ and $\hat{E}/[1-(T/T_c)]^{\beta\delta}$ was determined from a log-log plot which showed clearly a well-defined asymptotic behavior for the small and large "scaled" field. Comparison of the scaled data with the results from the mean-field theory showed good agreement. A phenomenological expression for the equation of state which matches all the empirical and homogeneity requirements has been formulated. Evidence for the validity of this equation of state for other transitions for which accurate data are available is discussed.

INTRODUCTION

THE second-order transition in ferroelectric triglycine sulfate (TGS) seems to be a very good test case for the mean-field theory. Previous work[1,2] has shown that the behavior of the dielectric constant and the spontaneous polarization is in agreement with the mean-field predictions. The present investigation, partially reported in a previous letter,[3] aimed at a more complete analysis of the order-disorder cooperative transition by means of a detailed study of the variation of polarization with electrical field as well as with temperature near the critical point. Accurate data of P versus E near T_c allow the determination of critical exponents through log-log graphic representations. In addition, once the two fundamental parameters β and δ, defined by $(P_s)_{T\sim T_c}=\text{const}\times[1-(T/T_c)]^\beta$ and $(P)_{T\sim T_c}=\text{const}\times E^{1/\delta}$, are determined, the way is open for the search of a "law of corresponding states" in terms of the properly "scaled" variables.

EXPERIMENTAL

The sample preparation and experimental procedure were described[1,3] in previous communications. The determination of the Curie temperature was done in two different ways. First, a plot of the squared spontaneous polarization (P_s^2) versus temperature was made which showed an almost perfect linear behavior yielding T_c by extrapolation to $P_s^2=0$. Alternatively, the method described by Kouvel and Fisher[4] was used, yielding the same result within experimental accuracy. It has been noted by Reese[5] that corrections due to the electrocaloric effect should be considered. While the accurate determination of these corrections near T_c is not easy, reasonable estimates indicate that our results would not be substantially altered by them. It may also be noted that perfect compensation of the P-versus-E hysteres is loops very near T_c could not be fully achieved with the Sawyer-Tower circuit, possibly due to a field dependence of the conductivity of the crystal. This behavior actually set limits of $\Delta T=T_c-T$ at $+0.02$ and -0.04°C within which a reliable determination of P for very small E was not possible.

As is well known, an increase of the amplitude of the ac field applied to the sample for displaying the P-versus-E curve, produces a relatively small increase of the absolute value of the polarization with respect to the corresponding values for lower ac amplitude. However, the relative variation of the polarization as a function of temperature was checked for various field amplitudes and it was found to be the same, the absolute values being different only by a constant factor. This effect might be attributed to a consistently partial switching of the ferroelectric domains at low ac amplitudes. The constant ac field amplitude chosen in our case, $E=190$ V/cm, was relatively low, which helps to keep down the electrocaloric effect in the vicinity of T_c.

RESULTS

The experimental results below T_c as described in a previous letter,[3] were shown to yield the value of the critical exponents δ and β, along with four other exponents indicating the field and temperature dependence of both derivatives of the polarization with respect to field and temperature. The experimental values obtained from log-log plots of the data are given in Table I and compared with those calculated from the mean-field model, using the expression[6]

$$\hat{E}=(1-t)\tanh^{-1}(\hat{P})-\hat{P},$$
$$[\hat{E}=E/E_0,\ t=1-(T/T_c),\ P=P/N\mu]$$

* Operated by the University of Puerto Rico for the U. S. Atomic Energy Commission.
[1] J. A. Gonzalo, Phys. Rev. **144**, 662 (1966).
[2] P. P. Craig, Phys. Letters **20**, 140 (1966).
[3] J. A. Gonzalo, Phys. Rev. Letters **21**, 749 (1968).
[4] J. S. Kouvel and M. E. Fisher, Phys. Rev. **136**, A1626 (1964).
[5] W. Reese (private communication). It should be noted that precisely because of the method we have used to determine T_c, in spite of the fact that the corrections due to the electrocaloric effect can affect the absolute value of T, the associate correction in (T_c-T) should go to *zero* as one approaches T_c, the temperature at which the hysteresis loops are recorded to practically disappear. Recently, very accurate experiments by E. Nakamura *et al.*, [Proceedings of the Second International Meeting on Ferroelectricity, Kyoto, Japan, 1969 (unpublished)] free from electrocaloric effects, establish conclusively the classical values previously reported for the exponents γ and β, the former even at temperatures as close to T_c as $T_c-T\sim0.01$°C.
[6] J. A. Gonzalo and J. R. López-Alonso, J. Phys. Chem. Solids **25**, 303 (1964).

TABLE I. Experimental critical exponents from TGS compared with mean-field theory predictions.

Defining relationship	Experimental value	Mean-field relationship	Theoretical value
$(P)_{t=0} \sim e^{\gamma_3}$	$\gamma_3 = \frac{1}{\delta} = 0.32 \pm 0.02$	$(p)_{t=0} = (3e)^{1/3}$	$\gamma_3 = \frac{1}{\delta} = \frac{1}{3}$
$(p)_{e=0} \sim t^{\gamma_4}$	$\gamma_4 = \beta = 0.50 \pm 0.03$	$(p)_{e=0} = (3t)^{1/2}$	$\gamma_4 = \beta = \frac{1}{2}$
$\left(\frac{\partial p}{\partial e}\right)_{t=0} \sim e^{\gamma_7}$	$\gamma_7 = -0.66 \pm 0.05$	$\left(\frac{\partial p}{\partial e}\right)_{t=0} = \frac{(1-p^2)}{p^2}$	$\gamma_7 = -\frac{2}{3}$
$\left(\frac{\partial p}{\partial e}\right)_{e=0} \sim t^{\gamma_8}$	$\gamma_8 = -\gamma' = -0.95 \pm 0.10$	$\left(\frac{\partial p}{\partial e}\right)_{e=0} = \frac{(1-p^2)}{(t-p^2)}$	$\gamma_8 = -1$
$\left(\frac{\partial p}{\partial t}\right)_{t=0} \sim e^{\gamma_9}$	$\gamma_9 = -0.33 \pm 0.05$	$\left(\frac{\partial p}{\partial t}\right)_{t=0} = \frac{\tanh^{-1} p (1-p^2)}{p^2}$	$\gamma_9 = -\frac{1}{3}$
$\left(\frac{\partial p}{\partial t}\right)_{e=0} \sim t^{\gamma_{10}}$	$\gamma_{10} = -0.45 \pm 0.10$	$\left(\frac{\partial p}{\partial t}\right)_{e=0} = \frac{-\tanh^{-1} p (1-p^2)}{(t-p^2)}$	$\gamma_{10} = -\frac{1}{2}$

where $E_0 \simeq 4.4 \times 10^6$ V/cm is the saturation internal field, and $N\mu \simeq 4.3$ $\mu C/cm^2$ is the saturation polarization. It is interesting to note that, as it should be expected (see Appendix), the ratio of the two critical exponents relating the same derivative of the free energy to field and temperature is constant.

In Table II a summary of data for polarization and field at various temperatures *below* and *above* T_c is given. These data were "scaled" to determine $p = \hat{P}/t^{\beta}$ and $e = \hat{E}/t^{\beta\delta}$, and plotted using a log-log scale. It can be seen from Fig. 1 that the scaling of the data is quite good, giving evidence of the existence of a law of corresponding states. From this, the sequence of critical exponents, found directly and reported in the previous short communication,[3] results in an automatic fashion. What is more important, however, is the fact that this log-log representation, which shows the critical behavior over a wide range of three decades in the reduced field e, shows clearly the asymptotic behavior of the equation of state for both small and large e, above and below T_c. This asymptotic character is in complete analogy with the observations of Green *et al.*[7] for liquid-vapor transitions in a good number of systems. We have also recently examined very accurate data[8,9] from ferromagnetic transitions, and the asymptotic trend for small and large scaled magnetic field is seen again to be fully analogous.

The asymptotic behavior can be summarized as

TABLE II. Polarization versus field for TGS from hysteresis loops in the vicinity of the Curie temperature, $T_c \simeq 322.50$°K.

Below Curie temperature			Above Curie temperature		
$P(\mu C/cm^2)$	$\Delta T(\times 10^{-2}\ °C)$	E(V/cm)	$P(\mu C/cm^2)$	$\Delta T(\times 10^{-2}\ °C)$	E(V/cm)
0.217	4.2	49.7	0.1085	2.3	12.7
0.217	10.7	27.7	0.1085	8.7	21.1
0.217	17.1	9.3	0.1085	15.2	30.3
0.244	4.2	73.0	0.1085	21.7	40.2
0.244	10.7	46.9	0.163	2.3	33.1
0.244	17.1	24.5	0.163	8.7	49.3
0.271	4.2	101.8	0.163	15.2	60.6
0.271	10.7	72.4	0.163	21.7	71.9
0.271	17.1	46.4	0.217	2.3	69.8
0.271	30.1	7.0	0.217	8.7	88.1
0.298	4.2	137.9	0.217	15.2	104.3
0.298	10.7	104.5	0.217	21.7	117.7
0.298	17.1	74.6	0.298	2.3	131.8
0.298	30.1	25.4	0.298	8.7	148.7
0.326	4.2	182.7	0.298	15.2	166.4
0.326	10.7	144.5	0.298	21.7	186.1
0.326	17.1	104.6			
0.326	30.1	53.8			
0.326	43.0	9.9			

[7] M. S. Green, M. Vicentini-Missoni, and J. M. H. Levelt Sengers, Phys. Rev. Letters **18**, 1113 (1967).
[8] J. S. Kouvel and J. B. Comly, Phys. Rev. Letters **20**, 1237 (1968).
[9] J. T. Ho and J. D. Litster, Phys. Rev. Letters **22**, 603 (1969).

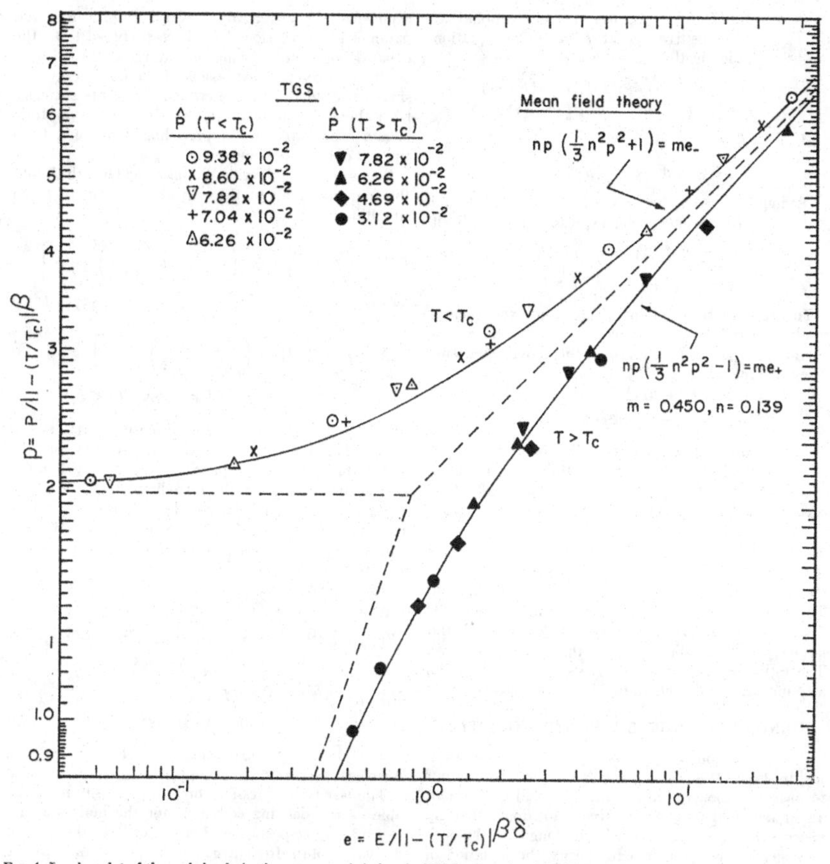

FIG. 1. Log-log plot of the scaled polarization versus scaled electric field for ferroelectric TGS near the Curie temperature. Full line is renormalized mean-field equation of state with $m=0.450$, $n=0.139$.

follows:

$$\text{Below } T_c: p \underset{e\to 0}{\sim} \text{const}, \tag{1a}$$

$$p \underset{e\to\infty}{\sim} \text{const}\times e^{1/\delta}, \tag{1b}$$

$$\text{Above } T_c: p \underset{e\to 0}{\sim} \text{const}\times e \tag{2a}$$

$$p \underset{e\to\infty}{\sim} \text{const}\times e^{1/\delta}. \tag{2b}$$

The implications of these expressions are obvious: (1a) means that for $\hat{E}\ll t^{\beta\delta}$, below T_c, we are approaching the coexistence curve, i.e., $P_s=\text{const}\times t^{\beta}$; (1b) and (2b) mean that for $\hat{E}\gg t^{\beta\delta}$ we are approaching the critical isotherm, i.e., $P=\text{const}\times E^{1/\delta}$; finally, (2a) means that for $\hat{E}\ll t^{\beta\delta}$, above T_c, we are approaching $\Delta P/\Delta E=\text{const}\times t^{-\gamma}$, with $-\gamma=\beta-\beta\delta=-\gamma'$ which in our case merely expresses the Curie-Weiss law.

MEAN-FIELD MODEL EQUATION OF STATE

The basic expression of the mean-field model for ferroelectrics can be written as

$$\hat{E}_{\mp}=(1\mp t)\tanh^{-1}(\hat{P})-\hat{P},$$

where $t = |1-(T/T_c)|$; the negative sign applies at $T<T_c$ and the positive sign at $T>T_c$. This equation can be expanded in the following way:

$$\hat{E}_{\mp} = (\hat{P}+\tfrac{1}{3}\hat{P}^3+\tfrac{1}{5}\hat{P}^5\cdots)\mp t(\hat{P}+\tfrac{1}{3}\hat{P}^3+\tfrac{1}{5}\hat{P}^5\cdots)-\hat{P}$$
$$=\mp t\hat{P}+\tfrac{1}{3}\hat{P}^3+(\tfrac{1}{5}\hat{P}^5+\cdots)\mp t(\tfrac{1}{3}\hat{P}^3+\tfrac{1}{5}\hat{P}^5\cdots).$$

Putting $e \equiv \hat{E}/t^{3/2}$ and $p=\hat{P}/t^{1/2}$,

$$e_{\mp} = \mp p + \tfrac{1}{3}p^3 + t[\mp\tfrac{1}{3}p^3+\tfrac{1}{5}p^5(1\mp t)+\cdots].$$

Obviously, if $t \ll 1$ (for instance, $1.0\times10^{-5} \leqslant t \leqslant 3.0 \times 10^{-3}$ in our experiment) this expression reduces itself to

$$e_- = p(\tfrac{1}{3}p^2-1) \quad \text{for} \quad T<T_c, \tag{3a}$$

$$e_+ = p(\tfrac{1}{3}p^2+1) \quad \text{for} \quad T>T_c. \tag{3b}$$

To check these equations against the experimental results it has been found necessary to introduce proportionality factors for e and p in the above equations. They became

$$me_- = np(\tfrac{1}{3}n^2p^2-1),$$
$$me_+ = np(\tfrac{1}{3}n^2p^2+1).$$

To introduce these proportionality factors is equivalent to modifying the normalization parameters in such a way that $(N\mu)/n$ replaces $(N\mu)$, and E_0/m replaces E_0. The best fit to the data is obtained with

$$m=0.450, \quad n=0.139.$$

Figure 1 shows a plot of the mean-field equation of state in scaling form, along with the experimental data. The agreement is very good except for a few points for $T<T_c$ in the intervening region between small and large e, which fall slightly above the theoretical curve. The estimated experimental errors go from 5 to 1% as p increases and from 10 to 2% as e increases.

PHENOMENOLOGICAL EQUATION OF STATE

The realization that the asymptotic behavior specified by Eqs. (1a)–(2b) is not only characteristic of our ferroelectric cooperative transition, but also of liquid-vapor and magnetic cooperative transitions, strongly suggests the convenience of using it along with Widom's homogeneity requirements to specify the equation of state for the system under consideration throughout T_c and in its vicinity.

Since the formulation of the homogeneity assumption[10] for the free energy of a cooperative system undergoing a second-order phase transition, considerable progress has been made in the understanding of the critical phenomena.[11] Griffiths[12] has studied the problem of constructing explicit analytic expressions for the equation of state relating the scaled extensive variable (polarization, magnetization, volume, etc.) to the intensive variable (electric field, magnetic field, pressure, etc., respectively) for the case of rational critical exponents. Very recently, several empirical[9,13] and parametric[14] expressions have been proposed to fit the equation of state of some real systems. We wish to construct a compact expression of the free energy in a simple way, matching the critical exponents sequence both above T_c and below it, as well as the asymptotic behavior indicated in the preceding paragraph, from the law of corresponding states.

Let us assume, following Widom[10] and Griffiths,[12] that the free energy about the critical point can be given simultaneously by

$$F(\hat{X},t)=\hat{X}^{(\delta+1)/\delta}\left[a_1+a_2\left(\frac{t}{\hat{X}^{1/\beta\delta}}\right)+a_3\left(\frac{t}{\hat{X}^{1/\beta\delta}}\right)^2+\cdots\right]$$
$$\text{for } \hat{X}^{1/\beta\delta} \gg t \quad (T \lesseqgtr T_c) \tag{4a}$$

$$F(\hat{X},t)=t^{\beta(\delta+1)}\left[b_1+b_2\left(\frac{\hat{X}}{t^{\beta\delta}}\right)+b_3\left(\frac{\hat{X}}{t^{\beta\delta}}\right)^2+\cdots\right]$$
$$\text{for } t^{\beta\delta} \gg \hat{X} \quad (T<T_c). \tag{4b}$$

Here $\hat{X}=X/X_0$ is the reduced intensive variable, and $t=1-(T/T_c)$ the reduced temperature. [It may be noted that while the expansion (4b) is very familiar in the literature since the introduction of the homogeneity assumption, relatively less attention has been paid to the complementary expansion (4a); Ho and Litster have made use of the latter in their recent work[9] on $CrBr_3$.]

These expansions ensure a sequence of critical exponents of the expected form

$$(F)_{t=0} \sim \hat{X}^{(\delta+1)/\delta}, \quad S=(\partial F/\partial t)_{t=0} \sim \hat{X}^{(\delta+1)/\delta-(1/\beta\delta)},$$
$$C=(\partial^2F/\partial t^2)_{t=0} \sim \hat{X}^{(\delta+1)/\delta-2(1/\beta\delta)}\cdots, \tag{5a}$$

$$(F)_{x=0} \sim t^{\beta(\delta+1)}, \quad \hat{Y}=(\partial F/\partial\hat{X})_{x=0} \sim t^{\beta(\delta+1)-\beta\delta},$$
$$\hat{Y}_x'=(\partial^2F/\partial\hat{X}^2)_{x=0} \sim t^{\beta(\delta+1)-2\beta\delta}\cdots, \tag{5b}$$

where the "gap" exponents are $(1/\Delta)=1/\beta\delta$ and $\Delta=\beta\delta$, respectively.

The last two expressions of (5b) are easily recognizable as the defining equations for the indices β and $-\gamma=\beta-\Delta$, respectively. Let us call $\hat{Y}$ the partial derivative of the free energy with respect to the variable $\hat{X}$. Its meaning will be, of course, that of the respective extensive variable in the various cases (polarization, magnetization, volume, etc.). From (4a) we obtain

$$\hat{Y}=\frac{\partial F}{\partial\hat{X}}=\hat{X}^{1/\delta}\left[a_1+a_2\left(\frac{t}{\hat{X}^{1/\beta\delta}}\right)+a_3\left(\frac{t}{\hat{X}^{1/\beta\delta}}\right)^2+\cdots\right]$$
$$+\hat{X}^{1/\delta}\left[a_2+2a_3\left(\frac{t}{\hat{X}^{1/\beta\delta}}\right)+3a_3\left(\frac{t}{\hat{X}^{1/\beta\delta}}\right)^2+\cdots\right]$$
$$\times\left(-\frac{1}{\beta\delta}\right)\left(\frac{t}{\hat{X}^{1/\beta\delta}}\right), \tag{6a}$$

[10] B. Widom, J. Chem. Phys. **43**, 3898 (1965).
[11] See, for example, M. E. Fisher, Rept. Progr. Phys. **30**, 615 (1967).
[12] R. B. Griffiths, Phys. Rev. **158**, 176 (1967).
[13] M. Vicentini-Missoni, J. M. H. Levelt Sengers, and M. S. Green, Phys. Rev. Letters **22**, 389 (1969).
[14] P. Schofield, Phys. Rev. Letters **22**, 606 (1969).

$$\hat{Y}=\frac{\partial F}{\partial \hat{X}}=t^{\beta}\left[b_2+2b_3\left(\frac{\hat{X}}{t^{\beta\delta}}\right)+3b_4\left(\frac{\hat{X}}{t^{\beta\delta}}\right)^2+\cdots\right], \quad (6b)$$

or in other words

$$y=\hat{Y}/t^{\beta}=x^{1/\delta}[c_1+c_2(x^{-1/\beta\delta})+c_3(x^{-1/\beta\delta})^2+\cdots] \quad \text{for } x\equiv\hat{X}/t^{\beta\delta}\gg 1 \quad (T\lesseqgtr T_c) \quad (7)$$

$$y=[d_1+d_2(x)+d_3(x)^2+\cdots] \quad \text{for } x\ll 1 \quad (T<T_c). \quad (8)$$

Our aim is to get a single expression for y which combines Eqs. (7) and (8) at $T<T_c$, approaching each of them for the limiting cases of $x\gg 1$ and $x\ll 1$. One can write formally

$$y=x^{1/\delta}\psi_1(x^{-1/\beta\delta})+\psi_2(x) \quad \text{for} \quad T<T_c, \quad (9)$$

$$y=x^{1/\delta}\psi_1(x^{-1/\beta\delta}) \quad \text{for} \quad T<T_c, \quad (10)$$

where ψ_1 and ψ_2 stand for the factors within brackets in Eqs. (7) and (8). This is our equation of state which already involves the correct sequence of critical exponents throughout the series expansion of ψ_1 in powers of $x^{-1/\beta\delta}$ and of ψ_2 in powers of x. Below T_c, according to Eqs. (4a) and (4b), $\psi_1(x^{-1/\beta\delta})$ should predominate for $x\gg 1$, and $\psi_2(x)$ for $x\ll 1$. Above T_c, it is clear that the spontaneous order ceases to be nonzero for $x=0$, so it is reasonable to eliminate the contribution from $\psi_2(x)$. At this point, the empirical asymptotic behavior indicated in the preceding paragraph should be incorporated. Below T_c, for $x\ll 1$, $\psi_2(x)$ should approach a constant, and for $x\gg 1$, $\psi_1(x^{-1/\beta\delta})$ should also approach a constant, according to (1a) and (1b), respectively. Above T_c, $\psi_2(x)$ does not exist and $\psi_1(x^{-1/\beta\delta})$ should approach a value proportional to $x^{1-(1/\delta)}$ for $x\ll 1$, remaining the same as below T_c for $x\gg 1$. One could try different functional expressions for ψ_2 and ψ_1, all of them susceptible to being expanded in the power series of the required form. In principle, a logarithm, a binomial, or an exponential would meet this requirement. However, after testing these three forms against the experimental data, not only for TGS but for magnetic and liquid-vapor systems, one comes to the conclusion that the logarithm changes too slowly with x and the exponential, on the other hand, too rapidly, in order to satisfy the asymptotic behavior indicated. On this ground, only the binomial form is left as a satisfactory one. The simplest binomial forms one can think of, meeting the above mentioned requirements, are

$$\psi_1(x^{-1/\beta\delta})=A\left[1+\left(\frac{x}{x_1}\right)^{-1/\beta\delta}\right]^{-\beta\delta(1-1/\delta)}, \quad (11a)$$

$$\psi_2(x)=B\left[1+\left(\frac{x}{x_2}\right)\right]^{-(1/\beta\delta)(1-\beta)}. \quad (11b)$$

The exponents $-\beta\delta(1-1/\delta)$ in Eq. (11a), and $-(1/\beta\delta)(1-\beta)$ in Eq. (11b), are the simplest ones which keep the homogeneity of Eq. (9) from $x/x_1\ll 1$ to $x/x_2\gg 1$ through the whole range in x. Also, the former is automatically required by the condition (2b).

As a check of the "phenomenological" equation of state obtained, the principal critical exponents may be calculated. Below T_c, from Eqs. (6)–(10), one obtains

$$x\ll 1, \quad \hat{Y}\simeq\psi_2 t^{\beta}\simeq Bt^{\beta}, \quad (12)$$

$$x\gg 1, \quad \hat{Y}\simeq\psi_1\hat{X}^{1/\delta}\simeq A\hat{X}^{1/\delta}, \quad (13)$$

$$x\ll 1, \quad \partial\hat{Y}/\partial\hat{X}=\hat{X}^{1/\delta}\partial\psi_1/\partial\hat{X}+(1/\delta)X^{(1/\delta)-1}\psi_1 +t^{\beta}\partial\psi_2/\partial\hat{X}=C_1t^{\beta-\beta\delta}=C_1t^{-\gamma'}. \quad (14)$$

Similarly, above T_c,

$$x\ll 1, \quad \hat{Y}\simeq\psi_1\hat{X}^{1/\delta}\simeq 0, \quad (15)$$

$$x\gg 1, \quad \hat{X}\simeq\psi_1\hat{X}^{1/\delta}\simeq A\hat{X}^{1/\delta}, \quad (16)$$

$$x\ll 1, \quad \partial\hat{Y}/\partial\hat{X}=\hat{X}^{1/\delta}\partial\psi_1/\partial X+(1/\delta)\hat{X}^{(1/\delta)-1}\psi_1 =C_2t^{\beta-\beta\delta}=C_2t^{-\gamma}. \quad (17)$$

The constants which appear in Eqs. (14) and (17) are, respectively,

$$C_1=\frac{A}{x_1^{1-(1/\delta)}}-\frac{1-\beta}{\beta\delta}\frac{B}{x_2} \quad \text{and} \quad C_2=\frac{A}{x_1^{1-(1/\delta)}}.$$

It is interesting to note that $\gamma=\gamma'$, also supported by available experimental evidence.

COMPARISON WITH EXPERIMENTAL RESULTS

Figure 2 shows that the use of Eqs. (11a) and (11b) in the equation of state given by (9) and (10), leads to excellent agreement with the experimental data for ferroelectric TGS, below T_c as well as above T_c; in this case, $y=p$ (scaled polarization) and $x=e$ (scaled electric field). Only four dimensionless numerical parameters have been used, their values being

$$A=2.12, \quad x_1=e_1=0.907,$$
$$B=1.87, \quad x_2=e_2=1.425. \quad (18)$$

For various liquid-vapor systems, it was earlier reported by Green *et al.*[7] that the data suggest a scaling law asymptotic behavior as that indicated by Eqs. (1)–(2). Using Green's *et al.* critical exponents, $\beta=0.35$ and $\delta=5.0$, one could try to fit Eqs. (9) and (10) to the data, given in the chemical potential-density representation, in order to determine A, x_1, B, and x_2. Since the scattering of the experimental points (collected from many authors on many different systems and temperature intervals) is fairly high, it does not seem to be justified. However, accurate data[15] for He^4 are available. By using the tabulated data of Roach,[15] one can calculate the fundamental critical exponents in the pressure-volume representation. Since the transition occurs at very low temperature, it is not surprising that the asymetry of the experimental data is considerable. It is convenient to bypass this difficulty

[15] P. R. Roach, Phys. Rev. **170**, 213 (1968).

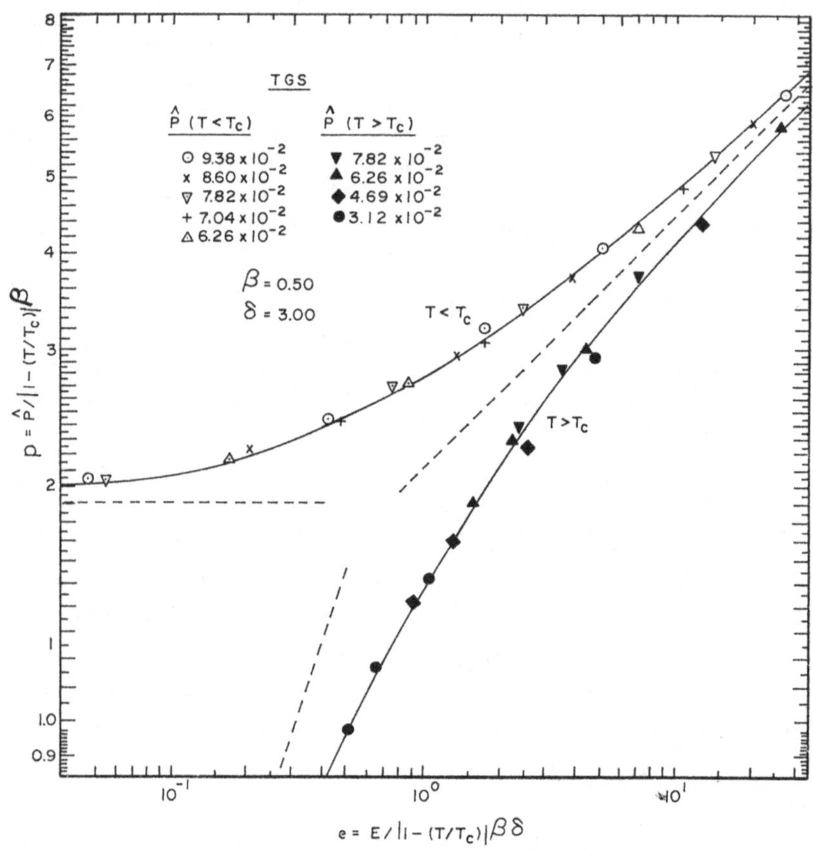

Fig. 2. Log-log plot of the scaled polarization versus scaled electric field for ferroelectric TGS near T_c. Full line is phenomenological equation of state with $A=2.12$, $B=1.87$, $x_1=0.907$, $x_2=1.425$.

by using the following definitions:

$$(V_{gas}-V_{liq})\propto \Delta T^{\beta} \text{ (along coexistence curve)}, \quad (19)$$

$$V(-\Delta P)-V(+\Delta P)\propto \Delta P^{1/\delta} \text{ (along critical isotherm)}, \quad (20)$$

where V is volume, T is temperature, and P is pressure.

In this way, a nicely defined straight line for two decades up to the vicinity of the critical point, is obtained in the log-log plot, which yields β. The analog plot for δ is only approximately linear in the last decade up to the vicinity of the critical point and since the trend suggests an increasing value of δ we extrapolate to the closest value which does not violate Griffith's inequality, taking $\alpha'\simeq 0$. This results in

$$\beta=0.411 \quad \text{and} \quad \delta=3.84. \quad (21)$$

These numerical values are somewhat different from those obtained by Roach[15] and Vicentini-Missoni[18] but it should be taken into account that they used the density instead of the volume and neglected the asymmetry. By using the exponents given by Eq. (21) one can scale the data corresponding to several isotherms above and below T_c. The results are seen in Fig. 3(a)

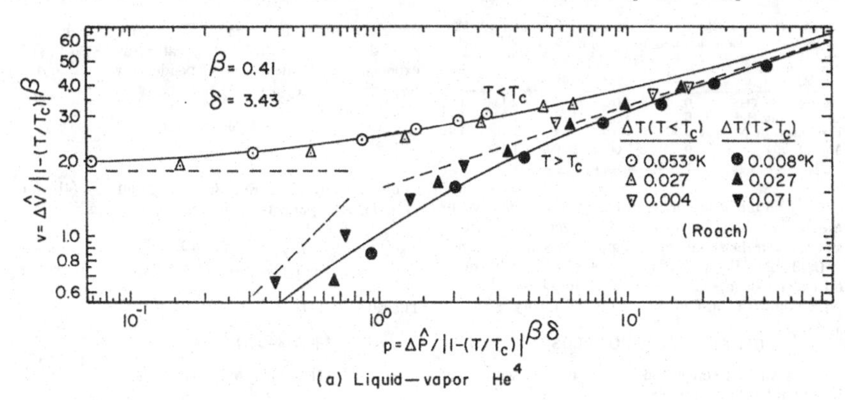

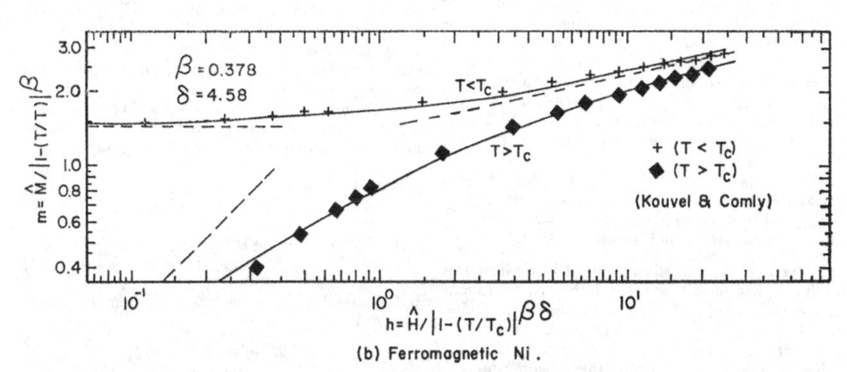

FIG. 3. Log-log plots of scaled quantities for (a) liquid-vapor He^4, volume versus pressure; (b) ferromagnetic Ni, magnetization versus magnetic field (averaged data). Full lines are phenomenological equations of state with (a) $A=1.92$, $B=1.83$, $x_1=7.57$, $x_2=11.3$; and (b) $A=1.18$, $B=1.42$, $x_1=0.334$, $x_2=0.360$.

where the phenomenological equation of state is represented, with

$$y=v=\frac{\Delta V/V_c}{t^\beta}\quad\text{and}\quad x=p=\frac{\Delta P/P_c}{t^{\beta\delta}},$$

where $V=14.49$ cc (per gram), $T_c=5.193$°K, $P_c=1710.0$ Torr. The equation that best fits the data has been obtained by using the dimensionless parameters

$$\begin{aligned}A&=1.92, & x_1&=p_1=7.57,\\ B&=1.83, & x_2&=p_2=11.3.\end{aligned}\tag{22}$$

An estimate of the experimental errors from Raach's data indicates that they go from 7 to 2% as v increases, and from 20 to 4% as p increases.

Finally, the accurate data for the ferroparamagnetic transition in Ni, by Kouvel and Comly,[8] have also been examined. The result is shown in Fig. 3(b). Again, the corresponding phenomenological equation of state is plotted with

$$y=m=\frac{M/M_0}{t^\beta}\quad\text{and}\quad x=h=\frac{H/H_0}{t^{\beta\delta}},$$

where $M_0=58.6$ emu/g, $H_0=kT_c/\mu_0=15.2\times10^6$ Oe. The equation that best fits the data is obtained with

$$\begin{aligned}A&=1.183, & x_1&=h_1=0.334,\\ B&=1.421, & x_2&=h_2=0.360.\end{aligned}\tag{23}$$

TABLE III. Comparison of critical exponents and coefficients (see text) for various real systems and the mean-field model.

System	β	δ	$\beta\delta$	A	B
Ferroelectric TGS	0.50	3.0	1.50	2.12	1.87
Liquid-vapor He^4	0.41	3.8	1.57	1.92	1.83
Ferromagnetic Ni	0.38	4.6	1.74	1.18	1.42
Mean field	0.500	3.00	1.500	1.44	1.73

Since no tables are given in Ref. 8 we cannot give an estimate of the relative errors. However, they must be small, given the small scattering of points in the graphs.

In Table III, a comparison is made of the main critical exponents and coefficients for TGS, He^4 and Ni, along with those of the mean-field theory.

SUMMARY AND CONCLUSIONS

The principal conclusions of the present work may be stated briefly as follows:

(1) The scaled data for P versus E from TGS in the vicinity and at both sides of T_c satisfy very approximately the mean-field theory predictions.

(2) A phenomenological equation of state has been constructed, based on the homogeneity assumption, which is simpler than previous proposals and reflects in a natural way the different asymptotic behavior for low- and high-scaled intensive variable (electric field in the case of TGS).

(3) The application of this phenomenological equation of state to representative ferroelectric, liquid-vapor, and ferromagnetic transitions shows fair agreement with the data in all three cases.

In the case of TGS, the agreement is excellent and improves that of the mean-field theory. All experimental points are within the estimated error limits. In the case of He^4 there is appreciable scattering of points at low p above T_c, and at high p below T_c. However, taking into account the experimental uncertainties, the agreement is fair. In the case of Ni, for which very accurate data are available, the agreement is also fair, but some small systematic deviations seem to be present, especially for $T>T_c$.

ACKNOWLEDGMENTS

I wish to thank Dr. P. Heller and Professor M. Fisher for valuable suggestions, and Dr. B. C. Frazer and Dr. G. Shirane for many stimulating discussions.

APPENDIX

From inspection of Eqs. (3a) and (3b), it is readily seen that the two critical exponents from the same derivative of the free energy, $F_{n,m}=\partial^{n+m}F/\hat{X}^n\partial t^m$, are related in a simple way. Let us call γ_k and γ_{k+1} to the exponents which define the dependence with $\hat{X}$ and t, respectively. They are

$$\gamma_k=(\delta+1)/\delta-m(1/\beta\delta)-n\,,\qquad \gamma_{k+1}=\beta(\delta+1)-n\beta\delta-m\,, \tag{A1}$$

as obtained by using Eq. (4a) for γ_k and Eq. (4b) for γ_{k+1}. These exponents define the relationships

$$F_{m,n}\sim\hat{X}^{\gamma_k},\quad (\hat{X}\gg t^{1/\beta\delta}),\qquad F_{m,n}\sim t^{\gamma_{k+1}},\quad (t^{\beta\delta}\gg\hat{X}). \tag{A2}$$

The ratio is then

$$\frac{\gamma_k}{\gamma_{k+1}}=\frac{(\delta+1)/\delta-m(1/\beta\delta)-n}{\beta\delta(\delta+1)/\delta-n\beta\delta-m}=\frac{1}{\beta\delta}=\text{const.} \tag{A3}$$

This general *equality* is quite useful, and is already implicit in earlier theoretical work (see Fisher's[12] work and references therein.) By using Eq. (A3), one can easily construct equalities relating triads of critical exponents. Let us take, for instance, the four most commonly used exponents, i.e., δ(critical isotherm), β(coexistence curve), $-\gamma'$ (compressibility versus temperature), and $-\alpha'$ (specific heat versus temperature). Four triads can be made in the following way:

$$[(1/\delta)+1]/(-\alpha'+2)=1/\beta\delta,\quad \text{i.e., } \alpha'+\beta(1+\delta)=2, \tag{A4}$$

$$[(1/\delta)-1]/(-\gamma')=1/\beta\delta,\quad \text{i.e., } +\gamma'+\beta(1-\delta)=0, \tag{A5}$$

and eliminating successively β and δ from Eqs. (A4) and (A5),

$$\alpha'+2\delta-\gamma'-\delta(\alpha'+\gamma')=2, \tag{A6}$$

$$\alpha'+\gamma'+2\beta=2. \tag{A7}$$

The expressions (A4), (A5), and (A7) can be recognized as the equality form of relationships introduced by Griffiths, Widom (also referred to as Kouvel-Rodbell relation), and Fisher-Rooshbrook. By using (A3), any desired relationship between three arbitrary critical exponents may be easily obtained.

The series of experimental exponents available from TGS enable a direct experimental check on the constancy of the exponent ratio specified by Eqs. (A3). Table I shows that the critical exponents associated with $F_{1,0}=P$, $F_{2,0}=\partial P/\partial E$, and $F_{1,1}=\partial P/\partial T$ satisfy the expected ratio $\gamma_k/\gamma_{k+1}=\frac{3}{2}$.

Reprinted from *Acta Crystallographica*, Vol. A 29, Part 4, July 1973

PRINTED IN DENMARK

Acta Cryst. (1973). A**29**, 423

X-ray Structural Damage of Triglycine Sulphate (TGS)

BY C. ALEMANY, J. MENDIOLA, B. JIMENEZ AND E. MAURER

Centro de Investigaciones Fisicas 'L. Torres Quevedo' Serrano 144, *Madrid* 6, *Spain*

(*Received* 8 *February* 1973; *accepted* 12 *February* 1973)

The complete spectrum emitted by a conventional X-ray diffraction Cu target produces damage in a small TGS crystal which is evident from the variation of integrated intensities of X-ray reflexions with irradiation time. An interpretation of the data is proposed which assumes that the trapping of irradiation products causes an exponential decrease of mosaic-block diameters. An empirical correction of the Zachariasen extinction factor for crystals belonging to type II is suggested, since the TGS crystal is of this type.

Introduction

Structure analysis requires the accurate determination of both integrated intensities and structure factors. It is well known that X-rays used to collect data have undesirable effects on the crystals. These effects are ascribed to material instability, or to defects produced by X-ray damage. Young (1969) and Milledge (1969) have made detailed reviews of the different problems concerning the precise determination of integrated X-ray reflexion intensities, emphasizing how to avoid the effects of damage rather than explaining the production mechanism.

There is a copious bibliography covering the topic of irradiation damage with special reference to X-ray diffraction effects. Examples are Lonsdale, Nave & Stephens (1966), Kolontsova & Telegina (1969), Krueger, Cook, Sartain & Yockey (1963), Telegina & Kolontsova (1970), Larson & Young (1972) and Baldwin & Dunn (1972). Work on X-irradiated TGS crystals has been carried out by Petroff (1971), who detected planar defects, and Mendiola & Alemany (1970), who pointed out large variations in the intensities of X-ray reflexions with cumulative doses.

In this paper we study the variation of integrated intensities of X-ray reflexions with time when a single crystal is irradiated by the complete spectrum emitted from a conventional X-ray diffraction Cu target. For reflexions with $F \geq 16$, a continuous increase in intensity is observed until a maximum is reached, followed by a decrease; but for reflexions with $F < 16$ the intensities diminish from the beginning. This behaviour is fairly well explained by an empirical correction to the Zachariasen extinction factor for crystals belonging to type II, as we suggest for TGS crystals, and assuming an exponential decrease of mosaic-block radius as well.

The TGS lattice parameters and the observed and calculated structure factors of the reflexions used in this paper are the early ones reported by Hoshino, Okaya & Pepinsky (1959). The results obtained there are substantially the same as those deduced from the fractional atomic coordinates x,y,z and temperature factors B_{ij} given by Itoh & Mitsui (1971). However we do not follow the latter paper because no extinction correction is made. In this paper we show that the inclusion of an extinction correction is necessary.

Theory

The concepts that are used in discussing the results have been exhaustively developed by Zachariasen (1967*a,b*; 1968*a,b,c,d*; 1969). According to his theory, the integrated intensity of a reflexion from a symmetrically shaped crystal of volume v, assumed to consist of nearly spherical domains of radius r, is given by

$$I = I_k \cdot y \qquad (1)$$

for unpolarized X-rays, where

$$I_k \text{ (the kinematical value)} = I_0 vAQ_0 \left(\frac{1+\cos^2 2\theta}{2}\right) \quad (2)$$

$$y \text{ (the extinction factor)} = (1+2x)^{-1/2} \quad (3)$$

$$x = \frac{1+\cos^4 2\theta}{1+\cos^2 2\theta} Q_0\lambda^{-1}Tr^* = Z \,.\, r^* \quad (4)$$

$$Q_0\lambda^{-1} = \left[\frac{e^2\lambda F}{mc^2V}\right]^2 \cdot \frac{1}{\sin 2\theta} \quad (5)$$

$$T = -\frac{1}{A}\frac{dA}{d\mu} \quad (6)$$

$$r^* = r\left[1 + \left(\frac{r}{\lambda g}\right)^2\right]^{-1/2}. \quad (7)$$

I_0 is the incident intensity, v the volume of the specimen, A the transmission factor and g the factor determining the misorientation of the perfect domains in the crystal. The other symbols have their usual meanings. Zachariasen (1967*b*) shows that very large extinction effects cannot occur in crystals of type I (for which $r/\lambda g \gg 1$) whilst they are possible for crystals of type II (*i.e.* $x \gg 1$ and $r/\lambda g \ll 1$, whence $r^* = r$).

The equation $y = (1+2x)^{-1/2}$ prevails over the two other theoretical expressions of the extinction factor (Zachariasen, 1967*b*) as shown in Table I from Zachariasen (1968*a*) where the calculated values of y are equal to or less than 0·5. Nevertheless, there is no experimental evidence of $y = (1+2x)^{-1/2}$ for y values close to unity (*i.e.* $x \ll 1$) since weak reflexions of small structure factor have never been corrected for extinction.

Let us restrict our discussion to crystals of type II ($r^* = r$). Now if r tends to zero through an external agent such as irradiation damage, x moves toward zero and y towards unity; therefore it is not necessary to make any extinction correction according to Zachariasen. However, as r diminishes, the blocks contain progressively fewer unit cells and in the limit $r = 0$ (*i.e.* $x = 0$), the crystal periodicity is completely lost and consequently so is the diffracted beam. Thus we propose a correction to Zachariasen's extinction factor for $x \ll 1$ in order to obtain a function that matches qualitatively the full-line curve in Fig. 1. Now, we may write for such a function

$$y = (1+2x)^{-1/2} \,.\, F(x) \quad (8)$$

where

$$F(x) = 0 \quad \text{for} \quad x = 0$$
$$F(x) \simeq 1 \quad \text{for} \quad x \geq 1 \,. \quad (9)$$

The last boundary condition comes from the experimental verification of Zachariasen's extinction factor for $x \geq 1$. Our present extinction function y shows a maximum with coordinates x_{max}, y_{max}.

Experimental

The crystals were grown by evaporation from a water solution of triglycine sulphate obtained through chemical reaction of pure products. In the first stages, needle-like crystals along [001] were obtained, from which a small specimen 0·2 × 0·3 × 0·5 mm was chosen. In order to make a correction for absorption, according to equation (6), the sample has been regarded as a sphere of diameter $T = 2{\cdot}3 \times 10^{-2}$ cm.

The crystal was centred along [001] on a goniometer head in a G.E. XRD-6 diffractometer at a distance of 5·73 in from the target. An unfiltered circular beam from a conventional Cu X-ray tube GE/CA-8S operating at 40 kV and 17 mA and equivalent to a radiation rate of 400 R/min was used to irradiate the sample, producing continuous damage. Meanwhile the sample was successively set at the Bragg angles of several reflexions for λ(Cu $K\alpha$), and the diffracted beams recorded by means of a proportional counter after being β-filtered, using a 2θ scanning speed of 0·4° min^{-1}, a range of ±0·6° about each Bragg angle, and a receiving slit of 0·2°. With this procedure the dead time between successive irradiations was eliminated, avoiding the possible recovery of the crystal (Mendiola & Alemany, 1970). Correction for thermal diffuse scattering has been disregarded.

The integrated intensities before irradiation, I_0, were obtained with nickel-filtered radiation, which it was assumed does not cause any defects. The experimental attenuation factor was 2·165, the sample being maintained at room temperature.

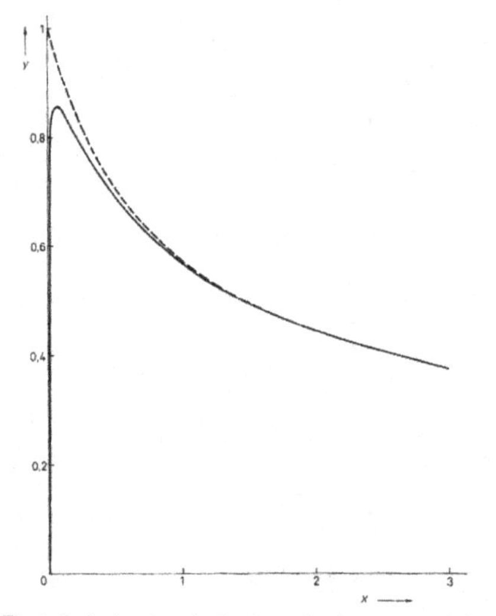

Fig. 1. Zachariasen's extinction factor (broken curve) and the proposed factor (solid curve).

Results and discussion

The data summarized in Fig. 2 show that intensity increases to a maximum and then decreases, except for the 051 reflexion. In Table 1 the reflexions considered are shown together with the observed and calculated structure factors, the Z-function values [as defined in equation (4)], the irradiation times, and the intensity ratios at the maximum.

A separate irradiation experiment was conducted with a new sample in the paraelectric phase at 68 °C where the behaviour of the integrated intensity of the 040 reflexion was studied. This intensity is compared in Fig. 3 with that obtained at room temperature. From this Figure it follows that the large intensity change upon irradiation is independent of whether the crystals are below or above their Curie temperature. On the other hand, there is a small intensity decrease during the first few minutes of irradiation which does not appear when the crystal is in the paraelectric phase. A sharp decrease of electrical capacitance has been found to occur during this short initial period (Alemany, Mendiola, Jimenez & Maurer, 1973) which suggests a probable relationship between the initial minimum in the intensities and the ferroelectricity, and hence an improvement of the crystal structure connected to the trap of domain walls; since there are no domain walls in the paraelectric phase, the initial minimum in the integrated intensities does not appear. Krueger, Cook, Sartain & Yockey (1963) have also observed an increase in perfection of γ-irradiated Rochelle salt resulting from low levels of radiation.

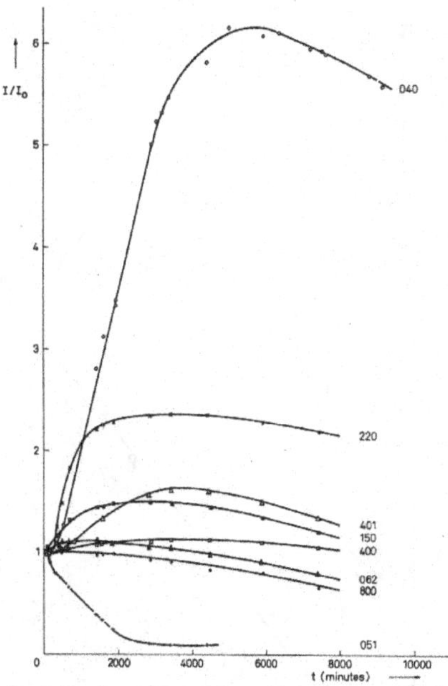

Fig. 2. Experimental X-ray integrated intensities *versus* irradiation time.

From Table 1 and equation (1), it follows that: for the 220 reflexion,

$$\frac{I_{\max}}{I_0} = \frac{y_{\max}}{y_0} = 2{\cdot}34;$$

for the 040 reflexion,

$$\frac{I_{\max}}{I_0} = \frac{y_{\max}}{y_0} = 6{\cdot}15\,.$$

That is to say, these reflexions have very large extinction effects and therefore TGS crystals may be assigned to type II which according to Zachariasen (1967*b*) fulfil $r^* = r$.

Gilleta, Taurel & Lauginie (1969) observed dislocations in TGS crystals produced by irradiation. Assuming these dislocations intermingle with the Frank network which forms the mosaicity of real crystals (Hedges & Mitchell, 1953), a continous decrease of the mean radius of mosaic blocks with time should be obtained. Let us suppose the decrease of radius is

$$r = r_0 \exp\left(-\frac{t}{\tau}\right) \qquad (10)$$

where r_0 is the mean radius of mosaic blocks in the undamaged crystal.

From (4) we can put

$$x = Zr_0 \exp\left(-\frac{t}{\tau}\right). \qquad (11)$$

Table 1. *Reflexions considered, together with observed and calculated structure factors, Z-function values, irradiation times and intensity ratios at the maxima*

Reflexion	F_o	F_c	Z_o(cm^{-1})	Z_c(cm^{-1})	$t_{\max}$(min)	$I_{\max}/I_0$
051	5·53	9·62	43·36	131·20	—	—
800	16·58	16·26	290·75	279·64	20	1·02
062	24·45	26·24	641·39	738·75	1000	1·11
400	31·72	27·39	1369·80	1021·35	3000	1·13
150	47·07	51·27	3351·17	3975·89	3200	1·48
401	57·30	62·94	3770·14	4548·85	3500	1·61
220	73·47	74·21	12709·01	12966·31	4500	2·34
040	102·22	117·94	21065·78	28043·23	5500	6·15

As x diminishes, the representative points of the reflexions move along the full curve of Fig. 1 according to the previous time dependence. If $x > x_{max}$, this latter value will be reached at t_{max}, when the maximum intensity is obtained. This value can then be written

$$x_{max} = Zr_0 \exp\left(-\frac{t_{max}}{\tau}\right)$$

$$\frac{t_{max}}{\tau} = \ln Z - \ln \frac{x_{max}}{r_0}. \qquad (12)$$

In Fig. 4, the experimental t_{max} values *versus* $\ln Z$ are shown to lie on a straight line, thus confirming the hypothesis of exponential decrease of the radius of mosaic blocks. From the ordinate intercept and slope of the graph we obtain

$$\frac{x_{max}}{r_0} = 266 \text{ cm}^{-1} \qquad (13)$$

$$\tau = 1180 \text{ min}.$$

Finally, by determining x_{max}, we can calculate the mean radius of the blocks in the undamaged crystal.

As (8) shows, x_{max} will depend on the function $F(x)$, but the exact form of this function is not known at this stage. We have approached the problem empirically, trying several forms of $F(x)$ to fit the curves of I_t/I_0 *versus* radiation time for the eight reflexions studied. The following forms were used:

$$F(x) = \exp\left(-\frac{a}{x^n}\right) \quad \text{for} \quad n = \tfrac{1}{2}, \tfrac{1}{3}, \tfrac{1}{4}, 1 \text{ and } 2$$

and

$$F(x) = 1 - \exp\left[-a(2x)^n\right] \quad \text{for} \quad n = 1, \tfrac{1}{2}, \tfrac{1}{3}, \text{and } \tfrac{1}{4} \quad (14)$$

as well as

$$y = (1+2x)^{-1/2} - (1+2x)^{-a}.$$

The best fit was obtained by putting

$$F(x) = 1 - \exp\left[-a(2x)^{1/4}\right] \qquad (15)$$

to give

$$y = (1+2x)^{-1/2}\{1 - \exp\left[-a(2x)^{1/4}\right]\}. \qquad (16)$$

The (x_{max}, y_{max}) coordinates are given as a function of the a parameter. By using equations (1) and (16), Fig. 5 shows the experimental and theoretical I_{max}/I_0 values *versus* Z. An a-curve family is then obtained, those for $a = 4$ fitting best the experimental points. Only the 040 reflexion does not lie on the empirical curve, probably because the structure-factor error has been exaggerated by X-ray damage.

Putting $a = 4$ in equation (16) we obtain the curve shown in Fig. 1 with $x_{max} = 0{\cdot}067$ and $y_{max} = 0{\cdot}855$. Using this value of x_{max} in equation (13) a value for $r_0 = 2{\cdot}52 \times 10^{-4}$ cm is obtained. This is the mean radius of the blocks for the undamaged crystal and agrees satisfactorily with the expected order of magnitude for a real crystal. At the same time we note that r_0 is slightly dependent on $F(x)$ provided that the parameter a has

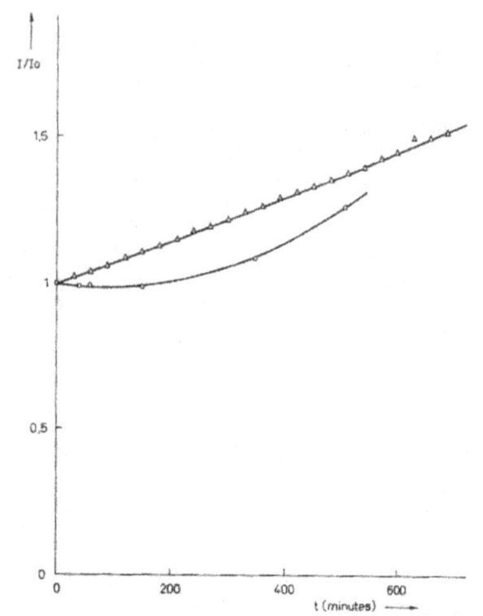

Fig. 3. Irradiation-time dependence of the 040 X-ray integrated intensity (○: crystal at 22 °C; △: crystal at 68 °C).

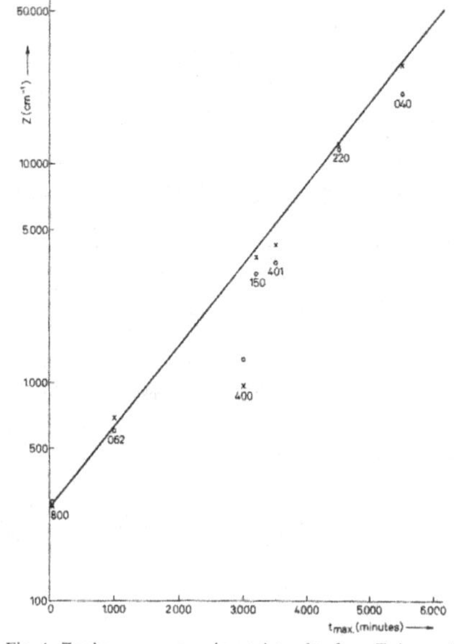

Fig. 4. Z values *versus* experimental t_{max} (○: from F observed; ×: from F calculated).

been chosen according to the limitations set by equation (9). Thus we can represent each reflexion by the function $I_t/I_0=y_t/y_0$ since we know all the parameters.

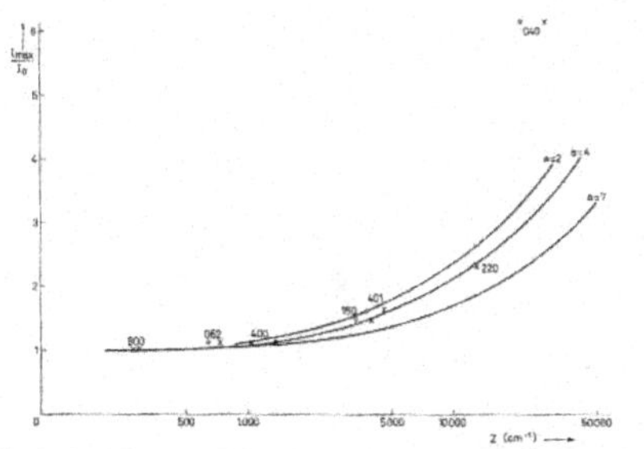

Fig. 5. I_{max}/I_0 *versus* Z. The curves are the theoretical values corresponding to different selected a parameters; ○ and × are the experimental points obtained when observed or calculated F are used to get Z.

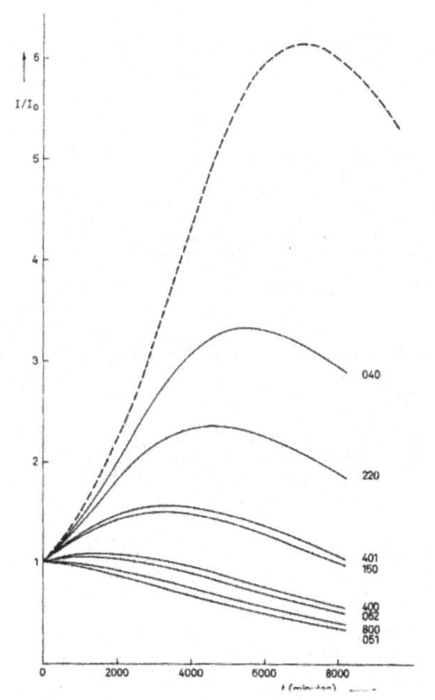

ig. 6. Calculated X-ray integrated intensities obtained by using F_c *versus* irradiation time. The broken curve corresponds to 040 when the proposed $F=223$ is used.

as shown in Fig. 6. In order to obtain the best agreement, a new value of the structure factor for 040 is also proposed in Fig. 6.

Disagreement between the experimental and calculated values could be due to large errors in Z resulting from inaccurate knowledge of the TGS crystal structure and also because of the simplicity of Zachariasen's mosaic-crystal model which only considers spherical blocks.

Conclusions

Defects created in TGS by large X-ray exposures perturb the crystal periodicity and therefore the X-ray diffraction extinction factor.

The proposed model assumes an exponential decrease, due to a defect-trapping process of lattice dislocation, in the mean radius of spherical blocks as the irradiation time increases, and a correction for the Zachariasen extinction factor for $x \ll 1$ in his theory of real crystals.

The authors are indebted to J. Tapia and A. B. Clarke for correcting the original manuscript.

References

ALEMANY, C., MENDIOLA, J., JIMENEZ, B. & MAURER, E. (1973). *Ferroelectrics*. In the press.

BALDWIN, T. O. & DUNN, J. E. (1972). Paper presented at 9th Int. Congr. Crystallogr. Kyoto. *Acta Cryst.* **A28**, S157.

GILLETA, F., TAUREL, L. & LAUGINIE, P. (1969). *Proc. Eur. Meet. Ferroelectr. Saarbrücken*, p. 225.

HEDGES, J. M. & MITCHELL, J. W. (1953). *Phil. Mag.* **44**, 223–224.

HOSHINO, K., OKAYA, Y. & PEPINSKY, R. (1959). *Phys. Rev.* **115**, 323–330.

ITOH, K. & MITSUI, T. (1971). *Ferroelectrics*, **2**, 225–226.

KOLONTSOVA, E. V. & TELEGINA, I. V. (1969). Paper presented at 8th Int. Congr. Crystallogr., New York. *Acta Cryst.* **A25**, S39.

KRUEGER, H. H. A., COOK, W. R., SARTAIN, C. C. & YOCKEY, H. P. (1963). *J. Appl. Phys.* **34**, 218–224.

LARSON, B. C. & YOUNG, F. W. (1972). Paper presented at 9th Int. Congr. Crystallogr. Kyoto. *Acta Cryst.* **A28**, S155.

LONSDALE, K., NAVE, E. & STEPHENS, J. F. (1966). *Phil. Trans.* **A261**, 1–31.

MENDIOLA, J. & ALEMANY, C. (1970). *Electron. Fis. Apl.* **13**, 237–242.

MILLEDGE, H. J. (1969). *Acta Cryst.* **A25**, 173–180.

PETROFF, J. F. (1971). 2nd Eur. Meet. Ferroelectr. Dijon, Abstract 22aB1.

TELEGINA, I. V. & KOLONTSOVA, E. V. (1970). *Kristallografiya*, **15**, 195–196.

YOUNG, R. A. (1969). *Acta Cryst.* **A25**, 55–66.

ZACHARIASEN, W. H. (1967*a*). *Phys. Rev. Lett.* **18**, 195–196.

ZACHARIASEN, W. H. (1967*b*). *Acta Cryst.* **23**, 558–564.

ZACHARIASEN, W. H. (1968*a*). *Acta Cryst.* **A24**, 212–216.

ZACHARIASEN, W. H. (1968*b*). *Acta Cryst.* **A24**, 324–325.

ZACHARIASEN, W. H. (1968*c*). *Acta Cryst.* **A24**, 421–424.

ZACHARIASEN, W. H. (1968*d*). *Acta Cryst.* **A24**, 425–427.

ZACHARIASEN, W. H. (1969). *Acta Cryst.* **A25**, 102.

PHYSICAL REVIEW B

CONDENSED MATTER

THIRD SERIES, VOLUME 53, NUMBER 6 — 1 FEBRUARY 1996-II

RAPID COMMUNICATIONS

Rapid Communications are intended for the accelerated publication of important new results and are therefore given priority treatment both in the editorial office and in production. A Rapid Communication in **Physical Review B** *may be no longer than four printed pages and must be accompanied by an abstract. Page proofs are sent to authors.*

Equation of state for the pressure- and temperature-induced transition in ferroelectric telluric acid ammonium phosphate

José R. Fernández-del-Castillo, Janusz Przeslawski,* and Julio A. Gonzalo

Departamento de Física de Materiales, C-IV, Universidad Autónoma de Madrid, 28049 Madrid, Spain

(Received 4 August 1995)

A careful experimental investigation of the ferroelectric-paraelectric transition in uniaxial ferroelectric telluric acid ammonium phosphate induced by pressure and by temperature in the vicinity of the phase transition has been carried out. Dielectric-constant and hysteresis-loop data allow the determination of the combined equation of state as a function of both temperature and pressure, relating accurately polarization, field, pressure, and temperature in the neighborhood of the transition.

Solid-state pressure-induced phase transitions are similar to temperature-induced phase transitions in many respects. The analogy between the role played by hydrostatic pressure in the former case, with the role played by thermal energy, in the latter, can be exploited to obtain a compact equation of state describing simultaneously the pressure and the temperature dependence of the relevant pair of conjugated variables (polarization and electric field in a ferroelectric). To make a proper quantitative investigation of the equation of state in the vicinity of the transition, accurate data, in terms of the hydrostatic pressure as well as in terms of temperature, are needed. Uniaxial ferroelectric telluric acid ammonium phosphate[1–8] (TAAP) chemical formula $|Te(OH)_6|\cdot|2NH_4H_2PO_4|\cdot|(NH_4)_2PO_4|$, is a good candidate for investigation, because it has (for ambient pressure, 1 bar) a second-order phase transition[2] to the paraelectric state at $T_c(0)\approx 320$ K, conveniently above room temperature, and also because it possesses a negative pressure coefficient,[4,5] $dT_c(p)/dp\approx -3.58$ °C/kbar, which allows investigation of pressure-induced transitions at constant temperatures between room temperature and 320 K. This crystal is monoclinic at room temperature ($Pn, Z=2$). At the ferroelectric transition the change in space group is $P2/n \leftrightarrow Pn$. The main feature of the atomic arrangement is the presence of two distinct anionic groups: TeO_6 and PO_4. Two types of planes perpendicular to the [101] direction can be distinguished. In the first type one can find PO_4 groups and NH_4O_n polyhedra, while in the second one, one has TeO_6 groups in addition to PO_4 groups and NH_4O_n polyhedra.[6] Four types of hydrogen bonds N-H-O linking PO_4 tetrahedra are present. All of them are asymmetric and the protons are *ordered* in the ferroelectric phase.[7] This ordering was confirmed also by the increase of T_c upon deuteration and Raman studies,[8] which showed that the groups of TeO_6, PO_4, and NH_4 do not undergo changes that could trigger the phase transition.

In this work we present and analyze high-resolution data ($\Delta p\approx 1.5$ bar steps, in an interval of about 3 kbar; $\Delta T \approx 0.01$ °C steps, in intervals of about 5 °C) on the dielectric constant and hysteresis loops for TAAP single crystals, undertaken with the aim of determining a combined pressure and temperature equation of state.

The samples were small platelets, $3\times 2\times 1.2$ mm^3 in size, cut from a larger single crystal grown from water solution, coated with silver paste electrodes deposited on the main surfaces. The sample holder, in which pressure and temperature could be accurately controlled, as mentioned before, was an adapted LC10 Unipress cell. Measurements of dielectric constant and loss factor were performed with a HP Precission LCR Meter 4384A ($E_{meas}=1$ V/cm, $f=1$ kHz). Automatic data, together with temperature and pressure readings were collected by means of a top desk computer. A Diamant-Drenck-Pepinsky bridge and a NICOLET 310 digital oscilloscope were used to do the digitalized hysteresis loops measurements ($E_{meas}=1.8$ kV/cm, $f=50$ Hz). A computer program was used to correct for internal bias in the hysteresis loops.

0163-1829/96/53(6)/2903(4)/$06.00 — 53 — R2903

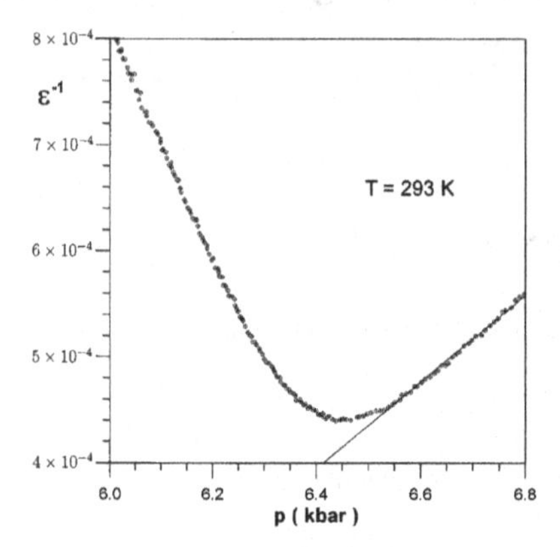

FIG. 1. Inverse dielectric constant ε^{-1} as a function of hydrostatic pressure at $T=293$ K for TAAP. Straight line indicates the Curie-Weiss behavior at $p>p_c=5.13$ kbar (note that vertical scale starts at $\varepsilon^{-1}=4\times10^{-4}$).

Figure 1 shows dielectric-constant data as a function of pressure through the transition at room temperature. The data can be fitted well with a Curie-Weiss-type law with pressure.

Figure 2 gives similar dielectric-constant data as a function of temperature through the transition at several pressures, which also follow well a Curie-Weiss law. It may be noted that $\varepsilon_{\max}$ becomes smaller for lower transition temperatures.

Figures 3(a) and 3(b) display normalized square spontaneous polarization as a function of temperature (a) at ambient pressure (1 bar), and pressure (b) at constant temperature $T=306.65$ K. It may be noted that, because of the lack of

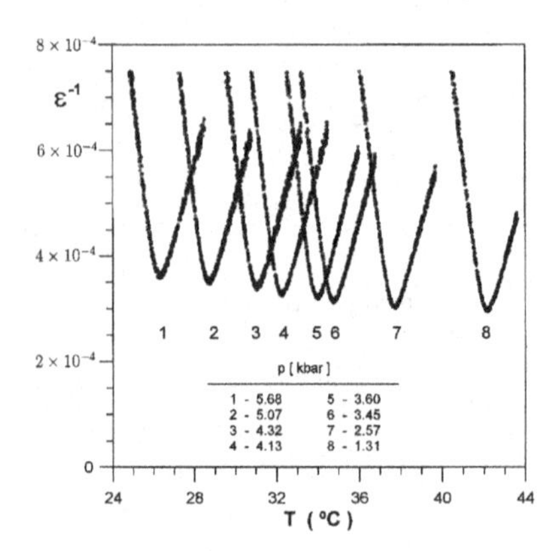

FIG. 2. Isobars of the temperature dependence of the inverse dielectric constant ε^{-1} for TAAP.

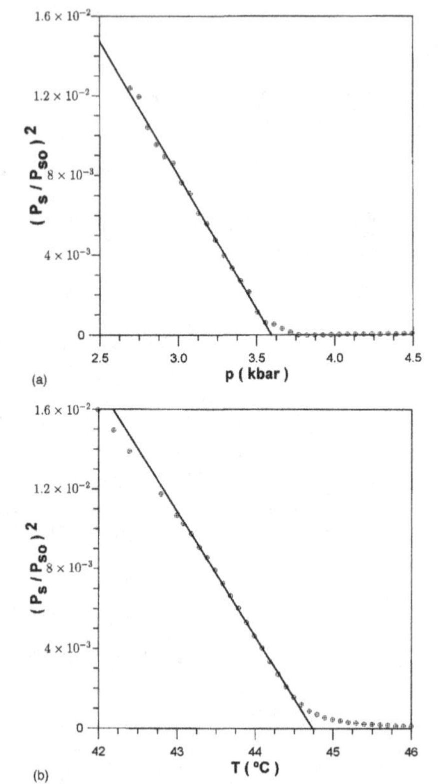

FIG. 3. Normalized spontaneous polarization square $(P_s/P_{so})^2$ vs pressure at $T=306.65$ K (a), and vs temperature at ambient pressure (1 bar) (b). $P_{so}=2.12$ μC/cm^2 is the saturation spontaneous polarization.

perfect compensation in the hysteresis loops, T_c and p_c are slightly overestimated.

Two sets of hysteresis-loop data, one at constant pressure (ambient pressure, i.e., 1 bar), and another at constant temperature ($T=306.65$ K, i.e., near and below the ambient pressure Curie temperature) are represented in Fig. 4 in scaled form. They fit well a combined equation of state, discussed below, which is easily deduced from basic considerations. It must be noted that we are not aware of previous descriptions of observed pressure and temperature behavior near a ferroelectric transition by means of a single equation of state.

In an order-disorder ferroelectric phase the existence of microscopic dipole moments reversible under the action of an external electric field can be associated with the presence of a double potential minimum along the polar axis of the

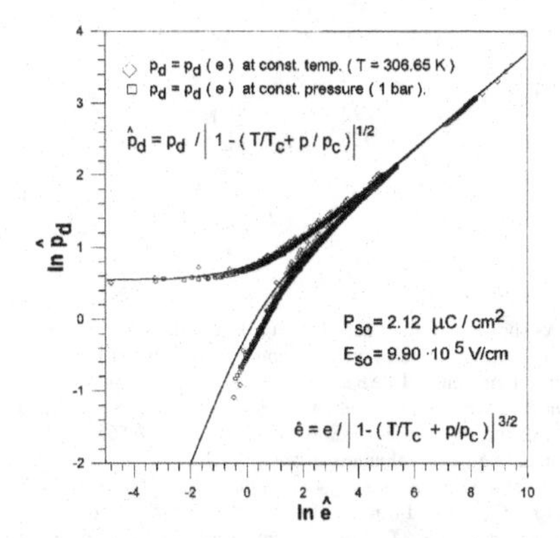

FIG. 4. Simultaneous scaling equation of state for the pressure-induced and the temperature-induced ferroelectric transition in TAAP. The continuous line is the theoretical scaling equation of state $\hat{e} \cong \pm \hat{p}_d + \frac{1}{3}\hat{p}_d^3$, with a − sign for $T<T_c$ and for $p<p_c$, and a + sign for $T>T_c$ and for $p>p_c$. $T_c(p=1\text{ bar})=317.25$ K, $p_c(T=0$ K$)=76.87$ kbar; $P_{so}=2.12$ μC/cm^2 the saturation spontaneous polarization, and $E_{so}=9.90\times10^5$ V/cm, the saturation spontaneous field, give the best fit to the theoretical equation of state.

unit cell for specific constituent ions. In the paraelectric, i.e., fully disordered, phase the jump probability per unit cell and unit time of those specific ions between the two symmetric minima is usually given by $\pi(\phi)=\nu e^{-\phi/k_BT}$, where $\nu\approx k_B\Theta_D/h$ is the attempt frequency, and ϕ the potential barrier height between the two minima. If the barrier height is increased by a small amount $d\phi$, the jump probability should decrease in such a way that

$$d\pi_0(\phi)=-(1/\phi^*)\pi_0 d\phi, \tag{1}$$

and then

$$\pi_0(\phi)=\pi_0(0)e^{-\phi/\phi^*}. \tag{2}$$

Here ϕ^* can be given in terms of the critical barrier height at which the ferroelectric-paraelectric transition, induced either by rising temperature or rising hydrostatic pressure, takes place. In a transition induced by temperature only (i.e., under ambient pressure) we have[9]

$$\phi_{\text{eff}}(T_c,0)=\beta N\mu^2=k_BT_c, \tag{3}$$

where T_c is the transition temperature and $E_{so}\mu=\beta N\mu^2$ is the electrostatic energy of the spontaneous field acting on the unit dipole μ. In a transition induced by pressure only (at 0 K) also

$$\phi_{\text{eff}}(0,p_c)=\beta N\mu^2=\Delta v_c p_c, \tag{4}$$

where p_c is the critical hydrostatic pressure and Δv_c the associated change in unit cell volume v_c needed to bring about the pressure-induced transition.

Consequently, in the general case of $T<T_c$, $p<p_c$ we may take

$$\phi^*=\phi_{\text{eff}}(T,p)=\beta N\mu^2\left(\frac{T}{T_c}+\frac{p}{p_c}\right). \tag{5}$$

This expression is used in the following to get in a straightforward manner a combined pressure/temperature equation of state for the ferroelectric transition.

In equilibrium under zero external field at the paraelectric phase, where the double potential minimum is symmetric, the number of dipoles per unit-volume pointing in the + direction (N_2) and in the − direction (N_1) are related by $N_2\pi_{21}=N_1\pi_{12}$, and since $\pi_{12}=\pi_{21}=\pi_0$, one has $N_1=N_2=N/2$. In general, however, i.e., under an effective field

$$E_{\text{eff}}=E+\beta P_d+\gamma P_d^3+\delta P_d^5+\cdots \tag{6}$$

one has

$$\pi_{21}=\pi_0(0)e^{-(E+\beta P_d+\cdots)\mu/\phi^*},$$

$$\pi_{12}=\pi_0(0)e^{(E+\beta P_d+\cdots)\mu/\phi^*}, \tag{7}$$

which, in equilibrium ($N_2\pi_{21}=N_1\pi_{12}$) leads to

$$\frac{P_d}{N\mu}=\frac{(N_2-N_1)\mu}{(N_2+N_1)\mu}=\tanh\left(\frac{(E+\beta P_d+\cdots)\mu}{\phi^*}\right) \tag{8}$$

and, using Eq. (5), to the combined equation of state

TABLE I. H=heating; C=cooling; D=decreasing of pressure. $T_c=T_c(p=0)=317.25$ K and $p_c=p_c(T=0)=76.87$ kbar were used for all data. $T_c(p\neq0)$ is the actual transition temperature for a specific pressure $p\neq0$; $p_c(T\neq0)$ is likewise the actual transition pressure for a specific temperature $T\neq0$.

	$T_c(p)$ (K)	p (kbar)	$T_c(p)/T_c$	(p/p_c)	$\{[T_c(p)/T_c]+(p/p_c)\}$
H	299.56	5.688	0.944	0.074	1.018
C	299.56	5.688	0.944	0.074	1.018
H	302.00	5.071	0.951	0.066	1.017
C	301.90	5.071	0.951	0.066	1.017
H	304.36	4.452	0.959	0.058	1.017
C	304.22	4.327	0.958	0.056	1.015
H	304.61	4.452	0.960	0.058	1.018
C	304.30	4.452	0.959	0.058	1.017
H	305.58	4.074	0.963	0.053	1.016
C	305.42	4.137	0.962	0.054	1.016
H	307.28	3.601	0.968	0.047	1.015
C	307.14	3.601	0.968	0.047	1.015
H	308.12	3.394	0.971	0.044	1.015
C	307.94	3.452	0.970	0.045	1.015
H	310.98	2.556	0.980	0.033	1.013
C	310.83	2.578	0.979	0.033	1.013
H	315.35	1.310	0.994	0.017	1.011
C	315.30	1.315	0.993	0.017	1.010
	T (K)	$p_c(T)$ (kbar)	(T/T_c)	$p_c(T)/p_c$	$\{(T/T_c)+[p(T)/p_c]\}$
D	293.15	5.13	0.924	0.067	0.991

$$\frac{P_d}{N\mu}=\tanh\left\{\left(\frac{T}{T_c}+\frac{p}{p_c}\right)^{-1}\left(\frac{E+\beta P_d+\cdots}{\beta N\mu}\right)\right\}, \quad (9)$$

or

$$\frac{E}{\beta N\mu}=\left(\frac{T}{T_c}+\frac{p}{p_c}\right)\tanh^{-1}\left(\frac{P_d}{N\mu}\right)$$
$$-\beta P_d\left[1+\left(\frac{\gamma}{\beta}\right)P_d^2+\left(\frac{\delta}{\beta}\right)P_d^4+\cdots\right], \quad (10)$$

which can be written in dimensionless form as

$$e=\left(\frac{T}{T_c}+\frac{p}{p_c}\right)\tanh^{-1}p_d-p_d(1+gp_d^2+hp_d^4+\cdots), \quad (11)$$

where p=pressure and p_d=reduced dipolar polarization must not be confused. $T_c=T_c(p=0)$ is the transition temperature at $p=0$, and $p_c=p_c(T=0)$ is the transition pressure at $T=0$ K.

For a continuous second-order transition,[10] close to the ordinary critical point ($p_d \ll 1$), Eq. (11) can be put in scaled form as

$$\hat{e}\cong\pm\hat{p}_d+\frac{1}{3}\hat{p}_d^3, \quad (12)$$

where

$$\hat{e}\equiv e\Big/\left|1-\left(\frac{T}{T_c}+\frac{p}{p_c}\right)\right|^{3/2},$$

$$\hat{p}_d\equiv p_d\Big/\left|1-\left(\frac{T}{T_c}+\frac{p}{p_c}\right)\right|^{1/2}. \quad (13)$$

As shown in Fig. 4, Eqs. (12)–(13) describe simultaneously the pressure-induced and the temperature-induced ferroelectric transition of TAAP simply and fairly accurately. From p_c and T_c we estimated the unit cell volume change associated with the phase transition as $\Delta v_c=k_BT_c/p_c=0.565$ Å^3 (the actual unit cell volume is $v_c=922.73$ Å^3).

Table I illustrates the good fit of data for $T_c(p)$ and $p_c(T)$ from Figs. 1 and 2 to Eq. (9), in which $(T/T_c+p/p_c)$ playsthe same role as T/T_c in a transition at zero pressure and as p/p_c in a transition at zero temperature.

We acknowledge the financial support of CICyT (Grant No. PB93-1253). One of us (J.P.) thanks DGICyT for financial support for a sabbatical (SAB94-0084) at the Ferroelectric Materials Laboratory, UAM.

*On leave from the Institute of Experimental Physics, University of Wroclaw, 50-205 Wroclaw, Poland.

[1] R. F. Weinland and H. Prause, Z. Anorg. Chem. **28**, 49 (1901).
[2] S. Guillot Gauthier, J. C. Peuzin, M. Olivier, and G. Rolland, Ferroelectrics **52**, 293 (1984).
[3] R. Sobiestianskas, J. Grigas, and Z. Czapla, Phase Trans. **37**, 157 (1992).
[4] M. N. Shashikala, H. L. Bhat, and P. S. Narayanan, J. Phys. Condens. Matter **2**, 5403 (1990).
[5] S. Haussühl and Y. F. Nicolau, Z. Phys. B **61**, 85 (1985).
[6] M. T. Averbuch-Pouchot and A. Durif, Ferroelectrics **52**, 271 (1984).
[7] J. Gaillard, J. Gloux, P. Gloux, B. Lamotte, and G. Rins, Ferroelectrics **54**, 81 (1984).
[8] M. N. Shashikala, B. Raghunata Chary, H. L. Bhat, and P. S. Narayanan, J. Raman Spectrosc. **20**, 351 (1989).
[9] See, e.g., J. A. Gonzalo, *Effective Field Approach to Phase Transitions and Some Applications to Ferroelectrics* (World Scientific, Singapore, 1991).
[10] T. Iglesias, B. Noheda, B. Gallego, J. R. Fernández del Castillo, G. Lifante, and J. A. Gonzalo, Europhys. Lett. **28**, 91 (1994).

PHYSICAL REVIEW B VOLUME 61, NUMBER 13 1 APRIL 2000-I

ARTICLES

Tetragonal-to-monoclinic phase transition in a ferroelectric perovskite: The structure of $PbZr_{0.52}Ti_{0.48}O_3$

B. Noheda* and J. A. Gonzalo
Departamento de Fisica de Materiales, UAM, Cantoblanco, 28049 Madrid, Spain

L. E. Cross, R. Guo, and S.-E. Park
Materials Research Laboratory, The Pennsylvania State University, Pennsylvania 16802-4800

D. E. Cox and G. Shirane
Department of Physics, Brookhaven National Laboratory, Upton, New York 11973-5000
(Received 11 October 1999; revised manuscript received 27 December 1999)

The perovskitelike ferroelectric system $PbZr_{1-x}Ti_xO_3$ (PZT) has a nearly vertical morphotropic phase boundary (MPB) around $x=0.45$–0.50. Recent synchrotron x-ray powder diffraction measurements by Noheda *et al.* [Appl. Phys. Lett. **74**, 2059 (1999)] have revealed a monoclinic phase between the previously established tetragonal and rhombohedral regions. In the present work we describe a Rietveld analysis of the detailed structure of the tetragonal and monoclinic PZT phases on a sample with $x=0.48$ for which the lattice parameters are, respectively, $a_t=4.044$ Å, $c_t=4.138$ Å, at 325 K, and $a_m=5.721$ Å, $b_m=5.708$ Å, $c_m=4.138$ Å, $\beta=90.496°$, at 20 K. In the tetragonal phase the shifts of the atoms along the polar [001] direction are similar to those in $PbTiO_3$ but the refinement indicates that there are, in addition, local disordered shifts of the Pb atoms of ~ 0.2 Å perpendicular to the polar axis. The monoclinic structure can be viewed as a condensation along one of the $\langle 110 \rangle$ directions of the local displacements present in the tetragonal phase. It equally well corresponds to a freezing-out of the local displacements along one of the $\langle 100 \rangle$ directions recently reported by Corker *et al.*[J. Phys.: Condens. Matter **10**, 6251 (1998)] for rhombohedral PZT. The monoclinic structure therefore provides a microscopic picture of the MPB region in which one of the "locally" monoclinic phases in the "average" rhombohedral or tetragonal structures freezes out, and thus represents a bridge between these two phases.

I. INTRODUCTION

Perovskitelike oxides have been at the center of research on ferroelectric and piezoelectric materials for the past fifty years because of their simple cubic structure at high temperatures and the variety of high symmetry phases with polar states found at lower temperatures. Among these materials the ferroelectric $PbZr_{1-x}Ti_xO_3$ (PZT) solid solutions have attracted special attention since they exhibit an unusual phase boundary which divides regions with rhombohedral and tetragonal structures, called the morphotropic phase boundary (MPB) by Jaffe *et al.*[1] Materials in this region exhibit a very high piezoelectric response, and it has been conjectured that these two features are intrinsically related. The simplicity of the perovskite structure is in part responsible for the considerable progress made recently in the determination of the basic structural properties and stability of phases of some important perovskite oxides, based on *ab initio* calculations (see, e.g., Refs. 2–9). Recently, such calculations have also been used to investigate solid solutions and, in particular, PZT, where the effective Hamiltonian includes both structural and compositional degrees of freedom.[10–12]

The PZT phase diagram of Jaffe *et al.*,[1] which covers only temperatures above 300 K, has been accepted as the basic characterization of the PZT solid solution. The ferroelectric region of the phase diagram consists mainly of two different regions: the Zr-rich rhombohedral region, (F_R) that contains two phases with space groups $R3m$ and $R3c$, and the Ti-rich tetragonal region (F_T), with space group $P4mm$.[13] The two regions are separated by a boundary that is nearly independent of temperature, the MPB mentioned above, which lies at a composition close to $x=0.47$. Many structural studies have been reported around the MPB, since the early 1950's, when these solid solutions were first studied,[13,14] since the high piezoelectric figure of merit that makes PZT so extraordinary is closely associated with this line.[1,15] The difficulty in obtaining good single crystals in this region, and the characteristics of the boundary itself, make good compositional homogeneity essential if single phase ceramic materials are to be obtained. Because of this, the MPB is frequently reported as a region of phase coexistence whose width depends on the sample processing conditions.[16–19]

Recently, another feature of the morphotropic phase boundary has been revealed by the discovery of a ferroelectric monoclinic phase (F_M) in the $Pb(Zr_{1-x}Ti_x)O_3$ ceramic system.[20] From a synchrotron x-ray powder diffraction study of a composition with $x=0.48$, a tetragonal-to-monoclinic phase transition was discovered at ~ 300 K. The monoclinic

0163-1829/2000/61(13)/8687(9)/$15.00 PRB 61 8687

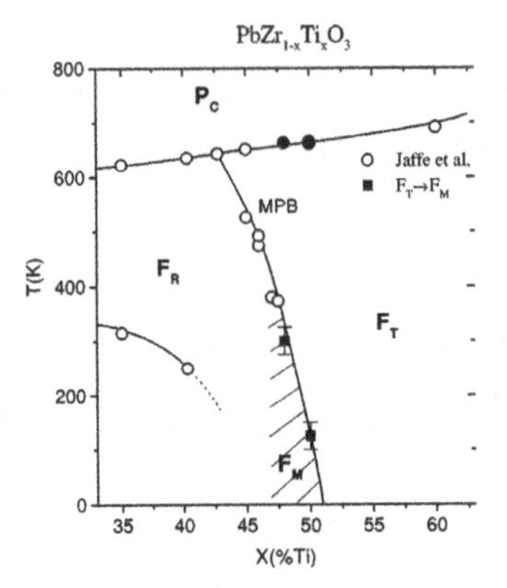

FIG. 1. Preliminary modification of the PZT phase diagram. The data of Jaffe *et al.* (Ref. 1) are plotted as open circles. The F_T-F_M and P_C-F_T transition temperatures for $x=0.48$ and $x=0.50$ are plotted as solid symbols. The F_T-F_M transition for $x=0.50$ is reported in Ref. 22.

unit cell is such that a_m and b_m lay along the tetragonal $[\bar{1}\bar{1}0]$ and $[1\bar{1}0]$ directions ($a_m \approx b_m \approx a_t\sqrt{2}$), and c_m deviates slightly from the [001] direction ($c_m \approx c_t$).[21] The space group is Cm, and the temperature dependence of the monoclinic angle β gives immediately the evolution of the order parameter for the tetragonal-monoclinic (F_T-F_M) transition. The polar axis of the monoclinic cell can in principle be directed along any direction within the ac mirror plane, making necessary a detailed structural study to determine its direction.

In the present work we present such a detailed structure determination of the monoclinic phase at 20 K and the tetragonal phase at 325 K in PZT with $x=0.48$. The results show that the polarization in the monoclinic plane lies along a direction between the pseudocubic $[001]_c$ and $[111]_c$ directions, corresponding to the first example of a species with $P_x^2=P_y^2\neq P_z^2$. A tentative phase diagram is presented in Fig. 1, which includes data for the $x=0.48$ composition together with those of the recently studied $x=0.50$ composition.[22] The most striking finding, however, is that the monoclinic cation displacements found here correspond to one of the three locally disordered sites reported by Corker *et al.*[23] for rhombohedral compositions in the region $x=0.1-0.4$, and thus provide a microscopic model of the rhombohedral-to-monoclinic phase transition. This, together with the fact that the space group of the new phase, Cm, is a subgroup of both $P4mm$ and $R3m$, suggests that F_M represents an intermediate phase connecting the well-known F_T and F_R PZT phases.

II. EXPERIMENTAL

A PZT sample with $x=0.48$ was prepared by conventional solid-state reaction techniques using appropriate amounts of reagent-grade powders of lead carbonate, zirconium oxide, and titanium oxide, with chemical purities better than 99.9%. Pellets were pressed and heated to 1250 °C at a ramp rate of 10 °C/min, held at this temperature in a covered crucible for 2 h, and furnace cooled. During sintering, $PbZrO_3$ was used as a lead source in the crucible to minimize volatilization of lead.

High-resolution synchrotron x-ray powder diffraction measurements were made at beam line X7A at the Brookhaven National Synchrotron Light Source. In the first set of measurements, an incident beam of wavelength 0.6896 Å from a Ge(111) double-crystal monochromator was used in combination with a Ge(220) crystal and scintillation detector in the diffraction path. The resulting instrumental resolution is about 0.01° on the 2θ scale, an order of magnitude better than that of a laboratory instrument. Data were collected from a disk in symmetric flat-plate reflection geometry over selected angular regions in the temperature range 20–736 K. Coupled θ-2θ scans were performed over selected angular regions with a 2θ step interval of 0.01°. The sample was rocked 1–2° during data collection to improve powder averaging.

Measurements above room temperature were performed with the disk mounted on a BN sample pedestal inside a wire-wound BN tube furnace. The furnace temperature was measured with a thermocouple mounted just below the pedestal and the temperature scale calibrated with a sample of CaF_2. The accuracy of the temperature in the furnace is estimated to be within 10 K, and the temperature stability about 2 K. For low-temperature measurements, the pellet was mounted on a Cu sample pedestal and loaded in a closed-cycle He cryostat, which has an estimated temperature accuracy of 2 K and stability better than 0.1 K. The diffracted intensities were normalized with respect to the incident beam monitor.

For the second set of measurements aimed at the detailed determination of the structure, a linear position-sensitive detector was mounted on the 2θ arm of the diffractometer instead of the crystal analyzer, and a wavelength of 0.7062 Å was used. This configuration gives greatly enhanced counting rates which make it feasible to collect accurate data from very narrow-diameter capillary samples in Debye-Scherrer geometry, with the advantage that systematic errors due to preferred orientation or texture effects are largely eliminated. A small piece of the sintered disk was carefully crushed and sealed into a 0.2 mm diameter glass capillary. The latter was loaded into a closed-cycle cryostat, and extended data sets were collected at 20 and 325 K while the sample was rocked over a 10° range. With this geometry the instrumental resolution is about 0.03° on the 2θ scale. Because lead is highly absorbing, the data were corrected for absorption effects[24] based on an approximate value of $\mu r=1.4$ determined from the weight and dimensions of the sample.

III. PHASE TRANSITIONS

The evolution of the lattice parameters with temperature was briefly summarized in Ref. 20, and a more complete analysis is presented below. The results of the full structure analysis are described later.

A transition from the cubic to the tetragonal phase was observed at ~660 K, in agreement with the phase diagram

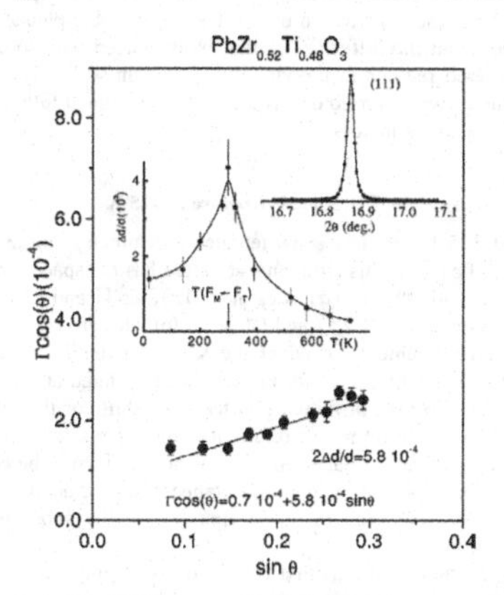

FIG. 2. The Williamson-Hall plot for PZT ($x=0.48$) derived from the measured diffraction peak widths in the cubic phase ($T=736$ K). Particle size and microstrain are estimated from a linear fit (solid line). The plot for the (111) reflection in the cubic phase demonstrates the excellent quality of the ceramic sample (peak width $\sim 0.02°$). The plot of $\Delta d/d$ vs temperature is also shown as an inset.

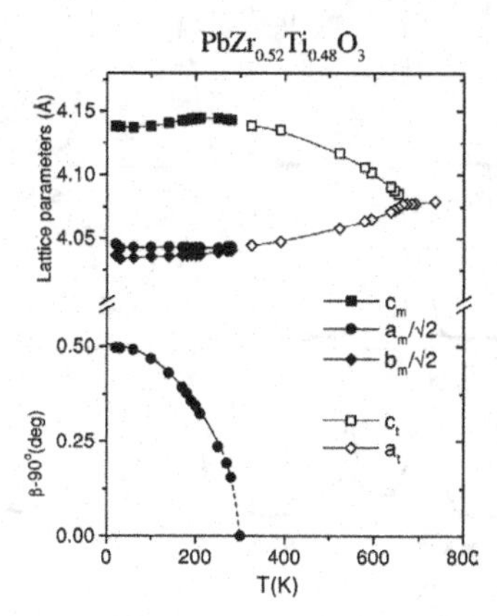

FIG. 3. Lattice parameters versus temperature for PZT ($x=0.48$) over the whole range of temperatures from 20 to 750 K showing the evolution from the monoclinic phase to the cubic phase via the tetragonal phase.

shown in Fig. 1. The measurements made on the pellet in the cubic phase at 736 K demonstrate the excellent quality of the sample, which exhibits diffraction peaks with full-widths at half-maximum (FWHM) ranging from 0.01° to 0.03° as shown for the (111) reflection plotted as the upper-right inset in Fig. 2. The FWHM's (Γ) for several peaks were determined from least-squares fits to a pseudo-Voigt function with the appropriate corrections for asymmetry effects,[25] and corrected for instrumental resolution. The corrected values are shown in Fig. 2 in the form of a Williamson-Hall plot[26]

$$\Gamma \cos\theta = \lambda/L + 2(\Delta d/d)\sin\theta, \qquad (3.1)$$

where λ is the wavelength and L is the mean crystallite size. From the slope of a linear fit to the data, the distribution of d spacings, $\Delta d/d$, is estimated to be $\sim 3\times 10^{-4}$, corresponding to a compositional inhomogeneity Δx of less than ± 0.003. From the intercept of the line on the ordinate axis the mean crystallite size is estimated to be ~ 1 μm.

A tetragonal-to-monoclinic phase transition in PZT with $x=0.48$ was recently reported by Noheda *et al.*[20] Additional data have been obtained near the phase transition around 300 K which have allowed a better determination of the phase transition to be made, as shown by the evolution of the lattice parameters as a function of temperature in Fig. 3. The tetragonal strain c_t/a_t increases as the temperature decreases from the Curie point ($T\approx 660$ K), to a value of 1.0247 at 300 K, below which peak splittings characteristic of a monoclinic phase with $a_m \approx b_m \approx a_t\sqrt{2}, \beta \neq 90°$, are observed (Fig. 3). As the temperature continues to decrease down to 20 K, a_m (which is defined to lie along the $[\bar{1}\bar{1}0]$ tetragonal direction) increases very slightly, and b_m (which lies along the $[1\bar{1}0]$ tetragonal direction) decreases. The c_m lattice parameter reaches a broad maximum value of 4.144 Å between 240–210 K and then reaches a shallow minimum value of 4.137 Å at 60 K. Over the same temperature region there is a striking variation of $\Delta d/d$ determined from Williamson-Hall plots at various temperatures, as shown in the upper-left inset in Fig. 2. $\Delta d/d$ increases rapidly as the temperature approaches the F_T-F_M transition at 300 K, in a similar fashion to the tetragonal strain, and then decreases rapidly below this temperature in the monoclinic region. Thus the microstrain responsible for the large increase in $\Delta d/d$ is an important feature of the phase transition, which may be associated with the development of local monoclinic order, and is very likely responsible for the large electromechanical response of PZT close to the MPB.[1]

The deviation of the monoclinic angle β from 90° is an order parameter of the F_T-F_M transition, and its evolution with temperature is also depicted in Fig. 3. This phase transition presents a special problem due to the steepness of the phase boundary (the MPB in Fig. 1). As shown in the previous section, the compositional fluctuations are quite small in these ceramic samples ($\Delta x \approx \pm 0.003$) but, even in this case, the nature of the MPB implies an associated temperature uncertainty of $\Delta T \approx 100$ K. There is, therefore, a rather wide range of transition temperatures instead of a single well-defined transition, so that the order parameter is smeared out as a function of temperature around the phase change, thereby concealing the nature of the transition.

Scans over the $(220)_c$ region for several different temperatures are plotted in Fig. 4, which shows the evolution of phases from the cubic phase at 687 K (upper-left plot) to the monoclinic phase at 20 K (lower-right plot), passing through the tetragonal phase at intermediate temperatures. With decreasing temperature, the tetragonal phase appears at ~ 660 K and the development of the tetragonal distortion can be observed on the left side of the figure from the splitting of

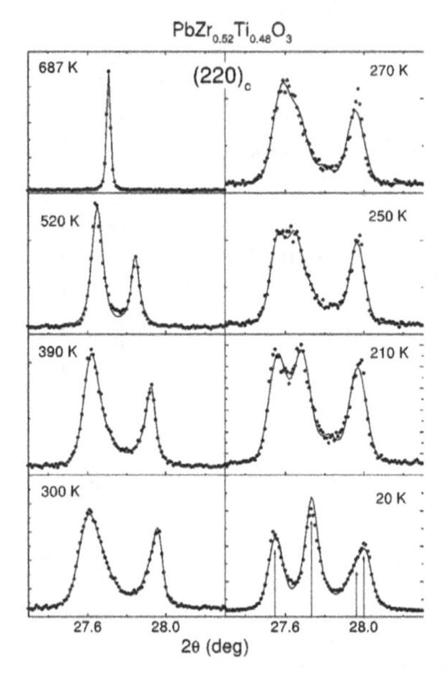

FIG. 4. Temperature evolution of the pseudocubic (220) peak from the cubic (top left) to the monoclinic (bottom right) phase.

the $(202)_t$ and $(220)_t$ reflections. On the right side of the figure, the evolution of the monoclinic phase, which appears below ~300 K, is shown by the splitting into the $(22\bar{2})_m$, $(222)_m$, $(400)_m$, and $(040)_m$ monoclinic reflections. It is quite evident from Fig. 4 that the $(202)_t$ peak is much broader than the neighboring $(220)_t$ peak, for example, and this "anisotropic" peak broadening is a general feature of the diffraction data for both phases. Another feature of the patterns is the presence of additional diffuse scattering between neighboring peaks, which is particularly evident between tetragonal $(00l)$ and $(h00)$ pairs, and the corresponding monoclinic $(00l)$ and $(hh0)$ pairs.

IV. STRUCTURE DETERMINATION

A detailed analysis of the 325 K tetragonal and 20 K monoclinic structures of $PbZr_{0.52}Ti_{0.48}O_3$ was carried out by Rietveld refinement using the GSAS program package.[27] The pseudo-Voigt peak shape function option was chosen[25] and background was estimated by linear interpolation between fixed values. An important feature of the refinements was the need to allow for the anisotropic peak broadening mentioned above. This was accomplished by the use of the recently incorporated generalized model for anisotropic peak broadening proposed by Stephens,[28] which is based on a distribution of lattice parameters. It was also necessary to take into account some additional diffuse scattering by modeling with a second, cubic, phase with broad, predominately Gaussian, peaks. A similar strategy has been adopted by Muller *et al.*[29] in a recent study of $PbHf_{0.4}Ti_{0.6}O_3$. Although in principle this could represent a fraction of untransformed cubic phase, we suspect that the diffuse scattering is associated with locally disordered regions in the vicinity of domain walls. The refinements were carried out with the atoms assigned fully ionized scattering factors.

A. Tetragonal structure at 325 K

At 325 K the data show tetragonal symmetry similar to that of $PbTiO_3$. This tetragonal structure has the space group $P4mm$ with Pb in $1(a)$ sites at $(0,0,z)$; Zr/Ti and O(1) in $1(b)$ sites at $(1/2,1/2,z)$ and O(2) in 2(c) sites at $(1/2, 0, z)$. For the refinement we adopt the same convention as that used in Refs. 30 and 31 for $PbTiO_3$, with Pb fixed at (0,0,0). However, instead of thinking in terms of shifts of the other atoms with respect to this origin, it is more physically intuitive to consider displacements of Pb and Zr/Ti from the center of the distorted oxygen cuboctahedra and octahedra, respectively. We shall take this approach in the subsequent discussion.

The refinement was first carried out with individual isotropic (U_{iso}) temperature factors assigned. Although a reasonably satisfactory fit was obtained ($R_{F^2}=8.9\%$), U_{iso} for O(1) was slightly negative and U_{iso} for Pb was very large, 0.026 Å^2, much larger than U_{iso} for the other atoms. Similarly high values for Pb(U_{iso}) in Pb-based perovskites are well known in the literature, and are usually ascribed to local disordered displacements, which may be either static or dynamic. Refinement with anisotropic temperature factors[32] (U_{11} and U_{33}) assigned to Pb (Table I, model I) gave an improved fit ($R_{F^2}=6.1\%$) with $U_{11}(=U_{22})$ considerably larger than U_{33} (0.032 and 0.013 Å^2, respectively) corresponding to large displacements perpendicular to the polar [001] axis. A further refinement based on local displacements of the Pb from the $1(a)$ site to the $4(d)$ sites at $(x,x,0)$, with isotropic temperature factors assigned to all the atoms, gave a small improvement in the fit ($R_{F^2}=6.0\%$) with $x\simeq 0.033$, corresponding to local shifts along the $\langle 110\rangle$ axes, and a much more reasonable temperature factor (Table I, model II). In order to check that high correlations between the temperature factor and local displacements were not biasing the result of this refinement, we have applied a commonly used procedure consisting of a series of refinements based on model II in which Pb displacements along $\langle 110\rangle$ were fixed but all the other parameters were varied.[34,35] Figure 5 shows unambiguously that there is well-defined minimum in the R factor for a displacement of about 0.19 Å, consistent with the result in Table I. A similar minimum was obtained for shifts along $\langle 100\rangle$ directions with a slightly higher R factor. Thus, in addition to a shift of 0.48 Å for Pb along the polar [001] axis towards four of its O(2) neighbors, similar to that in $PbTiO_3$,[30,31,36] there is a strong indication of substantial local shifts of ~0.2 Å perpendicular to this axis. The Zr/Ti displacement is 0.27 Å along the polar axis, once again similar to the Ti shift in $PbTiO_3$. Attempts to model local displacements along $\langle 110\rangle$ directions for the Zr/Ti atoms were unsuccessful due to the large correlations between these shifts and the temperature factor. Further attempts to refine the z parameters of the Zr and Ti atoms independently,

TABLE I. Structure refinement results for tetragonal $PbZr_{0.52}Ti_{0.48}O_3$ at 325 K, space group $P4mm$, lattice parameters $a_t=4.0460(1)$ Å, $c_t=4.1394(1)$ Å. Fractional occupancies N for all atoms taken as unity except for Pb in model II, where $N=0.25$. Agreement factors, R_{wp}, R_{F^2}, and χ^2 are defined in Ref. 33.

	Model I anisotropic lead temperature factors				Model II local ⟨110⟩ lead shifts			
	x	y	z	U(Å^2)	x	y	z	U_{iso} (Å^2)
Pb	0	0	0	$U_{11}=0.0319(4)$ $U_{33}=0.0127(4)$	0.0328(5)	0.0328(5)	0	0.0127(4)
Zr/Ti	0.5	0.5	0.4517(7)	$U_{iso}=0.0052(6)$	0.5	0.5	0.4509(7)	0.0041(6)
O(1)	0.5	0.5	−0.1027(28)	$U_{iso}=0.0061(34)$	0.5	0.5	−0.1027(28)	0.0072(35)
O(2)	0.5	0	0.3785(24)	$U_{iso}=0.0198(30)$	0.5	0	0.3786(24)	0.0197(30)
R_{wp}			4.00%				3.99%	
R_{F^2}			6.11%				6.04%	
χ^2			11.4				11.3	

as Corker *et al.* were able to do,[23] were likewise unsuccessful, presumably because the scattering contrast for x rays is much less than for neutrons.

From the values of the atomic coordinates listed in Table I, it can be inferred that the oxygen octahedra are somewhat more distorted than in $PbTiO_3$, the O(1) atoms being displaced 0.08 Å towards the O(2) plane above. The cation displacements are slightly larger than those recently reported by Wilkinson *et al.*[37] for samples close to the MPB containing a mixture of rhombohedral and tetragonal phases, and in excellent agreement with the theoretical values obtained by Bellaiche and Vanderbilt[38] for PZT with $x=0.50$ from first principles calculations. As far as we are aware no other structural analysis of PZT compositions in the tetragonal region has been reported in the literature.

Selected bond distances for the two models are shown in Table II. For model I, Zr/Ti has short and long bonds with O(1) of 1.85 and 2.29 Å, respectively, and four intermediate-length O(2) bonds of 2.05 Å. There are four intermediate-length Pb-O(1) bonds of 2.89 Å , four short Pb-O(2) bonds of 2.56 Å and four much larger Pb-O(2) distances of 3.27 Å. For model II, the Zr/Ti-O distances are the same, but the Pb-O distances change significantly. A Pb atom in one of the four equivalent $(x,x,0)$ sites in Table I now has a highly distorted coordination, consisting of two short and two intermediate Pb-O(2) bonds of 2.46 and 2.67 Å, and one slightly longer Pb-O(1) bond of 2.71 Å (Table II). The tendency of Pb^{+2}, which has a lone *sp* electron pair, to form short covalent bonds with a few neighboring oxygens is well documented in the literature.[23,39–41]

The observed and calculated diffraction profiles and the difference plot are shown in Fig. 6 for a selected 2θ range between 7° and 34° (upper figure). The short vertical markers represent the calculated peak positions. The upper and lower sets of markers correspond to the cubic and tetragonal phases, respectively. We note that although agreement between the observed and the calculated profiles is considerably better when the diffuse scattering is modeled with a cubic phase, the refined values of the atomic coordinates are not significantly affected by the inclusion of this phase. The anisotropic peak broadening was found to be satisfactorily

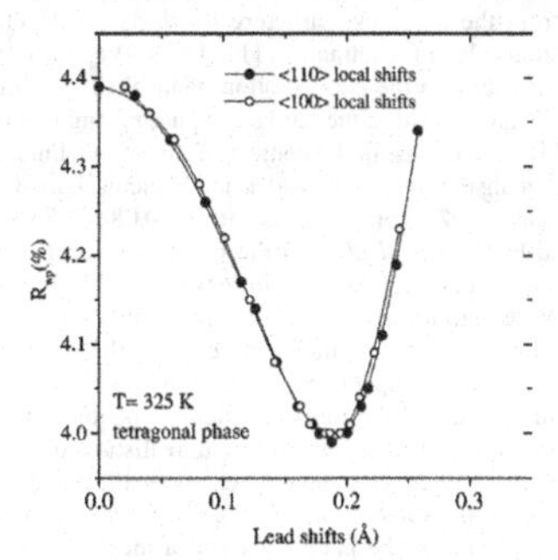

FIG. 5. Agreement factor R_{wp} as a function of Pb displacements for refinements with fixed values of x along tetragonal ⟨110⟩ and ⟨100⟩ directions as described in text. The well-defined minimum at $x\sim0.19$ Å confirms the result listed in Table I for model II.

TABLE II. Selected Zr/Ti-O and Pb-O bond lengths in the tetragonal and monoclinic structures. Models I and II refer to the refinements with anisotropic temperature factors and local ⟨110⟩ displacements for Pb, respectively (see Table I). The standard errors in the bond lengths are ~0.01 Å .

	Bond lengths (Å) tetragonal model I	tetragonal model II	monoclinic
Zr/Ti-O(1)	1.85×1	1.85×1	1.87×1
	2.29×1	2.29×1	2.28×1
Zr/Ti-O(2)	2.05×4	2.05×4	2.13×2
			1.96×2
Pb-O(1)	2.89×4	2.90×2	2.90×2
		2.71×1	2.60×1
Pb-O(2)	2.56×4	2.67×2	2.64×2
		2.46×2	2.46×2

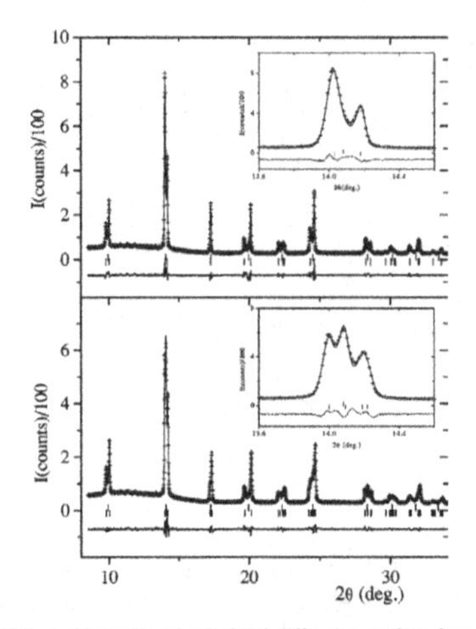

FIG. 6. Observed and calculated diffraction profiles from the Rietveld refinement of the tetragonal (top) and monoclinic (bottom) phases of PZT ($x=0.48$) at 325 and 20 K, respectively. The difference plots are shown below, and the short vertical markers represent the peak positions (the upper set correspond to the cubic phase as discussed in text). The insets in each figure highlight the differences between the tetragonal and the monoclinic phase for the pseudocubic (110) reflection, and illustrate the high resolution needed in order to characterize the monoclinic phase.

described by two of the four parameters in the generalized model for tetragonal asymmetry.[28]

B. Monoclinic structure at 20 K

As discussed above, the diffraction data at 20 K can be indexed unambiguously on the basis of a monoclinic cell with the space group Cm. In this case Pb, Zr/Ti, and O(1) are in $2(a)$ sites at $(x,0,z)$, and O(2) in $4(b)$ sites at (x,y,z). Individual isotropic temperature factors were assigned, and Pb was fixed at (0,0,0). For monoclinic symmetry, the generalized expression for anisotropic peak broadening contains nine parameters, but when all of these were allowed to vary the refinement was slightly unstable and did not completely converge. After several tests in which some of the less significant values were fixed at zero, satisfactory convergence was obtained with three parameters ($R_{wp}=0.036, \chi^2=11.5$). During these tests, there was essentially no change in the refined values of the atomic coordinates. A small improvement in the fit was obtained when anisotropic temperature factors were assigned to Pb ($R_{wp}=0.033, \chi^2=9.2$). The final results are listed in Table III, and the profile fit and difference plot are shown in the lower part of Fig. 6.

From an inspection of the results in Tables I and III, it can be seen that the displacements of the Pb and Zr/Ti atoms along [001] are very similar to those in the tetragonal phase at 325 K, about 0.53 and 0.24 Å, respectively. However, in the monoclinic phase at 20 K, there are also significant shifts

TABLE III. Structure refinement results for monoclinic $PbZr_{0.52}Ti_{0.48}O_3$ at 20 K, space group Cm, lattice parameters $a_m=5.72204(15)$ Å, $b_m=5.70957(14)$ Å, $c_m=4.13651(14)$ Å, $\beta=90.498(1)°$. Agreement factors $R_{wp}=3.26\%, R_{F^2}=4.36\%, \chi^2=9.3$.

	x_m	y_m	z_m	U_{iso}(Å^2)
Pb	0	0	0	0.0139[a]
Zr/Ti	0.5230(6)	0	0.4492(4)	0.0011(5)
O(1)	0.5515(23)	0	−0.0994(24)	0.0035(28)
O(2)	0.2880(18)	0.2434(20)	0.3729(17)	0.0123(22)

[a]For Pb, U_{iso} is the equivalent isotropic value calculated from the refined anisotropic values [$U_{11}=0.0253(7)$ Å^2, $U_{22}=0.0106(6)$ Å^2, $U_{33}=0.0059(3)$ Å^2, $U_{13}=0.0052(4)$ Å^2].

of these atoms along the monoclinic $[\bar{1}00]$, i.e., pseudocubic [110], towards their O(2) neighbors in adjacent pseudocubic (110) planes, of about 0.24 and 0.11 Å, respectively, which corresponds to a rotation of the polar axis from [001] towards pseudocubic [111]. The Pb shifts are also qualitatively consistent with the local shifts of Pb along the tetragonal ⟨110⟩ axes inferred from the results of model II in Table I, i.e., about 0.2 Å. Thus it seems very plausible that the monoclinic phase results from the condensation of the local Pb displacements in the tetragonal phase along one of the ⟨110⟩ directions.

Some selected bond distances are listed in Table II. The Zr/Ti-O(1) distances are much the same as in the tetragonal structure, but the two sets of Zr/Ti-O(2) distances are significantly different, 1.96 and 2.13 Å, compared to the single set at 2.04 Å in the tetragonal structure. Except for a shortening in the Pb-O(1) distance from 2.71 to 2.60 Å, the Pb environment is quite similar to that in the tetragonal phase, with two close O(2) neighbors at 2.46 Å, and two at 2.64 Å.

V. DISCUSSION

In the previous section, we have shown that the low-temperature monoclinic structure of $PbZr_{0.52}Ti_{0.48}O_3$ is derived from the tetragonal structure by shifts of the Pb and Zr/Ti atoms along the tetragonal [110] axis. We attribute this phase transition to the condensation of local ⟨110⟩ shifts of Pb which are present in the tetragonal phase along one of the four ⟨110⟩ directions. In the context of this monoclinic structure it is instructive to consider the structural model for rhombohedral PZT compositions with $x=0.08-0.38$ recently reported by Corker *el al.*[23] on the basis of neutron powder diffraction data collected at room temperature. In this study and also an earlier study[42] of a sample with $x=0.1$, it was found that satisfactory refinements could only be achieved with anisotropic temperature factors, and that the thermal ellipsoid for Pb had the form of a disk perpendicular to the pseudocubic [111] axis. This highly unrealistic situation led them to propose a physically much more plausible model involving local displacements for the Pb atoms of about 0.25 Å perpendicular to the [111] axis and a much smaller and more isotropic thermal ellipsoid was obtained. Evidence for local shifts of Pb atoms in PZT ceramics has also been demonstrated by pair-distribution function analysis by Teslic and co-workers.[39]

TABLE IV. Comparison of refined values of atomic coordinates in the monoclinic phase with the corresponding values in the tetragonal and rhombohedral phases for both the "ideal" structures and those with local shifts, as discussed in text.

	tetragonal $x=0.48$, 325 K		monoclinic $x=0.48$, 20 K	rhombohedral (Ref. 43) $x=0.40$, 295 K	
	ideal	local shifts [a]	as refined	local shifts [b]	ideal
$x_{Zr/Ti}$	0.500	0.530	0.523	0.520	0.540
$z_{Zr/Ti}$	0.451	0.451	0.449	0.420	0.460
$x_{O(1)}$	0.500	0.530	0.551	0.547	0.567
$z_{O(1)}$	−0.103	−0.103	−0.099	0.093	−0.053
$x_{O(2)}$	0.250	0.280	0.288	0.290	0.310
$y_{O(2)}$	0.250	0.250	0.243	0.257	0.257
$z_{O(2)}$	0.379	0.379	0.373	0.393	0.433
a_m(Å)	5.722		5.722	5.787	
b_m(Å)	5.722		5.710	5.755	
c_m(Å)	4.139		4.137	4.081	
β(°)	90.0		90.50	90.45	

[a]Tetragonal local shifts of (0.03,0.03,0).
[b]Hexagonal local shifts of (−0.02,0.02,0).

We now consider the refined values of the Pb atom positions with local displacements for rhombohedral PZT listed in Table IV of Ref. 23. With the use of the appropriate transformation matrices, it is straightforward to show that these shifts correspond to displacements of 0.2–0.25 Å along the direction of the monoclinic [100] axis, similar to what is actually observed for $x=0.48$. It thus seems equally plausible that the monoclinic phase can also result from the condensation of local displacements perpendicular to the [111] axis.

The monoclinic structure can thus be pictured as providing a "bridge" between the rhombohedral and tetragonal structures in the region of the MPB. This is illustrated in Table IV, which compares the results for PZT with $x=0.48$ obtained in the present study with earlier results[43] for rhombohedral PZT with $x=0.40$ expressed in terms of the monoclinic cell.[44] For $x=0.48$, the atomic coordinates for Zr/Ti, O(1) and O(2) are listed for the "ideal" tetragonal structure (model I) and for a similar structure with local shifts of (0.03,0.03,0) in the first two columns, and for the monoclinic structure in the third column. The last two columns describe the rhombohedral structure for $x=0.40$ assuming local shifts of (−0.02,0.02,0) along the hexagonal axes and the as-refined "ideal" structure, respectively. It is clear that the condensation of these local shifts gives a very plausible description of the monoclinic structure in both cases. It is also interesting to note the behavior of the corresponding lattice parameters; metrically the monoclinic cell is very similar to the tetragonal cell except for the monoclinic angle, which is close to that of the rhombohedral cell.

Evidence for a tetragonal-to-monoclinic transition in the ferroelectric material $PbFe_{0.5}Nb_{0.5}O_3$ has also been reported by Bonny *et al.*[45] from single crystal and synchrotron x-ray powder diffraction measurements. The latter data show a cubic-tetragonal transition at ~376 K, and a second transition at ~355 K. Although the resolution was not sufficient to reveal any systematic splitting of the peaks, it was concluded that the data were consistent with a very weak monoclinic distortion of the pseudorhombohedral unit cell. In a recent neutron and x-ray powder study, Lampis *et al.*[46] have shown that Rietveld refinement gives better agreement for the monoclinic structure at 80 and 250 K than for the rhombohedral one. The resulting monoclinic distortion is very weak, and the large thermal factor obtained for Pb is indicative of a high degree of disorder.

The relationships between the PZT rhombohedral, tetragonal and monoclinic structures are also shown schematically in Fig. 7, in which the displacements of the Pb atom are shown projected on the pseudocubic (110) mirror plane. The four locally disordered ⟨110⟩ shifts postulated in the present paper for the tetragonal phase are shown superimposed on the [001] shift at the left [Fig. 7(a)] and the three locally disordered ⟨100⟩ shifts proposed by Corker *et al.*[23] for the rhombohedral phase are shown superimposed on the [111]

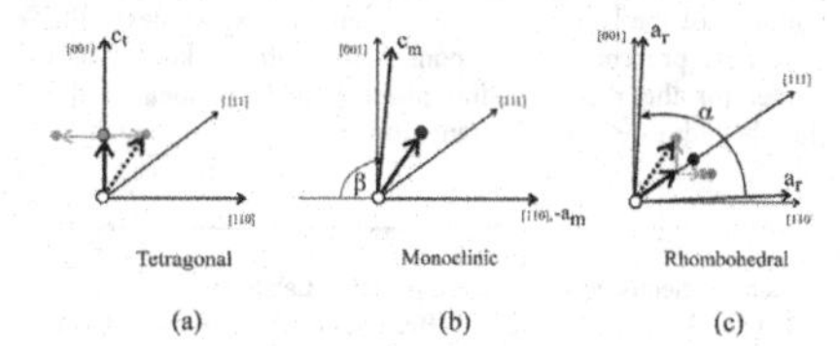

Tetragonal Monoclinic Rhombohedral

(a) (b) (c)

FIG. 7. Schematic illustration of the tetragonal (a), monoclinic (b), and rhombohedral (c) distortions of the perovskite unit cell projected on the pseudocubic (110) plane. The solid circles represent the observed shifts with respect to the ideal cubic structure. The light grey circles represent the four locally disordered ⟨100⟩ shifts in the tetragonal structure (a) and the three locally disordered shifts in the rhombohedral structure (c) described by Corker *et al.* (Ref. 23). The heavy dashed arrows represent the freezing-out of one of these shifts to give the monoclinic observed structure. Note that the resultant shifts in the rhombohedral structure can be viewed as a combination of a [111] shift with local ⟨100⟩ shifts, as indicated by the light grey arrows.

shift at the right [Fig. 7(c)]. It can be seen that both the condensation of the [110] shift in the tetragonal phase and the condensation of the [001] shift in the rhombohedral phase leads to the observed monoclinic shift shown at the center [Fig. 7(b)]. We note that although Corker *et al.* discuss their results in terms of ⟨100⟩ shifts and a [111] shift smaller than that predicted in the usual refinement procedure, they can be equally well described by a combination of shifts perpendicular to the [111] axis and the [111] shift actually obtained in the refinement, as is evident from Fig. 7(c).

We conclude, therefore, that the F_M phase establishes a connection between the PZT phases at both sides of the MPB through the common symmetry element, the mirror plane, and suggest that there is not really a morphotropic phase boundary, but rather a "morphotropic phase," connecting the F_T and F_R phases of PZT. In the monoclinic phase the difference vector between the positive and negative centers of charge defines the polar axis, whose orientation, in contrast to the case of the F_T and F_R phases, cannot be determined on symmetry grounds alone. According to this, from the results shown in Table III, the polar axis in the monoclinic phase is found to be tilted about 24° from the [001] axis towards the [111] axis. This structure represents the first example of a ferroelectric material with $P_x^2=P_y^2\neq P_z^2$, (P_x,P_y,P_z) being the Cartesian components of the polarization vector. This class corresponds to the so-called $m3m(12)A2Fm$ type predicted by Shuvalov.[47] It is possible that this new phase is one of the rare examples of a two-dimensional ferroelectric[48] in which the unit cell dipole moment switches within a plane containing the polar axis, upon application of an electric field.

This F_M phase has important implications; for example, it might explain the well-known shifts of the anomalies of many physical properties with respect to the MPB and thus help in understanding the physical properties in this region, of great interest from the applications point of view.[1] It has been found that the maximum values of d_{33} for rhombohedral PZT with $x=0.40$ are not obtained for samples polarized along the [111] direction but along a direction close to the perovskite [001] direction.[49] This points out the intrinsic importance of the [001] direction in perovskites, whatever the distortion present, and is consistent with Corker *et al.*'s model for the rhombohedral phase,[23] and the idea of the rhombohedral-tetragonal transition through a monoclinic phase.

It is also to be expected that other systems with morphotropic phase boundaries between two nonsymmetry-related phases (e.g., other perovskites or tungsten-bronze mixed systems) may show similar intermediate phases. In fact, an indication of symmetry lowering at the MPB of the PZN-PT system has been recently reported by Fujishiro *et al.*[50] From a different point of view, a monoclinic ferroelectric perovskite also represents a new challenge for first-principles theorists, until now used to dealing only with tetragonal, rhombohedral and orthorhombic perovskites.

A structural analysis of several other PZT compositions with $x=0.42-0.51$ is currently in progress in order to determine the new PZT phase diagram more precisely. In the preliminary version shown in Fig. 1 we have included data for a sample with $x=0.50$ made under slightly different conditions[22] at the Institute of Ceramic and Glass (ICG) in Madrid, together with the data described in the present work for a sample with $x=0.48$ synthesized in the Materials Research Laboratory at the Pennsylvania State University (PSU). As can be seen the results for these two compositions show consistent behavior, and demonstrate that the F_M-F_T phase boundary lies along the MPB of Jaffe *et al.* Preliminary results for a sample from PSU with $x=0.47$ show unequivocally that the monoclinic features are present at 300 K. However, measurements on an ICG sample with the same nominal composition do not show monoclinicity unambiguously, but instead a rather complex poorly defined region from 300–400 K between the rhombohedral and tetragonal phases.[22] The extension of the monoclinic region and the location of the F_R-F_M phase boundary are still somewhat undefined, although it is clear that the monoclinic region has a narrower composition range as the temperature increases. The existence of a quadruple point in the PZT phase diagram is an interesting possibility.

ACKNOWLEDGMENTS

We wish to gratefully acknowledge B. Jones for the excellent quality of the $x=0.48$ sample used in this work, and we thank L. Bellaiche, A. M. Glazer, E. Moshopoulou, C. Moure, and E. Sawaguchi for their helpful comments. Support by NATO (Grant No. R.C.G. 970037), the Spanish CICyT (Project No. PB96-0037), and the U.S. Department of Energy, Division of Materials Sciences (Contract No. DE-AC02-98CH10886) is also acknowledged.

*Visiting scientist at Brookhaven National Laboratory.

[1] B. Jaffe, W.R. Cook, and H. Jaffe, *Piezoelectric Ceramics* (Academic Press, London, 1971).
[2] R.E. Cohen, Nature (London) **358**, 136 (1992).
[3] R.D. King-Smith and D. Vanderbilt, Phys. Rev. B **49**, 5828 (1994).
[4] W. Zhong, D. Vanderbilt, and K. Rabe, Phys. Rev. Lett. **73**, 1861 (1994); Phys. Rev. B **52**, 6301 (1995).
[5] W. Zhong and D. Vanderbilt, Phys. Rev. Lett. **74**, 2587 (1995).
[6] A. Garcia and D. Vanderbilt, Phys. Rev. B **54**, 3817 (1996).
[7] K.M. Rabe and U.W. Waghmare, Phys. Rev. B **55**, 6161 (1997).
[8] K.M. Rabe and E. Cockayne, in *First-Principles Calculations for Ferroelectrics*, AIP Conf. Proc. No. 436, edited by R. Cohen (AIP, New York, 1998), p. 61.
[9] Ph. Ghosez, E. Cokayne, U.V. Waghmare, and K.M. Rabe, Phys. Rev. B **60**, 836 (1999).
[10] L. Bellaiche, J. Padilla, and D. Vanderbilt, Phys. Rev. B **59**, 1834 (1999).
[11] L. Bellaiche, J. Padilla, and D. Vanderbilt, *First-principles Calculations for Ferroelectrics: 5th Williamsburg Workshop*, edited by R. Cohen (AIP, Woodbury, 1998), p. 11.
[12] G. Soghi-Szabo and R. E. Cohen, Ferroelectrics **194**, 287 (1997).
[13] G. Shirane and K. Suzuki. J. Phys. Soc. Jpn. **7**, 333 (1952).
[14] E. Sawaguchi, J. Phys. Soc. Jpn. **8**, 615 (1953).
[15] Y. Xu, *Ferroelectric Materials and their Applications* (North Holland, Amsterdam, 1991).
[16] K. Kakewaga, O. Matsunaga, T. Kato, and Y. Sasaki, J. Am. Ceram. Soc. **78**, 1071 (1995).

[17] J.C. Fernandes, D.A. Hall, M.R. Cockburn, and G.N. Greaves, Nucl. Instrum. Methods Phys. Res. B **97**, 137 (1995).

[18] M. Hammer, C. Monty, A. Endriss, and M.J. Hoffmann, J. Am. Ceram. Soc. **81**, 721 (1998).

[19] W. Cao and L.E. Cross, Phys. Rev. B **47**, 4825 (1993).

[20] B. Noheda, D.E. Cox, G. Shirane, J.A. Gonzalo, L.E. Cross, and S-E. Park, Appl. Phys. Lett. **74**, 2059 (1999).

[21] The $[\overline{1}\overline{1}0]$ and $[1\overline{1}0]$ directions are chosen so that the monoclinic angle $\beta > 90°$ to conform with usual crystallographic convention.

[22] B. Noheda, J.A. Gonzalo, A.C. Caballero, C. Moure, D.E. Cox, and G. Shirane, cond-mat/9907286, Ferroelectrics (to be published).

[23] D.L. Corker, A.M. Glazer, R.W. Whatmore, A. Stallard, and F. Fauth, J. Phys.: Condens. Matter **10**, 6251 (1998).

[24] C.W. Dwiggins Jr., Acta Crystallogr., Sect. A: Cryst. Phys., Diffr., Theor. Gen. Crystallogr. **31**, 146 (1975).

[25] L.W. Finger, D.E. Cox, and A.P. Jephcoat, J. Appl. Crystallogr. **27**, 892 (1994).

[26] G.K. Williamson and W.H. Hall, Acta Metall. **1**, 22 (1953).

[27] A.C. Larson and R.B. Von Dreele (unpublished).

[28] P.W. Stephens, J. Appl. Crystallogr. **32**, 281 (1999).

[29] C. Muller, J.-L. Baudour, V. Madigou, F. Bouree, J.-M. Kiat, C. Favotto, and M. Roubin, Acta Crystallogr., Sect. B: Struct. Sci. **55**, 8 (1999).

[30] A.M. Glazer and S.A. Mabud, Acta Crystallogr., Sect. B: Struct. Crystallogr. Cryst. Chem. **34**, 1065 (1978).

[31] R.J. Nelmes and W.F. Kuhs, Solid State Commun. **54**, 721 (1985).

[32] The structure factor correction is defined in terms of the anisotropic u_{ij} thermal factors as $\exp\{-[2\pi^2(\Sigma i,j u_{ij} a_i^* a_j^*)]\}$, a_i^* being the lattice vectors of the reciprocal unit cell.

[33] L.B. McCusker, R.B. von Dreele, D.E. Cox, D. Louër, and P. Scardi, J. Appl. Crystallogr. **32**, 36 (1999).

[34] K. Itoh, L.Z. Zeng, E. Nakamura, and N. Mishima, Ferroelectrics **63**, 29 (1985).

[35] C. Malibert, B. Dkhil, J.M. Kiat, D. Durand, J.F. Berar, and A. Spasojevic-de Bire, J. Phys.: Condens. Matter **9**, 7485 (1997).

[36] G. Shirane, R. Pepinski, and B.C. Frazer, Acta Crystallogr. **9**, 131 (1956).

[37] A.P. Wilkinson, J. Xu, S. Pattanaik, and J.L. Billinge, Chem. Mater. **10**, 3611 (1998).

[38] L. Bellaiche and D. Vanderbilt, Phys. Rev. Lett. **83**, 1347 (1999).

[39] S. Teslic, T. Egami, and D. Viehland, J. Phys. Chem. Solids **57**, 1537 (1996); Ferroelectrics **194**, 271 (1997).

[40] D.L. Corker, A.M. Glazer, W. Kaminsky, R.W. Whatmore, J. Dec, and K. Roleder, Acta Crystallogr., Sect. B: Struct. Sci. **54**, 18 (1998).

[41] S. Teslic and T. Egami, Acta Crystallogr., Sect. B: Struct. Sci. **54**, 750 (1998).

[42] A.M. Glazer, S.A. Mabud, and R. Clarke, Acta Crystallogr., Sect. B: Struct. Crystallogr. Cryst. Chem. **34**, 1060 (1978).

[43] A. Amin, R.E. Newnham, L.E. Cross, and D.E. Cox, J. Solid State Chem. **37**, 248 (1981).

[44] The rhombohedral unit cell can be expressed in terms of a monoclinic one by $a_m = 2a_r\cos(\alpha/2)$, $b_m = 2a_r\sin(\alpha/2)$, $c_m = a_r$, $\beta = 180° - \phi$, where $\cos\phi = 1 - 2\sin^2(\alpha/2)/\cos(\alpha/2)$ and a_r and α are the $R3m$ cell parameters. Note that a_r in Ref. 43 refers to the doubled cell.

[45] V. Bonny, M. Bonin, P. Sciau, K.J. Schenk, and G. Chapuis, Solid State Commun. **102**, 347 (1997).

[46] N. Lampis, P. Sciau, and A.G. Lehmann, J. Phys.: Condens. Matter **11**, 3489 (1999).

[47] L.A. Shuvalov, J. Phys. Soc. Jpn. **28**, 38 (1970).

[48] S.C. Abrahams and E.T. Keve, Ferroelectrics **2**, 129 (1971).

[49] X-h Du, J. Zheng, U. Belegundu, and K. Uchino, Appl. Phys. Lett. **72**, 2421 (1998).

[50] K. Fujishiro, R. Vlokh, Y. Uesu, Y. Yamada, J-M. Kiat, B. Dkhil, and Y. Yamashita, Ferroelectrics (to be published).

APPLIED PHYSICS LETTERS VOLUME 77, NUMBER 22 27 NOVEMBER 2000

Formation and evolution of charged domain walls in congruent lithium niobate

V. Ya. Shur,[a] E. L. Rumyantsev, E. V. Nikolaeva, and E. I. Shishkin
Institute of Physics and Applied Mathematics, Ural State University, Ekaterinburg 620083, Russia

(Received 8 June 2000; accepted for publication 2 October 2000)

We present experimental evidence of the formation of stable charged domain walls (CDWs) in congruent lithium niobate during switching. CDW evolution under the action of field pulses was *in situ* visualized. CDW boundary motion velocity is about 60 μm/s at 20 kV/mm. Relief of CDW strongly depends on applied field. Dielectric response in the presence of CDW demonstrates the pronounced frequency dependence in the range 50–150 °C. We propose the mechanism of CDW self-maintained propagation governed by self-consistent electrostatic interaction between the wall's steps. © *2000 American Institute of Physics.* [S0003-6951(00)05048-8]

The creation of a variety of domain structures with controlled parameters in commercially available ferroelectric materials is the main objective of domain engineering. Such ability opens the opportunity to optimize device characteristics for application in nonlinear-optical[1–3] and electro-optical[4] devices. One of the most difficult and challenging problems is the formation of stable charged domain walls (CDWs). It is known that any counter domain wall possesses extremely high surface energy due to the bound charge on it. Therefore, their arising is assumed to be of low probability.[5] On the other hand, it is well known that polarization reversal is realized through formation and growth of spike-like domains with CDWs. Moreover, the typical as-grown (virgin) domain structure in real ferroelectrics contains CDWs.[6–8] A stable periodic laminar domain structure with CDWs has been observed in $LiNbO_3$ (LN) crystals with growth striations.[9] It was pointed out that during domain patterning in congruent LN using Al electrodes that domains did not grow through the sample and nonswitched layer remains at the Z^- surface.[10] Residual internal needle-like persistent domains were found in triglycine sulfate.[11] Thus, the counter domains, being far from equilibrium according to theory, nevertheless are stable.[6,12] In this letter, we study the field-induced formation of the CDW in LN and its evolution in electric field.

The formation and evolution of CDWs were investigated in optical-grade single-domain 0.2-mm-thick congruent LN wafers cut perpendicular to the polar axis with typical area 6×6 mm^2. Transparent In_2O_3:Sn electrodes (3 mm diam) were deposited on Z^+ and Z^- surfaces. Rectangular voltage pulses with amplitude up to 5 kV and duration ranging from 1 to 30 s were applied to the wafer at room temperature (RT). The domain evolution was *in situ* visualized by an optical microscope in crossed polarizers and recorded using a television camera. Static domain patterns were revealed by etching of the polar surfaces and cross sections with HF acid (during 5 min at RT) and examined by optical microscope.

Our observations demonstrated that in the experimental conditions used, the domains nucleated at the Z^+ surface, grew in the forward direction, and coalesced at Z^+, but did not reach the Z^- surface, thus forming large isolated domains bounded by the CDW (Fig. 1). Such an exotic dented domain pattern does not disappear after removing the external electric field. The electric field produced by the CDW induces variation of the refractive index due to the electrooptic effect,[13,14] which allows us to visualize the dynamic domain patterns during switching, and the static ones after switching off the field (Fig. 2).

Growth of the created stable CDW starts at the field about 17 kV/mm, which is lower than the threshold for switching from a single-domain state ($E_{th}\sim21$ kV/mm). The observations revealed that relief of the CDW essentially depends on the wall propagation velocity, which is determined by the applied field.

The successive positions of the CDW boundary during switching have been obtained by image processing of the sequence of instantaneous patterns [Fig. 3(a)]. CDW boundary propagation is uniform with rare pinning. It is seen that

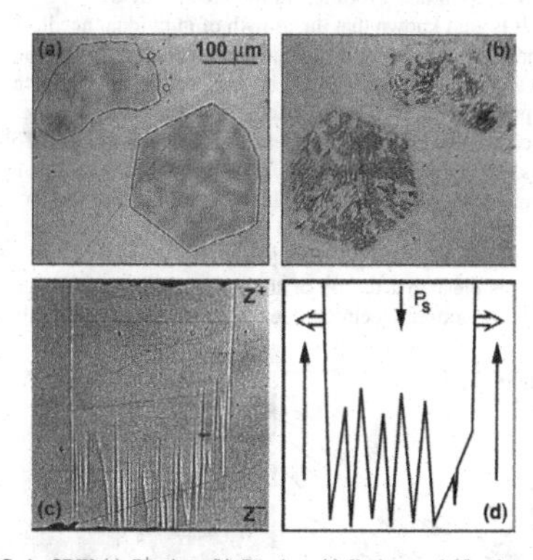

FIG. 1. CDW (a) Z^+ view, (b) Z^- view, (c) Y view, and (d) scheme (Y view, white arrows show the direction of boundary propagation), patterns revealed by etching and visualized by optical microscope; sample thickness, 0.2 mm.

[a]Electronic mail: vladimir.shur@usu.ru

0003-6951/2000/77(22)/3636/3/$17.00

Appl. Phys. Lett., Vol. 77, No. 22, 27 November 2000

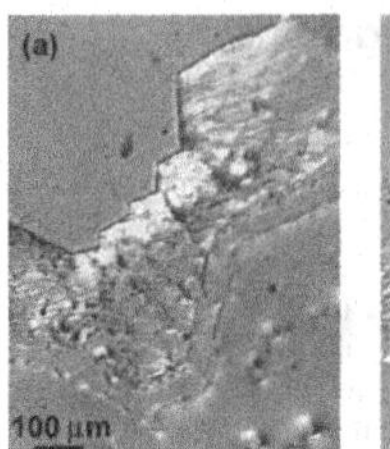
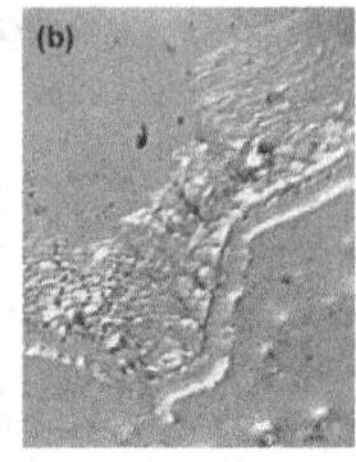

FIG. 2. CDW patterns visualized by optical microscope with crossed polarizers (a) in field, (b) without field; Z^+ view.

its orientation is in accord with the crystal symmetry [Fig. 3(b)]. The averaged value of the boundary velocity ranges from 40 to 80 μm/s for $E \sim 20$ kV/mm.

Consecutive etching allows us to study the variations of the CDW structure with depth in the surface layer of about several microns. The revealed patterns demonstrate the uniform switched areas at Z^+ [Figs. 1(a) and 1(c)] and oriented quasiregular structure of the needle-like domains in the vicinity of Z^- [Figs. 1(b) and 3(b)] in accord with *in situ* optical observations (Fig. 2).

Application of a reversed field up to 22.5 kV/mm does not move the boundary of the created CDW backward. At the same time, a pronounced change of the CDW relief is observed (Fig. 4). The domain patterns revealed by long-time etching (15 min) are visualized on both surfaces (Fig. 5), which show that during backswitching the needle-like domains have reached the Z^+ surface in contrast to the CDW produced by switching (compare with Fig. 1).

The fact of the formation of a stable CDW poses two main problems. First, why do the domains arisen at the Z^+ surface not reach Z^-? Second, what is the mechanism of CDW growth and evolution in the electric field?

It is well known that the growth of individual needle-like domains in the forward direction is a result of step generation at the domain walls at the Z^+ surface and their growth in the polar direction (Fig. 6, inset).[5] The growth rate is determined by the local electric field at the charged ends of the steps. Its value is the sum of external, depolarization and various screening fields.[15,16] The electrostatic interaction between a given step's end and its neighbors determines the growth velocity of a given step.

It is clear that the tip of the needle-like domain cannot reach the bottom electrode due to the existence of an intrinsic dielectric gap.[17,18] This gap makes complete external screening (redistribution of the charges at the electrodes accompanied by the current through the external circuit)[15,17] of the field produced by the terminated tip impossible. This field leads to the slowing and final termination of motion of the approaching neighboring step ends, thus preventing step annihilation. In turn, these terminated step ends stop their neighbors, and so on. Thus, the step concentration (the bound charge at the wall) increases because the steps cannot annihilate. Such a process in the case of constant step generation at Z^+ results in the formation of a CDW fragment abnormally deflected from the polar direction [Fig. 1(d)].

The high value of the depolarization field at that wall fragment can be decreased only by generation of additional spikes [Fig. 1(d)]. The spike tip propagates to Z^- and stops at the gap. This cyclic process occurs over and over, thus providing CDW growth. The distance between the CDW dents is determined by the thickness of the dielectric gap and the value of the applied field. The considered mechanism explains the observed formation of the self-organized dented structure at Z^- revealed by etching [Fig. 3(b)] and the correspondence between the orientation of quasiregular arrays and sequent positions of the CDW boundary (Fig. 3).

We have simulated the dependence of the local field at the approaching step on the distance from the immobile tip for different gap thicknesses (Fig. 6). It is seen that the "termination distance" is larger for thinner gaps. Thus, the observed absence of a CDW during switching by liquid electrolyte electrodes[2,3,14,19] can be explained if we suppose that the dielectric gap for liquid electrodes is thicker than for metal ones.

The evolution of CDW relief under a reversed field and the appearance of a quasiregular structure on Z^+ can be considered within the same approach as the forward growth of the spike-like domains towards Z^+. The termination of the tip motion at the Z^+ dielectric gap stabilizes the dent-like structure. Such a mechanism leads to the formation of the

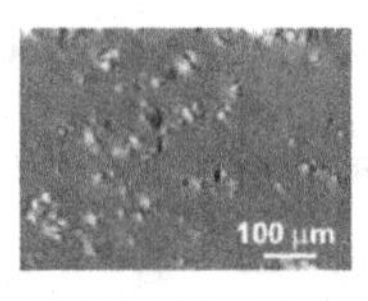
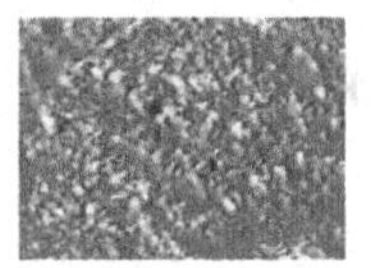

FIG. 4. Change of the relief of the CDW during backswitching (*in situ* optical observation, Z^+ view).

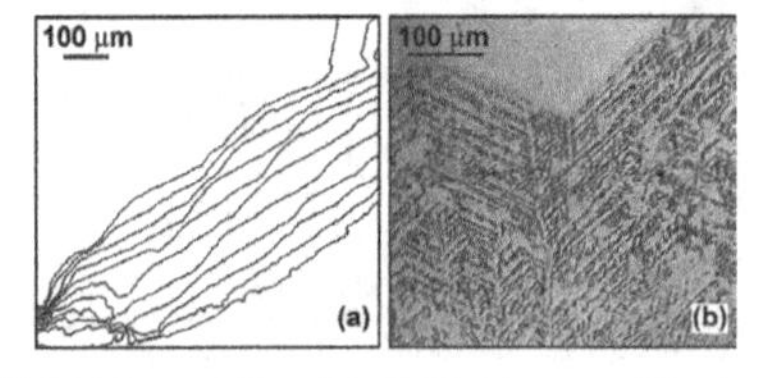

FIG. 3. (a) Sequent positions of the CDW boundary extracted from *in situ* optical observation of switching in constant external field (Z^+ view); (b) quasiregular domain pattern revealed by etching at Z^- after switching.

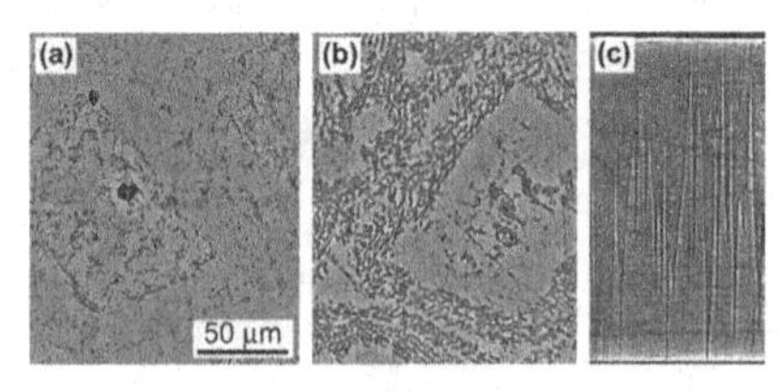

FIG. 5. Domain patterns after backswitching (a) Z^+ view, (b) Z^- view, and (c) Y view (revealed by long-time etching).

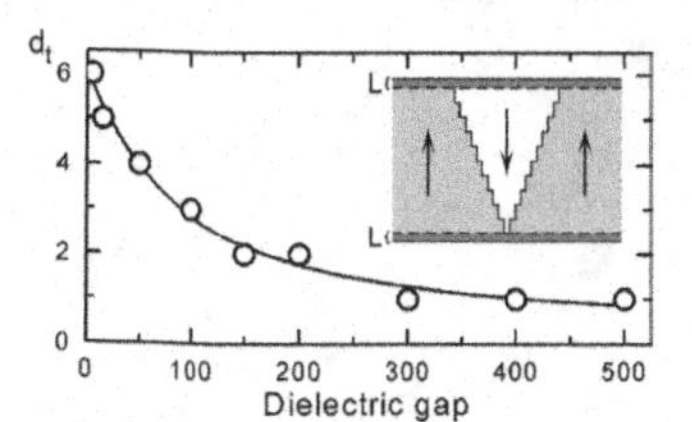

FIG. 6. Simulated dependence of the termination distance d_t of the step motion on the thickness of dielectric gap L. In the inset, the scheme of the forward growth of the needle-like domain, Y view.

persistent quasiregular CDW structure, which is similar at the top and bottom (Fig. 5).

The presence of a CDW drastically changes the dielectric behavior of LN in the temperature range 50–150 °C. The time dependence of the switching current during application of ac rectangular field pulses ($E = 50$ V/mm, $f = 0.05$ Hz) has been measured in the sample with the created stable CDW in the temperature range 20–220 °C. Comparison of the obtained data with the ones measured without the CDW allows us to extract the temperature dependence of the CDW input in the dielectric response at low frequencies ($f = 0.1$–20 Hz) (Fig. 7). The pronounced increase of this response is observed at low frequencies with the maximum at 135 °C.

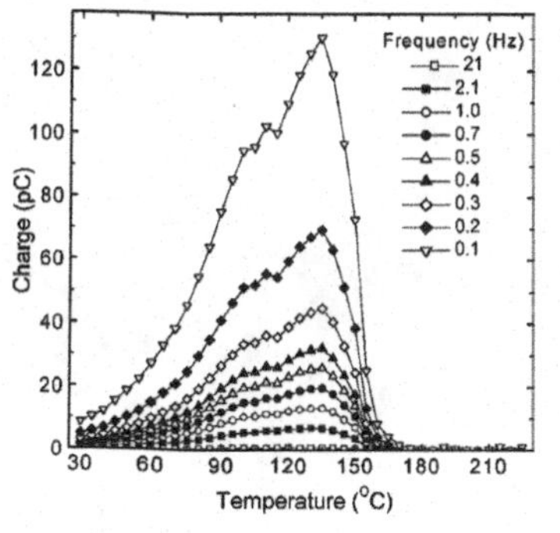

FIG. 7. Temperature dependence of the CDW dielectric response at low frequencies.

Within the discussed approach, the CDW response can be considered as a result of the reversible motion of the wall steps. The high value of the dielectric response is due to the extremely high step concentration. The observed temperature dependence can be explained as a result of competition between the temperature-induced increase of the step mobility and the decrease of the local switching field by bulk screening. The significant increase of the bulk screening is caused by the abrupt increase of the bulk conductivity usually observed in this temperature range.[13]

In summary, we investigated, using *in situ* methods, the process of CDW formation and field-induced evolution in congruent LN at room temperature. A strong dependence of CDW relief on the value and sign of the applied field was observed. We propose the mechanism of CDW boundary propagation and formation of the quasiregular dented structure. A strong frequency and temperature dependence of the CDW dielectric response was observed in the range of 50–150 °C.

The research was made possible in part by the Program "Basic Research in Russian Universities," by a grant from the Ministry of Education of the RF, by a grant from the RFBR, and by Award No. REC-005 of the U.S. Civilian Research and Development Foundation for Independent States of the Former Soviet Union.

[1] R. L. Byer, J. Nonlinear Opt. Phys. Mater. **6**, 549 (1997).
[2] R. G. Batchko, V. Y. Shur, M. M. Fejer, and R. L. Byer, Appl. Phys. Lett. **75**, 1673 (1999).
[3] V. Ya. Shur, E. L. Rumyantsev, E. V. Nikolaeva, E. I. Shishkin, D. V. Fursov, R. G. Batchko, L. A. Eyres, M. M. Fejer, and R. L. Byer, Appl. Phys. Lett. **76**, 143 (2000).
[4] M. Yamada, M. Saitoh, and H. Ooki, Appl. Phys. Lett. **69**, 3659 (1996).
[5] M. E. Lines and A. M. Glass, *Principles and Application of Ferroelectrics and Related Materials* (Clarendon, Oxford, 1977), p. 96.
[6] V. Ya. Shur, E. L. Rumyantsev, and A. L. Subbotin, Ferroelectrics **140**, 305 (1993).
[7] V. Ya. Shur and E. L. Rumyantsev, J. Korean Phys. Soc. **32**, S727 (1998).
[8] V. Ya. Shur, A. L. Gruverman, V. V. Letuchev, E. L. Rumyantsev, and A. L. Subbotin, Ferroelectrics **98**, 29 (1989).
[9] N. Ming, J. Hong, and D. Feng, J. Mater. Sci. **17**, 1663 (1982).
[10] L. E. Myers, Ph.D. thesis, Stanford University (1995).
[11] A. G. Chynoweth, Phys. Rev. **117**, 1235 (1960).
[12] L. E. Cross and T. W. Cline, Ferroelectrics **11**, 333 (1976).
[13] A. M. Prokhorov and Y. S. Kuzminov, *Physics and Chemistry of Crystalline Lithium Niobate* (Adam Hilger, Bristol, 1990).
[14] V. Gopalan, Q. X. Jia, and T. E. Mitchell, Appl. Phys. Lett. **75**, 2482 (1999).
[15] V. Ya. Shur, *Ferroelectric Thin Films: Synthesis and Basic Properties* (Gordon and Breach, New York, 1996), Vol. 10, Chap. 6.
[16] V. Ya. Shur, E. L. Rumyantsev, V. P. Kuminov, A. L. Subbotin, and V. L. Kozhevnikov, Phys. Solid State **41**, 269 (1999).
[17] V. Ya. Shur and E. L. Rumyantsev, Ferroelectrics **191**, 319 (1997).
[18] V. M. Fridkin, *Ferroelectrics Semiconductors* (Consulting Bureau, New York, 1980).
[19] G. D. Miller, R. G. Batchko, M. M. Fejer, and R. L. Byer, Proc. SPIE **2700**, 34 (1996).

Nanoscale backswitched domain patterning in lithium niobate

V. Ya. Shur,[a] E. L. Rumyantsev, E. V. Nikolaeva, E. I. Shishkin, and D. V. Fursov
Institute of Physics and Applied Mathematics, Ural State University, Ekaterinburg 620083, Russia

R. G. Batchko, L. A. Eyres, M. M. Fejer, and R. L. Byer
E.L. Ginzton Laboratory, Stanford University, Stanford, California 94305

(Received 29 July 1999; accepted for publication 11 November 1999)

We demonstrate a promising method of nanoscale domain engineering; which allows us to fabricate regular nanoscale domain patterns consisting of strictly oriented arrays of nanodomains (diameter down to 30 nm and density up to 100 μm^{-2}) in lithium niobate. We produce submicron domain patterns through multiplication of the domain spatial frequency as compared with the electrode one. The fabrication techniques are based on controlled backswitched poling. © *2000 American Institute of Physics.* [S0003-6951(00)01402-9]

In recent years, domain engineering, as a new branch of technology connected with fabrication of periodic ferroelectric domain structures with desirable parameters, is rapidly developing. Its advance is a critical step in the manufacturing of electro-optical[1] and nonlinear optical devices.[2–5] Engineerable nonlinear optical materials have permitted the development of a wide range of tunable coherent light sources based on quasiphase matching.[6] Lithium niobate is widely used due to its large electro-optical and nonlinear optical coefficients.[7] Application of an electric field through lithographically defined electrodes for domain patterning is one of the most promising methods.[8] However, high coercive electric fields and the strong effect of domain widening out of the electroded area impose limitations on the period of short-pitch domain patterns. Up to now, domain patterns with periods less than 1 μm have not been fabricated. In this work, we report an approach which allows to us overcome this obstacle.

Periodic domain structures were obtained in standard optical-grade single-domain 0.5-mm-thick $LiNbO_3$ wafers of congruent composition cut perpendicular to the polar axis. The wafers were photolithographically patterned with a periodic stripe metal–electrode structure (NiCr) deposited on a $\mathbf{Z}^+$ surface and oriented along the **Y** axis. The patterned surface was covered by a thin (about 0.5-mm-thick) insulating layer (photoresist) [Fig. 1(a)]. A high-voltage pulse producing an electric field greater than the coercive field for quasistatic switching (E_c=21.5 kV/mm) was applied to the structure through a fixture containing liquid electrolyte (LiCl).[9,10] The wave form for backswitched poling consisted of three levels of external field: "high field," "low field" and "stabilization field," [Fig. 1(b)]. The switching from the single-domain state took place at the "high field" and the backswitching (flip-back)[11–13] occurred at the "low field." The crucial parameters for backswitching kinetics were the duration of the "high-field" stage Δt_{sp} and the field-diminishing amplitude ΔE. The domain patterns obtained for different durations of the "low field" stage yielded information about the domain structure development during backswitching. For observation of the domain patterns after partial poling, the **Z** surfaces and polished **Y** cross sections were etched for 5–10 min by hydrofluoric acid without heating.[10] The obtained surface relief was visualized by optical microscope, scanning electron microscopy (SEM), and atomic force microscopy (AFM), techniques.

Distinguishable stages of domain evolution have been observed during backswitched poling.[11–13] The process starts with nucleation (arising of new domains) at the $\mathbf{Z}^+$ polar surface along the electrode edges [Fig. 2(a)]. During the second stage, these domains grow and propagate through the sample. As a result of merging, laminar domains with plane walls are formed. Pronounced domain broadening out of the electroded area is always observed [Fig. 2(b)]. After rapid decreasing of the poling field, the backswitching starts through shrinking of the laminar domains by the backward wall motion and nucleation along the electrode edges [Fig. 2(c)]. We have shown that various regular nanodomain structures can be produced by backswitched poling by presetting the voltage wave-form parameters. The types of domain patterns depend also on the value of the domain-wall shift out of the electrodes.

Backswitched domain frequency multiplication. For small domain-wall shift out of electrodes, we have revealed a backswitched effect—multiplication of the domain pattern

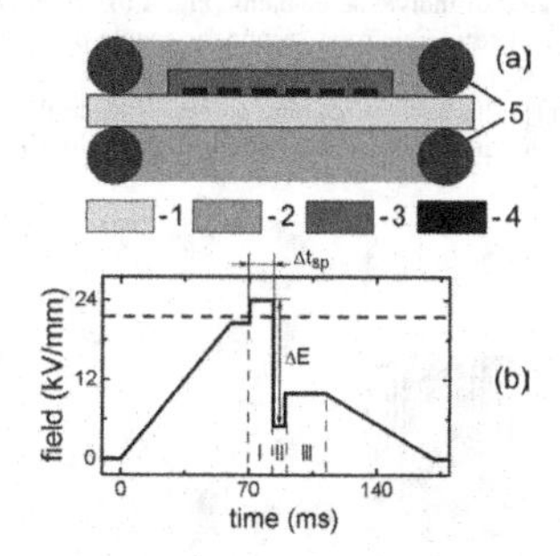

FIG. 1. (a) Scheme of the experimental setup and (b) backswitched poling voltage waveform.

[a]Electronic mail: vladimir.shur@usu.ru

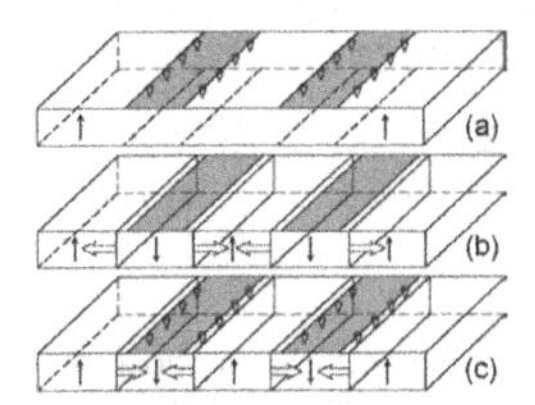

FIG. 2. Stages of domain evolution during backswitched poling. White arrows: directions of domain-wall motion.

frequency as compared to the electrode one. The mechanism of frequency multiplication is based on the nucleation along the electrode edges during backswitching, while the role of the wall motion in this case is negligible. For "frequency tripling" [Fig. 3(c)], the subsequent growth and merging of nucleated domains lead to formation of a couple of strictly oriented sub-micron-width domain stripes under each electrode. Their depth is about 20–50 μm [Fig. 3(d)]. It is clear that this structure can be produced only by using wide enough electrodes. For narrow electrodes these stripes merge and only the "frequency doubling" can be obtained [Fig. 3(a)]. The depth of these backswitched domain stripes is, typically, about 50–100 μm [Fig. 3(b)].

The backswitched domain cross sections reveal two distinct variants of domain evolution during frequency multiplication: "erasing" and "splitting." During "erasing," backswitched domains are formed in the earlier switched area without variation of the external shape of the switched laminar domains [Fig. 3(e)]. During "splitting," the backswitched domains cut the initial ones conserving its volume [Fig. 3(f)].

More complicated frequency multiplication is demonstrated for sufficient domain spreading out of electrodes during switching. For long switching pulse $\Delta t_{sp} \sim 15$ ms and large field-diminishing amplitude $\Delta E \sim 20$ kV/mm [Fig. 1(b)], the backswitching also starts with the formation of the couple of arrays under the electrode edges [Fig. 4(a)]. Then, the arrays turn into a pair of stripe domains through growth and merging of individual domains [Fig. 4(b)]. Surprisingly, after complete merging the secondary couple of arrays appear in the nonelectroded area parallel to the initial ones [Fig. 4(c)]. This self-maintaining process leads to the formation of periodic stripe domains oriented along the electrodes

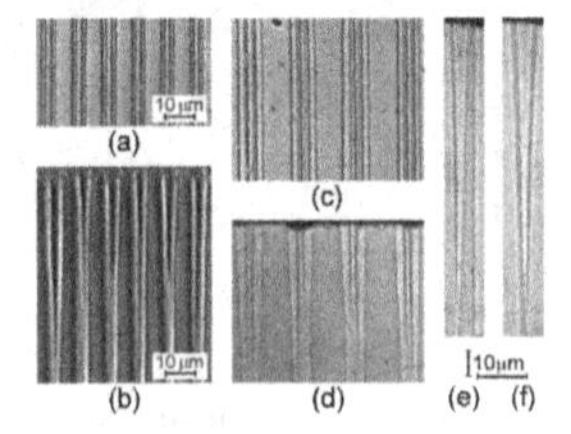

FIG. 3. Backswitched domain frequency multiplication. (a) and (b) "frequency doubling;" (c) and (d) "frequency tripling;" (e) "erasing;" (f) "splitting." (a) and (c) $\mathbf{Z}^+$ view; and (b), (d), (e), and (f) $\mathbf{Y}$ cross sections.

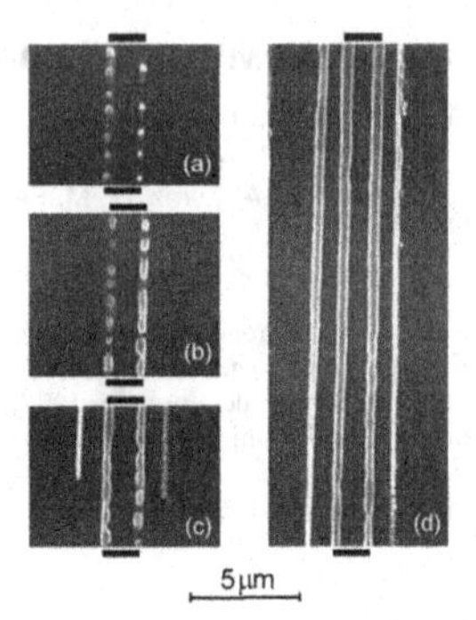

FIG. 4. SEM patterns demonstrating the stages of the formation of periodic stripe domains along the electrode edges. $\mathbf{Z}^+$ view. Electrodes cover the area between the black rectangles. Electrodes are oriented along the $\mathbf{Y}$ direction.

[Fig. 4(d)]. The distance between the secondary and initial stripes is about the thickness of the insulating layer.

Formation of the regular nanodomain structures. For sufficient domain spreading out of electrodes, but for short switching pulse duration $\Delta t_{sp} \sim 5$ ms and small field-diminishing amplitude $\Delta E \sim 2$ kV/mm, the backswitched domain evolution is changed drastically and represents self-maintained self-organized growth of oriented nanoscale domain arrays. The domain patterns revealed by etching and visualized by SEM demonstrate the array-assisted reversal motion of the existing domain walls through propagation of the highly organized quasiperiodical structure of domain arrays strictly oriented along crystallographic directions (Fig. 5). Each quasiregular array is comprised of nanodomains with a diameter of 30–100 nm and an average linear density exceeding 10^4 mm^{-1}.

Two variants of array orientation are obtained. In the usual case, all nanodomain arrays are strictly oriented along the $\mathbf{Y}^-$ direction at 60° to the electrode edges [Fig. 5(a)]. A similar oriented nucleation has been observed during polarization reversal in strong homogeneous fields.[14,15] In some cases the domain arrays are strictly oriented along the $\mathbf{X}^-$ or $\mathbf{X}^+$ directions [Fig. 5(b)], and individual nanodomains have a triangular shape. There are four equal $\mathbf{X}$ directions oriented at 30° to the electrode edges. So, the domain pattern consisting of regular 30° array fragments with different array orientations are obtained along one domain wall [Fig. 5(c)]. The array patterns growing along the $\mathbf{X}$ directions at 90° to the electrodes are different, as the fast growth of nanodomains along the electrodes leads to the formation of a periodic set of nanoscale stripe domains with a period less than 60 nm [Fig. 5(d)].

All effects observed during backswitching can be explained by the nonuniform distribution of the local backswitching field near the electrode edges and domain walls due to spatially nonuniform screening of the external and depolarization fields.[16] Our estimations reveal the field maximum at a distance about the thickness of the insulating layer from the domain wall or nanodomain array.[15] An array ag-

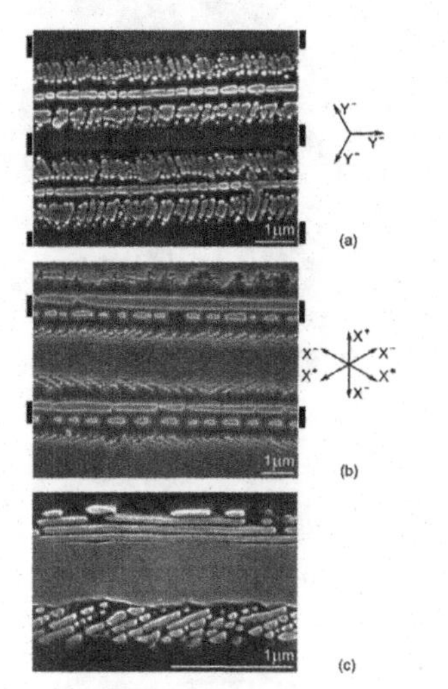

FIG. 5. SEM nanodomain patterns with arrays oriented along different axes: (a) **Y**$^-$ at 60° to the electrode edges, (b) **X** at 30° and 90° to the electrode edges, and (c) coexistence of different array orientations. **Z**$^+$ view. Electrodes cover the area between the black rectangles. Electrodes are oriented along the **Y** direction.

gregate existing at a given moment generates the new maximum of the backswitching field at a fixed distance in front of its boundary and triggers the arising of the new array. The developing correlated nucleation leads to the self-maintaining generation of the parallel arrays. The array-assisted propagation of the correlated domain structure is similar to the formation of the ''wide domain boundary'' discovered in $Pb_5Ge_3O_{11}$ during switching in ''super-strong'' fields.[17]

We demonstrate promising possibilities in domain engineering based on backswitched poling. The multiplication of the domain spatial frequency as compare with the electrode one allows us to produce domain structures with submicron periods. Quasiregular patterns of oriented arrays of nanoscale domains (with a density up to 100 μm^{-2}) were fabricated. The proposed method of domain engineering can be used in a wide range of applications and for manufacturing of various devices.

This material is based upon work partially supported by the Program ''Basic Research in Russian Universities'' under Grant No. 5563, by the EOARD, Air Force Office of Scientific Research, Air Force Research Laboratory, under Contract No. F61775-99-WE037, and by DARPA/ONR through the Center of Nonlinear Optical Materials (CNOM) at Stanford University under ONR Grant No. 00014-92-J-1903 and LLNL.

[1] M. Yamada, M. Saitoh, and H. Ooki, Appl. Phys. Lett. **69**, 3659 (1996).
[2] D. Feng, N. B. Ming, J. F. Hong, Y. S. Yang, J. S. Zhu, Z. Yang, and Y. N. Wang, Appl. Phys. Lett. **37**, 607 (1980).
[3] N. B. Ming, J. F. Hong, and D. Feng, J. Mater. Sci. **17**, 1663 (1982).
[4] A. Feisst and P. Koidl, Appl. Phys. Lett. **47**, 1125 (1985).
[5] R. L. Byer, J. Nonlinear Opt. Phys. Mater. **6**, 549 (1997).
[6] K. C. Burr, C. L. Tang, M. A. Arbore, and M. M. Fejer, Appl. Phys. Lett. **70**, 3341 (1997).
[7] A. M. Prokhorov and Y. S. Kuzminov, *Physics and Chemistry of Crystalline Lithium Niobate* (Adam Hilger, Bristol, 1990), p. 263.
[8] M. Yamada, N. Nada, M. Saitoh, and K. Watanabe, Appl. Phys. Lett. **62**, 435 (1993).
[9] L. E. Myers, R. C. Eckardt, M. M. Fejer, R. L. Byer, and W. R. Bosenberg, Opt. Lett. **21**, 591 (1996).
[10] G. D. Miller, R. G. Batchko, M. M. Fejer, and R. L. Byer, Proc. SPIE **2700**, 34 (1996).
[11] V. Ya. Shur, R. G. Batchko, E. L. Rumyantsev, G. D. Miller, M. M. Fejer, and R. L. Byer, Proc. 11th ISAF, Piscataway, IEEE, NJ (1999), pp. 399.
[12] V. Shur, E. Rumyantsev, R. Batchko, G. Miller, M. Fejer, and R. Byer, Ferroelectrics **221**, 157 (1999).
[13] R. G. Batchko, V. Ya. Shur, M. M. Fejer, and R. L. Byer, Appl. Phys. Lett. **75**, 1673 (1999).
[14] V. Ya. Shur, E. L. Rumyantsev, R. G. Batchko, G. D. Miller, M. M. Fejer, and R. L. Byer, Phys. Solid State **41**, 1681 (1999).
[15] V. Ya. Shur, E. L. Rumyantsev, E. V. Nikolaeva, E. I. Shishkin, R. G. Batchko, L. A. Eyres, M. M. Fejer, and R. L. Byer, Phys. Rev. Lett. (in press).
[16] V. Ya. Shur, E. V. Nikolaeva, E. L. Rumyantsev, E. I. Shishkin, A. L. Subbotin, and V. L. Kozhevnikov, Ferroelectrics **222**, 323 (1999).
[17] V. Ya. Shur, A. L. Gruverman, N. Yu. Ponomarev, and N. A. Tonkachyova, Ferroelectrics **126**, 371 (1992).

PHYSICAL REVIEW B VOLUME 40, NUMBER 1 1 JULY 1989

First-order phase transition in order-disorder ferroelectrics

C. L. Wang and Z. K. Qin
Department of Physics, Shandong University, Jinan, People's Republic of China

D. L. Lin
Department of Physics and Astronomy, State University of New York at Buffalo, Amherst, New York 14260
(Received 24 October 1988)

A four-spin coupling term is introduced to the pseudospin model Hamiltonian to account for the recently observed first-order phase transition in order-disorder ferroelectrics. The critical behavior is investigated and characteristic temperatures are calculated. We have also expressed the free energy in terms of microscopic variables and demonstrated that, in the limit of small mean spin $\langle S^z \rangle$, its behavior near the critical point is identical to what is expected in thermodynamical theory.

I. INTRODUCTION

A first-order phase transition in certain order-disorder ferroelectrics has been observed recently. The transition[1] of $C_4H_2O_4$ near 373 K is just one example. It can be well understood in terms of Landau's thermodynamic theory,[2] especially in the neighborhood of the critical point where the theory correctly predicts characteristic temperatures of the first-order phase transition: the paraelectric phase-stability-limit temperature T_0, the critical temperature T_c, and the ferroelectric phase-stability-limit temperature T_0^-. The thermodynamic theory, however, can never interpret the observed phenomena in terms of the microscopic atomic interactions.

A pseudospin model[3] is able to account for the order-disorder phase-transition phenomena. The theory cannot handle the first-order phase transition by itself even though it explains the second-order phase transition quite successfully. Further work on the coupling between the pseudospin and lattice has been suggested[4] to discuss the phase-transition problem of order-disorder-type ferroelectrics. Again, it does not yield satisfactory results for the first-order phase transition.

It appears, therefore, necessary to examine more carefully the structure of order-disorder–type ferroelectrics in order to understand the phase-transition properties. One of the typical order-disorder ferroelectrics is potassium dihydrogen phosphate[4] (KH_2PO_4, KDP), in which the group PO_4 forms a tetrahedron with P^{5+} at its center and the four oxygen ions at its vertices. Hydrogen bonds connect each of the vertices to other molecules on equal footing, and the potassium ion may be either above or below the tetrahedron as shown in Fig. 1. At higher temperatures, protons move back and forth along the equilibrium hydrogen bonds and the system is in a disordered state. When the phase transition occurs, the upper two protons move away from, and the lower two protons move toward, the tetrahedron. The system becomes ordered and a net dipole moment is produced along the c axis because the ion P^{5+} is pushed upward and K^+ moves downward to cause the spontaneous polarization.

The squaric acid[5] ($C_4H_2O_4$, H_2SQ) has a layered structure, as can be seen in Fig. 2. Each molecule is a square with oxygen ions at its four vertices linked by hydrogen bonds on equal footing. At room temperature the layers are ferroelectrically ordered and antiferroelectrically stacked. Protons move randomly along the hydrogen bonds above the transition temperature, and ordered motion results at critical temperature. The net dipole moment produced in this case lies in the plane of the square, a different situation from the KDP.

From the above discussion, it is clear that in a ferroelectric material of order-disorder type, the four hydrogen bonds usually appear as a group and every one is equivalent to another. This means that the four-body interaction in such structures is generally important. In fact, Deiningham and Mehring[5] have pointed out the existence of four-body interaction in the $C_4H_2O_4$ structure

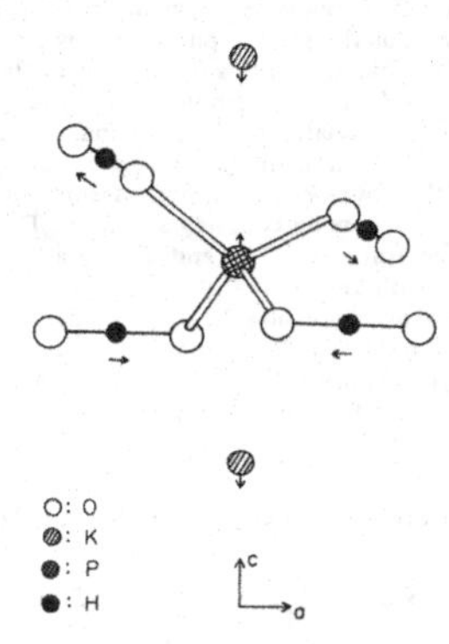

FIG. 1. Tetrahedronic structure of KH_2PO_4 crystal.

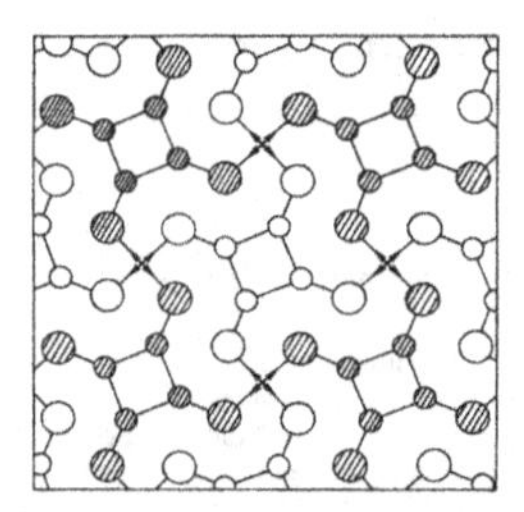

FIG. 2. Schematic diagram of the structure of squaric acid crystal cut across the layers. Different layers are denoted by open and shaded circles. Large and small circles represent O atoms and C atoms, respectively. Double arrows indicate the hydrogen bonds.

but neglected it in their actual calculation, hence, their result is still a second-order phase transition. A similar conclusion follows also from the Slater-Takagi model. Slater assumes that[6] only two protons can approach the tetrahedron, and his theory predicts only the second-order phase transition. On the other hand, Takagi[7] considers all possibilities and obtains the first-order phase transition. It is therefore necessary to include four-body interactions in the discussion of the first-order phase transition.

On the other hand, in the theoretical studies of the Ising model,[8–10] it has been shown that a first-order phase transition is possible only if the number of coupled spins is four or larger. But these investigations are carried out by means of either a renormalization group or finite-size scaling, and they do not involve dynamical properties.

On the basis of the pseudospin theory, we consider in this paper the four-spin interaction. The method of retarded Green's function is employed to study how the transition order is related to the coupling strength. It is found that the system exhibits a first-order phase transition when the four-body coupling strength $J'>4J/3$, where J is the two-body coupling strength. The characteristic temperatures T_0, T_0^-, and T_c are all calculated. To compare with the results of thermodynamic theory, we also derive an expression for the free energy. We find that in the limit of small mean spin $\langle S^z\rangle$, its behavior near the critical point is in complete agreement with predictions of the thermodynamic theory.

II. THEORY

The order-disorder system is described by the Hamiltonian

$$H=-\Omega\sum_i S_i^x-\tfrac{1}{2}\sum_{i,j}J_{ij}S_i^zS_j^z-\tfrac{1}{4}\sum_{i,j,k,l}J'_{ijkl}S_i^zS_j^zS_k^zS_l^z , \tag{1}$$

where the first two terms constitute the original pseudospin model and the third term represents the four-body interaction. S_i^x is called the tunneling operator, which measures the tunneling power of the proton between the hydrogen double well, Ω is the tunneling frequency, and S_i^z is half of the difference of occupation probabilities for the proton to be found in the two equilibrium positions of the hydrogen bond. The two-body coupling J_{ij} is the same for every pair of protons in KDP. In the case of H_2SQ, it represents the coupling between protons in neighboring layers as well as those in the same layer. The four-body coupling J_{ijkl} refers to the four hydrogen bonds in the PO_4 group in KDP, and in H_2SQ crystals it represents the interactions between the four hydrogen bonds in the C_4O_4 group.

The retarded Green's function for any two operators $A(t)$ and $B(t')$ is defined by

$$\langle\langle A(t)|B(t')\rangle\rangle=-i\Theta(t-t')\langle[A(t),B(t')]\rangle , \tag{2}$$

where $\Theta(t)$ is the step function and the commutator $[A,B]=AB-BA$. The Fourier component of the Green's function satisfies the equation of motion

$$\omega\langle\langle A|B\rangle\rangle_\omega=(2\pi)^{-1}\langle[A,B]\rangle+\langle\langle[A,H]|B\rangle\rangle , \tag{3}$$

where H is the Hamiltonian of the system. The correlation function $\langle BA\rangle$ is related to the Green's function by the spectral theorem

$$\langle BA\rangle=i\int_{-\infty}^{\infty}d\omega\, e^{-i\omega(t-t')}\times\frac{\langle\langle A|B\rangle\rangle_{\omega+i0^+}-\langle\langle A|B\rangle\rangle_{\omega-i0^+}}{e^{\beta\omega}-1} , \tag{4}$$

where $\beta=k_BT$. It is not difficult to show from the definition (2) that the Green's function remains unchanged when the operators are changed by constants. That is,

$$\langle\langle A+\gamma|B+\gamma'\rangle\rangle=\langle\langle A|B\rangle\rangle , \tag{5}$$

where γ and γ' are arbitrary constants. But the correlation function

$$\langle(B+\gamma')(A+\gamma)\rangle\neq\langle BA\rangle . \tag{6}$$

This is evidently inconsistent with the spectral theorem (4) which implies that the correlation function does not change when constants are added to the operators. This inconsistency has been resolved by the method of undetermined constant.[11]

Let us choose the operators $A=S_i^z,S_i^+,S_i^-$ and $B=S_j^z,S_j^x$ to form Green's functions. With the Hamiltonian (1), higher-order Green's functions result from the equation of motion (3). We apply the Tyablikov scheme[12] to decouple the Green's functions involving two pseudospins,

$$\sum_j J_{ij}\langle\langle S_m^zS_j^z|S_n^z\rangle\rangle\simeq\sum_j J_{ij}\langle S_j^z\rangle\langle\langle S_m^z|S_n^z\rangle\rangle =J\langle S^z\rangle\langle\langle S_m^z|S_n^z\rangle\rangle , \tag{7}$$

where we have assumed for simplicity $J=\sum_j J_{ij}$, and $\langle S_i^z\rangle=\langle S_j^z\rangle=\langle S^z\rangle$. The Green's functions involving four pseudospins are decoupled according to a scheme proposed in a recent paper,[13] that is,

$$\sum_{j,k,l} J'_{ijkl}\langle\langle S^z_m S^z_j S^z_k S^z_l | S^z_n\rangle\rangle \simeq \sum_{j,k,l} J'_{ijkl}\langle S^z_j\rangle\langle S^z_k\rangle\langle S^z_l\rangle\langle\langle S^z_m|S^z_n\rangle\rangle = J'\langle S^z\rangle^3\langle\langle S^z_m|S^z_n\rangle\rangle , \quad (8)$$

where $J'=\sum_{j,k,l} J'_{ijkl}$.

The Fourier components of the linearized Green's function satisfy the equation of motion

$$\begin{bmatrix} \omega & \Omega/2 & -\Omega/2 & & & \\ \Omega & \omega-K & 0 & & & \\ -\Omega & 0 & \omega+K & & & \\ & & & \omega & \Omega/2 & -\Omega/2 \\ & & & \Omega & \omega-K & 0 \\ & & & -\Omega & 0 & \omega+K \end{bmatrix} \begin{bmatrix} G^{zz}(\omega) \\ G^{+z}(\omega) \\ G^{-z}(\omega) \\ G^{zx}(\omega) \\ G^{+x}(\omega) \\ G^{-x}(\omega) \end{bmatrix} = \frac{1}{2\pi}\begin{bmatrix} 0 \\ -\langle S^+\rangle \\ \langle S^-\rangle \\ 0 \\ \langle S^z\rangle \\ -\langle S^z\rangle \end{bmatrix} , \quad (9)$$

where $K=J\langle S^z\rangle + J'\langle S^z\rangle^3$, and $G^{mn}(\omega)=\langle\langle S^m|S^n\rangle\rangle$, $m,n=x,y,z$. Following the procedures described in Ref. 11, we obtain, after some calculation, the following self-consistent equations:

$$[\langle S^x\rangle(J+J'\langle S^z\rangle^2)-\Omega]\langle S^z\rangle=0 , \quad (10a)$$

$$\frac{\langle S^x\rangle}{2[\langle S^x\rangle^2+\langle S^z\rangle^2]}=\frac{\Omega}{\omega_0}\coth\frac{\omega_0}{2k_BT} , \quad (10b)$$

where

$$\omega_0=[\Omega^2+(J+J'\langle S^z\rangle^2)^2\langle S^z\rangle^2]^{1/2} . \quad (11)$$

It can easily be seen that Eqs. (10) possess two sets of solutions. $\langle S^z\rangle=0$ corresponds to the paraelectric phase and $\langle S^z\rangle\neq 0$ corresponds to the ferroelectric phase. In the following sections, we shall discuss the properties of these phases on the basis of (10).

III. STABILITY-LIMIT TEMPERATURES

When the system is in the ferroelectric phase, or $\langle S^z\rangle\neq 0$, we have from (10a)

$$\langle S^x\rangle=\frac{\Omega}{J+J'\langle S^z\rangle^2} .$$

Substituting in (10b), we find

$$1=\frac{J+J'\langle S^z\rangle^2}{2\omega_0}\tanh\frac{\omega_0}{2k_BT} . \quad (12)$$

For simplicity we assume $\Omega=0$ without loss of generality; then (12) becomes, upon substitution of (11),

$$\langle S^z\rangle=\tfrac{1}{2}\tanh\frac{J\langle S^z\rangle+J'\langle S^z\rangle^3}{2k_BT} . \quad (13)$$

The variation of temperature with the mean spin value $\langle S^z\rangle$ for different ratios of the coupling strength J and J' is plotted in Fig. 3. To study the critical behavior, we have introduced a characteristic temperature $T_0=J/4k_B$ for the unit of temperature in the plot. The unit of $\langle S^z\rangle$ is its value $\langle S^z\rangle_0$ at $T=0$. The three curves are for the relative coupling strengths (*a*) $J'=J/3$, (*b*) $J'=4J/3$, and (*c*) $J'=7J/3$.

For $J'/J\leq 4/3$, or cases corresponding to curves *a* and *b*, the system has two sets of solutions. When $T\geq T_0$, there is only one solution with $\langle S^z\rangle=0$ corresponding to the paraelectric phase. When $T<T_0$, both solutions with $\langle S^z\rangle=0$ and $\langle S^z\rangle\neq 0$ are possible. As it will become clear in Sec. V, when we analyze the behavior of the free energy near T_0, the solution with $\langle S^z\rangle=0$ is unstable for $T<T_0$, and only $\langle S^z\rangle\neq 0$ is a stable solution. This means that the system changes continuously from the paraelectric phase ($\langle S^z\rangle=0$) to the ferroelectric phase ($\langle S^z\rangle\neq 0$) when its temperature decreases from above T_0 to below. Thus the system exhibits a second-order phase transition with the critical temperature T_0.

For $J'/J>4/3$ or the case corresponding to curve *c* in Fig. 3, the system has three sets of solutions. When $T\geq T_0^-$, it has a paraelectric solution with $\langle S^z\rangle=0$. When $T_0<T<T_0^-$, the equation has three solutions: one with $\langle S^z\rangle=0$ and two with $\langle S^z\rangle\neq 0$ corresponding to the solid and dashed lines of the curve. When $T<T_0$, there are two possible solutions with $\langle S^z\rangle=0$ and $\langle S^z\rangle\neq 0$.

It is observed from Fig. 3 that T_0^- is the limiting temperature below which the system can be found in the ferroelectric phase. The value of T_0^- can be determined by the condition

$$\frac{\partial T}{\partial\langle S^z\rangle}=0 . \quad (14)$$

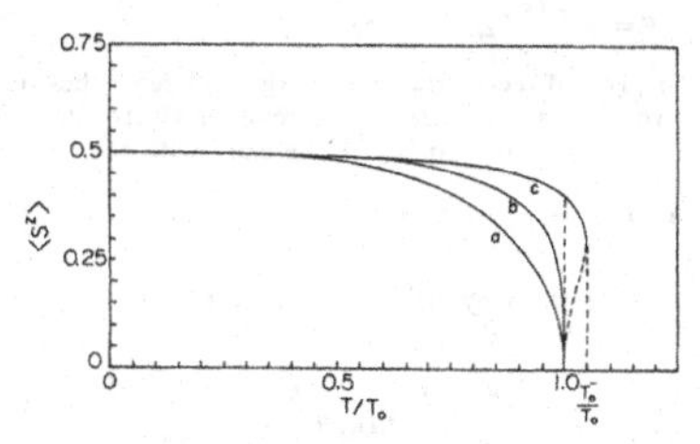

FIG. 3. Temperature dependences of $\langle S^z\rangle$. (*a*) $J'=J/3$, (*b*) $J'=4J/3$, (*c*) $J'=4J/3$.

Solving (14) we find the mean spin $\langle S^z\rangle_-$ at T_0^-. Combining (13) and (14), we obtain

$$4k_B T_0^- = (1-4\langle S^z\rangle_-^2)(J+3J'\langle S^z\rangle_-^2), \tag{15}$$

which, together with (13), yields the critical values for T_0^- and $\langle S^z\rangle_-$. An approximate expression can be obtained for sufficiently small $\langle S^z\rangle_-$. We expand (13) around $\langle S^z\rangle=0$ to the fourth order in $\langle S^z\rangle$ and then combine with (15) to find

$$T_0^- = T_0\left[1+\frac{3}{16}\frac{(J'/J-4/3)^2}{J'/J+16/15}\right]. \tag{16}$$

Again from the analysis of the free energy in Sec. V, the ferroelectric phase is metastable but the paraelectric phase is stable in the neighborhood of T_0^-. Hence, the system remains in the paraelectric phase when the temperature drops down to T_0^-. On the other hand, the paraelectric phase is metastable and the ferroelectric phase is stable at T_0+0^+, or the system remains in the ferroelectric phase as the temperature increases to T_0+0^+. This implies that T_0 is the limiting temperature for the stable paraelectric phase. Therefore, paraelectric and ferroelectric phases coexist when the temperature is within the interval $T_0 < T < T_0^-$. We then conclude that within this interval there exists a temperature T_c at which the stability of the two phases are equal. This is the Curie temperature of the first-order phase transition.

IV. DETERMINATION OF T_c

The critical temperature T_c and the order parameter $\langle S^z\rangle_c$ are determined by the condition that the system possesses the same free energy in both phases. Hence, we proceed to calculate the free-energy change when the system changes phase. The Helmholtz free energy F is related to the external electric field E by

$$E=\left[\frac{\partial F}{\partial p}\right]_T, \tag{17}$$

where p stands for the polarization intensity. For the present problem, we have $p=2N\mu\langle S^z\rangle$, where μ is the effective dipole moment and N is the number of hydrogen bonds per unit volume. Equation (17) can then be put in the integral form

$$\Delta F = N\int_0^{\langle S^z\rangle} 2\mu E\, d\langle S^z\rangle, \tag{18}$$

which gives directly the free-energy difference between the two phases. In order to express the electric field in terms of the mean spin $\langle S^z\rangle$, we start with the Hamiltonian

$$H(E)=H-2\mu E\sum_i S_i^z, \tag{19}$$

where H is given by (1). Repeating the procedures that lead to (13), we find

$$\langle S^z\rangle = \tfrac{1}{2}\tanh\frac{J\langle S^z\rangle + J'\langle S^z\rangle^3+2\mu E}{2k_B T}, \tag{20}$$

which can be rewritten as

$$2\mu E = -J\langle S^z\rangle - J'\langle S^z\rangle^3 + k_B T\ln\left[\frac{1+2\langle S^z\rangle}{1-2\langle S^z\rangle}\right]. \tag{21}$$

Substituting (21) in (18) and integrating over $\langle S^z\rangle$, we obtain

$$\frac{\Delta F}{N} = -\tfrac{1}{2}J\langle S^z\rangle^2 - \tfrac{1}{4}J'\langle S^z\rangle^4 + \tfrac{1}{2}k_B T\ln(1-4\langle S^z\rangle^2) + k_B T\ln\left[\frac{1+2\langle S^z\rangle}{1-2\langle S^z\rangle}\right]. \tag{22}$$

At the critical point, $\Delta F=0$ and $\langle S^z\rangle = \langle S^z\rangle_0$, we have

$$-\tfrac{1}{2}J\langle S^z\rangle_c^2 - \tfrac{1}{4}J'\langle S^z\rangle_c^4 + \tfrac{1}{2}k_B T_c\ln(1-4\langle S^z\rangle_c^2) + k_B T_c\ln\left[\frac{1+2\langle S^z\rangle_c}{1-2\langle S^z\rangle_c}\right]=0, \tag{23}$$

which, together with (13), can be solved for T_c and $\langle S^z\rangle_c$. If one assumes $\langle S^z\rangle_c$ to be small, these equations can be expanded around $\langle S^z\rangle_c=0$ up to the fourth order in $\langle s^z\rangle$ to find an approximate expression for the Curie temperature

$$T_c = T_0\left[1+\frac{1}{6}\frac{(J'/J+1/5)(J'/J-4/3)^2}{(J'/J+4/5)^2}\right]. \tag{24}$$

A comparison with (16) then yields $T_0^- > T_c > T_0$, as expected.

V. FREE ENERGY NEAR THE CRITICAL POINT

Landau's thermodynamic theory of phase transition is based on the expansion of the free energy in a power series of the polarization which serves as the order parameter of the system. Thus, the free energy for the ferroelectric phase can be written as

$$F = F_0 + \tfrac{1}{2}a(T-T_0)p^2 + \tfrac{1}{4}bp^4 + \tfrac{1}{6}cp^6, \tag{25}$$

where F_0 is the free energy for the paraelectric phase and the coefficients a, b, and c are assumed to be temperature independent. When $b>0$, the last term can be ignored and F may be employed to describe the second-order phase transition. When $b<0$, only $c>0$ yields a stable minimum free energy. Thus the last term must be included to explain the first-order phase transition.

In order to express the quantities a, b, c, and T_0 in terms of the microscopic parameters J and J', we expand (22) around the critical point up to and including $\langle S^z\rangle^6$:

$$\frac{\Delta F}{N} = \tfrac{1}{2}(4k_B T - J)\langle S^z\rangle^2 + \tfrac{1}{4}(\tfrac{16}{3}k_B T - J')\langle S^z\rangle^4 + \tfrac{32}{15}k_B T\langle S^z\rangle^6. \tag{26}$$

A direct comparison with (25) leads to $T_0 = J/4k_B$, which means that the Curie-Weiss temperature (the limiting temperature for the stable paraelectric phase) is directly proportional to the two-body coupling strength J. Since (26) is an expansion around the critical point, and since the coefficients of the second and third terms have no singularities at $T=T_0$, we can replace with good approximation $T\approx T_0=J/4k_B$ in these terms. Thus, Eq. (26) becomes

$$\frac{\Delta F}{N}=\tfrac{1}{2}(4k_BT-J)\langle S^z\rangle^2+\tfrac{1}{4}(4J/3-J')\langle S^z\rangle^4+\tfrac{8}{15}J\langle S^z\rangle^6 . \quad (27)$$

It is observed from (27) that the coefficient b in the thermodynamical expansion (25) is proportional to $4J/3-J'$. Therefore, a positive b implies $J'<4J/3$, and the free energy describes a second-order phase transition. In this case, the last term of the order of $\langle S^z\rangle^6$ can be neglected. When $b<0$, $J'>4J/3$, and F describes a first-order phase transition, exactly the same as what Landau's theory concludes. The critical strength $J'=4J/3$ for the onset of first-order phase transition is determined approximately here by the expansion around the critical point. It is, in fact, an exact result.[14]

To compare, in more detail, predictions near the critical point by the microscopic model and the macroscopic theory, we plot the reduced free energy $f=\Delta F/NJ$ calculated from (22) in Figs. 4 and 5. For the case of second-order phase transition or $J'<4J/3$, Fig. 4 illustrates the behavior of f at different temperatures. Above the critical temperature T_0, the reduced free energy has only one minimum at $\langle S^z\rangle=0$, as shown in 4(a), while it has two minima at $\pm\langle S^z\rangle_m$ and a maximum at $\langle S^z\rangle=0$ for $T<T_0$, as shown in 4(c). Therefore, the paraelectric phase is unstable below the critical temperature. When $T=T_0$, the free energy as can be seen in 4(b) still has a minimum at $\langle S^z\rangle=0$, but the curve has a flat bottom indicating a vanishing second-order derivative.

The case of first-order phase transition or $J'>4J/3$ is depicted in Fig. 5. The reduced free energy has two minima at $\pm\langle S^z\rangle_m$ for $T<T_0$ when the system is in the ferroelectric phase. This is illustrated in 5(a). Another

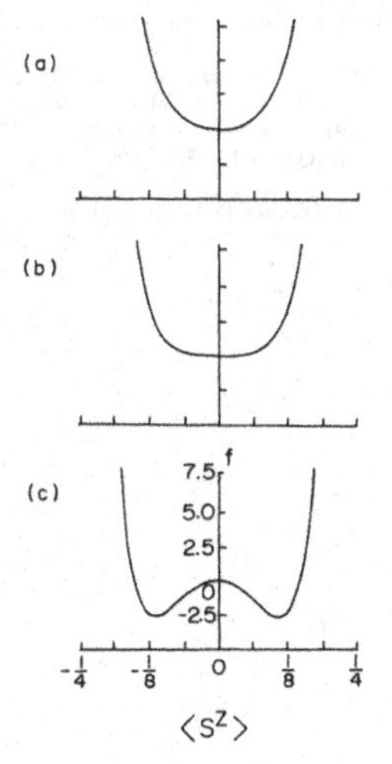

FIG. 4. Polarization dependence of the reduced free energy $f=\Delta F/NJ$ near the second-order phase transition temperature T_0. In our numerical calculations, however, we have introduced the dimensionless temperature $\Theta=k_BT/J$. (a) $T>T_0$ and $\Theta=0.250\,09$, (b) $T=T_0$ and $\Theta=0.25$, (c) $T<T_0$ and $\Theta=0.2498$.

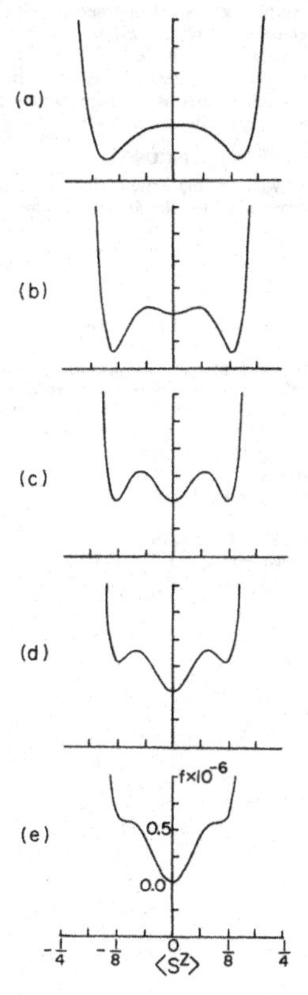

FIG. 5. Polarization dependence of the reduced free energy near the first-order phase transition temperature T_c. (a) $T<T_0$ and $\Theta=0.25$, (b) $T_c>T>T_0$ and $\Theta=0.250\,04$, (c) $T=T_c$ and $\Theta=0.250\,06$, (d) $T_0^->T>T_c$ and $\Theta=0.250\,08$, (e) $T=T_0^-$ and $\Theta=0.250\,09$.

minimum with larger free energy occurs at $\langle S^z \rangle = 0$ for temperatures within the range $T_c > T > T_0$, as shown in 5(b). This indicates that the system may be found in the paraelectric phase, which is far less stable than the ferroelectric phase. Figure 5(c) shows the situation for $T = T_c$ when three minima of equal free energy exist. Thus, the two phases can coexist with equal stability, as discussed in Sec. III. The order parameter changes abruptly from $\langle S^z \rangle_c$ to zero and the transition is of the first order. When the temperature increases to above T_c but still below T_0^-, the minimum at $\langle S^z \rangle = 0$ has lower free energy, implying a stable paraelectric phase and a metastable ferroelectric phase. Figure 5(d) illustrates this situation. When $T = T_0^-$, the ferroelectric phase starts to disappear, as is shown in 5(e). The system can only be in the paraelectric phase at this temperature, since it has a minimum free energy only at $\langle S^z \rangle = 0$.

VI. DISCUSSION

We first showed in this article that the four-spin interaction is responsible for the first-order phase transition observed in the order-disorder ferroelectrics. The critical four-body coupling strength $J' = 4J/3$ was then calculated by the method of retarded Green's function. By investigating the variation of the order parameter $\langle S^z \rangle$ with temperature, we found the conditions for the determination of the limiting temperatures T_0 and T_0^- for the stable paraelectric and ferroelectric phases, respectively. For the special case in which the tunneling frequency vanishes, approximate expressions for these limiting temperatures were derived.

On the basis of the above results, the free energy of the system was expressed in terms of the microscopic parameters, and the Curie temperature T_c and the corresponding order parameter $\langle S^z \rangle_c$ were calculated. Finally, we analyzed the critical behavior of the free energy at different temperatures. Our conclusion from this analysis is that, in the limit of small mean spin $\langle S^z \rangle$, the microscopic model completely predicts the same behavior of the free energy near the critical point as the macroscopic Landau theory.

[1]G. Fischer, J. Peterson, and D. Michel, Z. Phys. B **67**, 387 (1987).
[2]E. Fetuzzo and W. J. Merz, in *Ferroelectricity*, edited by E. P. Wohlfarth (North-Holland, Amsterdam, 1967).
[3]R. Blinc and B. Zeks, *Soft Modes in Ferroelectrics and Antiferroelectrics* (North-Holland, Amsterdam, 1974), p. 105.
[4]K. K. Kobayshi, J. Phys. Soc. Jpn. **24**, 497 (1968); Y. Ishibashi, *ibid.* **52**, 200 (1983).
[5]U. Deiningham and M. Mehring, Solid State Commun. **39**, 1257 (1981).
[6]J. C. Slater, J. Chem. Phys. **9**, 16 (1941).
[7]Y. Takagi, J. Phys. Soc. Jpn. **3**, 271 (1948).
[8]I. Turban, J. Phys. C **15**, L65 (1982).
[9]K. A. Penson, R. Jullien, and P. Pfeuty, Phys. Rev. B **26**, 6334 (1982).
[10]F. Igloi, D. V. Kapor, M. Skrinjar, and J. Solyom, J. Phys. A **19**, 1189 (1986).
[11]J. B. Zhang and Z. K. Qin, Phys. Rev. A **36**, 915 (1987).
[12]N. N. Bogolyubov and S. V. Tyablikov, Dokl. Akad. Nauk. SSSR **126**, 53 (1959) [Sov. Phys. Dokl. **4**, 604 (1959)].
[13]C. L. Wang, Z. K. Qin, and J. B. Zhang, Ferroelectrics **77**, 21 (1987).
[14]C. L. Wang, Z. K. Qin, and D. L. Lin (unpublished).

PHYSICAL REVIEW B VOLUME 55, NUMBER 10 1 MARCH 1997-II

O_3 tilt and the Pb/(Zr/Ti) displacement order parameters in Zr-rich $PbZr_{1-x}Ti_xO_3$ from 20 to 500 K

Noé Cereceda, Beatriz Noheda, Tomás Iglesias, José R. Fernández-del-Castillo, and Julio A. Gonzalo
Universidad Autónoma de Madrid, 28049 Madrid, Spain

Ning Duan and Yong Ling Wang
Shanghai Institute of Ceramics, Chinese Academy of Sciences, 1295 Ding Xi Road, Shanghai 200050, China

David E. Cox and Gen Shirane
Department of Physics, Brookhaven National Laboratory, Upton, New York 11973-5000
(Received 21 June 1996; revised manuscript received 24 September 1996)

Neutron diffraction and dielectric data have been collected from a polycrystalline sample of $PbZr_{1-x}Ti_xO_3$ ($x=0.035$) to determine the temperature dependence of the two-order parameters associated with the oxygen octahedral tilt and the Pb/(Zr/Ti) displacement, respectively, through the sequence of phase transitions F_{RL}-F_{RH}-P_C present in this mixed system. The weak coupling between tilt and polarization (displacements) is satisfactorily described within the framework of a two-order-parameters statistical theory in which the respective effective fields involve two higher-order terms in addition to a linear term. [S0163-1829(97)01709-8]

I. INTRODUCTION

Perovskite structures, which are cubic at high temperatures, frequently display a rich variety of phase transitions at lower temperatures[1] that usually involve cation shifts and/or oxygen octahedra tilts, with respect to the high-symmetry phase. These shifts and tilts are basically independent, in the sense that one can appear in the absence of the other, but they can couple weakly to each other. The mixed perovskite system $PbZr_{1-x}Ti_xO_3$ (generally denoted PZT) is especially interesting,[2] not only because it can be prepared over the entire compositional range ($0\leqslant x\leqslant 1$), with a variety of distorted structures at room temperature, e.g., tetragonal (F_T), zero-tilt rhombohedral high-temperature (F_{RH}), nonzero-tilt rhombohedral low-temperature (F_{RL}), and antiferroelectric orthorhombic (AF_O) phases, but also because[3] of its important technological applications.

The basic features of the phase diagram of PZT solid solutions were established by Shirane, Suzuki, and Takeda.[2] Barnett[4] pointed out the existence of an additional low-temperature phase change for Zr-rich compositions, later characterized as the F_{RL}-F_{RH} transition. Michel *et al.*[5] determined the space groups of the two rhombohedral phases as $R3c$ (F_{RL}) and $R3m$ (F_{RH}) by means of x-ray and neutron powder diffraction at room temperature using two compositions $x=0.10$ (F_{RL}, with two formula units per unit cell, involving oxygen octahedra tilts), and $x=0.42$ (F_{RH}, with one formula unit per unit cell and zero tilt). In summary, the sequence of phase transitions in Zr-rich PZT is F_{RL} (rhombohedral $R3c$)$\rightarrow F_{RH}$(rhombohedral $R3m$)$\rightarrow P_C$(cubic). Later, Glazer, Mabud, and Clarke[6] made a further neutron powder study on PZT with $x=0.10$, characterizing the F_{RL} phase in terms of the tilting of oxygen octahedra[7] and the expansion or contraction of the octahedra triangles neighboring the shifted B cation (Zr/Ti).

Glazer[7] has also made a general classification of tilted octahedra in perovskites. He pointed out that the cation displacements, which are directly related to the ferroelectric character of the perovskites, have only a small effect on the lattice parameters and that, generally, the overall symmetries follow those of the tilts.

Research has been done to describe the possible relations between the atomic displacements in these materials. Megaw and Darlington[8] described the perovskite structures in terms of four structural parameters, classifying them by their space groups to allow comparisons. They found no correlation between the cation displacements and the octahedra tilts in the space groups $R3c$ and $R3m$, to which rhombohedral PZT belongs. A study of such a correlation, if it exists, is important to determine which parameters are the cause of the deformations and which parameters are only consequences. Nevertheless, even if displacement and tilt are independent, they may be coupled, mainly because of the shared-corner linkage between octahedra, which is required by the packing.

We have recently reported some preliminary neutron diffraction measurements[9] on a ceramic sample of PZT with $x=0.035$ over a limited temperature region around the phase transition between the two ferroelectric rhombohedral phases, F_{RL} and F_{RH} which occurs at a temperature $T_{LH}\approx 323$ K. This work allowed a detailed description of the structural parameters which characterize the transition, namely, the cation displacements along the [111] axis, **s** (Pb) and **t** (Zr, Ti), the distortion of the oxygen octahedra **d**, and a parameter **e**, which describes the rotation of the oxygen octahedra around the [111] axis with a tilt angle, ω, in the (111) plane such that $\tan\omega=4\sqrt{3}\mathbf{e}$.[6] We have also reported some measurements on the dielectric response of Zr-rich PZT (Ref. 10) in the temperature region around the rhombohedral-cubic phase transition, F_{RH}-P_C, which occurs at ≈ 510 K and is shown to be first order.

The weak coupling between tilt and displacements has been recently studied by Dai, Lie, and Viehland[11] in PZT ($x=0.35$) by means of dielectric constant, hysteresis loops, and dilatometric and electron diffraction techniques. They proposed that the inability of the oxygen octahedra to rotate coherently within a rigid lattice generates random internal stresses which constrain the polarization. They found doubled hysteresis loops for this composition, which appear to be relaxed with La and Nb doping.

To analyze in depth the coupling between tilt and displacements, it is very convenient to use neutron diffraction data, especially to measure the tilt angle, which is difficult to determine by other techniques. In the present work, we investigate the temperature dependence of the two-order parameters associated with the octahedral tilting and the cation displacements over a wide temperature range spanning both the F_{RL}-F_{RH} and $F_{RH}-P_C$ transitions, with special emphasis on the weak coupling between polarization, directly associated with the cation shifts, and tilt in a Nb-doped $PbZr_{1-x}Ti_xO_3$ ceramic sample with $x=0.035$. Nb doping is important, among other things, to reduce the electrical conductivity and to facilitate dielectric constant and polarization measurements.

The squared structure factor of the first superstructure peak, $\frac{1}{2}(311)$, which can be expressed as

$$|F|^2=8[\sin^2(4\pi\mathbf{e})][1+\cos(24\pi\mathbf{d})]\approx 16\sin^2(4\pi\mathbf{e}),$$

is, in practice, directly proportional to the squared rotation parameter $\mathbf{e}^2$. This allows us to define the temperature dependence of the tilt order parameter as

$$\eta_s(T)=[I_{RL}(T)/I_{RL}(0)]^{1/2},$$

where I_{RL} is the integrated intensity of the superstructure $\frac{1}{2}(311)$ Bragg reflection. We have accurately measured $\eta_s(T)$ between 20 and 375 K by neutron diffraction.

Dielectric constant data, which show a small anomaly at T_{LH}, and pyroelectric charge measurements through T_{LH}, were used to characterize the small decrease in spontaneous polarization which accompanies the F_{RL}-F_{RH} transition. Measurements of hysteresis loops, which were previously used to characterize the behavior of the polarization order parameter $p_s(T)=P_s(T)/P_s(O)$ at the F_{RH}-P_C (ferroelectric-paraelectric) transition, were attempted at $T\leqslant T_{LH}$, but were unsuccessful because of the high values of the coercive field for the sample in this temperature range.

The combined neutron diffraction and dielectric data were used for a theoretical analysis of the weak tilt-polarization coupling. Prior work by Halemane *et al.*[12,13] made use of Landau's theory to describe the simultaneous temperature dependence of tilt and polarization in PZT with a different composition, $x=0.10$, but no detailed information on the temperature dependence of the tilt order parameter was available at this time. We describe below an analysis of our data for PZT with $x=0.035$ by means of a simple two-order-parameters statistical theory, in which the effective fields[14] for both tilt and polarization involve linear, cubic, and fifth power terms in the conjugate variable. In spite of its simplicity, this theoretical approach produces a simple relationship between tilt and polarization which is borne out quantitatively by the experimental data.

II. EXPERIMENT

The samples were high quality polycrystalline ceramics prepared at the Shanghai Institute of Ceramics, with nominal composition $Pb_{1-y/2}(Zr_{1-x}Ti_x)_{1-y}Nb_yO_3$, with $x=0.035$ and $y=0.039$, confirmed by chemical analysis. The constituent oxides were ground, pressed into pellets, and fired at 1350 °C for 2 h. Cylinders of about 1 cm in height and 1 cm in diameter were sintered and hot pressed for the neutron diffraction runs, and thin plates of about 1 mm in thickness and 1 cm in diameter were used for the dielectric measurements. Data were collected with neutrons of 14.7 meV energy for several low angle peaks, including the superstructure peak $\frac{1}{2}(311)$ (Ref. 9) at one of the triple-axis diffractometers at the Brookhaven National Laboratory HFBR between 20 K and $T_{LH}\cong 323$ K. The dielectric constant data, capacitance and dissipation factor, were measured at regular intervals of about 0.1 K by means of an automatic Hewlett-Packard Precision LCR Meter (Model 4284A) with a field amplitude of 8.3 V/cm to an accuracy better than 1 part in 10^4 at a frequency of 1 KHz. The rate of temperature change was ~20 K/h for both heating and cooling runs. The pyroelectric charge released through the F_{RL}-F_{RH} transition was measured with a Keithley Electrometer (Model 610C), and measurements of hysteresis loops data were made with a modified Diamant-Pepinsky-Drenck circuit with a Nicolet Digital Scope (Model NIC-310). The temperature for the electrical measurements was measured in all cases by a chromel-alumel thermocouple with a Keithley Digital Multimeter (Model 196).

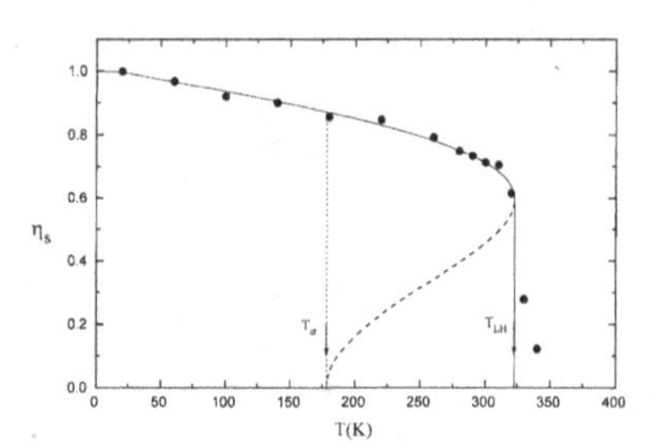

FIG. 1. Tilt order parameter vs temperature obtained from neutron diffraction measurements on a ceramic sample of Nb-doped $PbZr_{1-x}Ti_xO_3$ with $x=0.035$. The tilt order parameter is defined as $\eta_s=[I_{RL}(T)/I_{RL}(0)]^{1/2}$, where I_{RL} corresponds to the integrated intensity of the first superstructure peak $(hkl)=\frac{1}{2}(311)$ of the rhombohedral unit cell. Dashed line indicates calculated metastable region between low-temperature and high-temperature phases. The theoretical curves (full and dashed lines) are calculated from Eq. (7) with $x=0$.

III. RESULTS

Figure 1 depicts the tilt order parameter $\eta_s(T)$ obtained directly from the neutron diffraction data. It can be seen that $\eta_s(T)$ decreases gradually with increasing temperature, reaches a value of $\eta_s(T_{LH})\cong 0.59$, and then drops fairly abruptly to nearly zero for $T>T_{LH}\cong 323$ K. As shown below, a good fit of the data for $\eta_s(T)$ between 20 K$<T<T_{LH}$ is obtained with an extrapolated Curie temperature $T_{ct}\cong 178.2$ K, much lower than the transition temperature T_{LH}, and in-

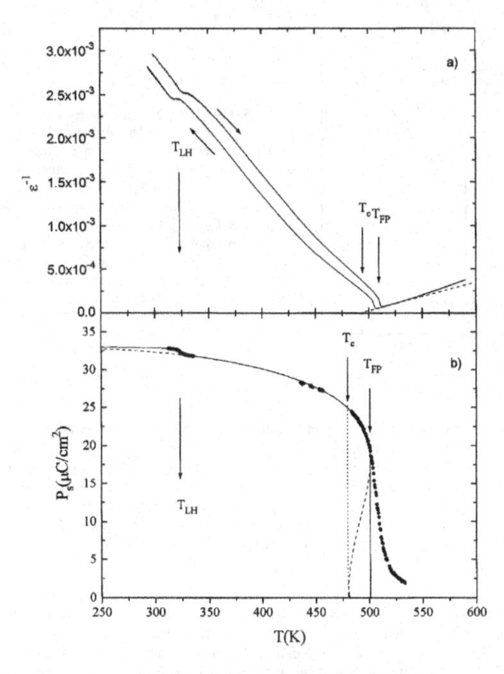

FIG. 2. (a) Inverse dielectric constant vs temperature (heating and cooling) for Nb-doped $PbZr_{1-x}Ti_xO_3$ ($x=0.035$) showing the $F_{RL}-F_{RH}$ transition associated with the O_3 tilt and the $F_{RH}-P_C$ transition to the cubic paraelectric phase. (b) Spontaneous polarization vs temperature for Nb-doped $PbZr_{1-x}Ti_xO_3$ ($x=0.035$). Since hysteresis loops data show an increasingly large coercive field as T decreases and approaches T_{LH}, making measurements near this temperature impossible, pyroelectric charge measurements were made. It may be noted that in ceramic rhombohedral perovskites, the saturation polarization is substantially lower than in single crystals, the ratio being P_s (s.c)$\approx 1.15\times P_s$ (ceramic) (see text). Dashed line indicates calculated metastable region. The theoretical curves (full and dashed lines) are calculated from Eq. (4) with $e=0$.

dicative of the pronounced first-order character of the transition. The fit is done by varying the tilt saturation value, $\eta_s(0)\equiv\eta_{s0}$, and the effective Curie temperature, T_{ct}, from estimated initial values to optimize the agreement between available experimental data and the equation of state arrived at in Sec. IV. It may be noted that, in our case, $\eta_s(0)\approx\eta(20$ K) is well defined beforehand, while $T_{ct}<T_{LH}=323$ K (transition temperature) is not, because it is inaccessible experimentally. In Sec. V the fitting process is described in more detail.

Figure 2(a) shows the inverse dielectric constant as a function of temperature over a wide range, from slightly below $T_{LH}=323$ K to well above $T_{FP}=509.6$ K, the ferroelectric-paraelectric transition temperature. In this case, the extrapolated Curie temperature, $T_c=489.9$ K, is relatively close to the transition temperature. A small anomaly in $\epsilon^{-1}(T)$ at $T\cong T_{LH}$ marks the onset of the tilting of the oxygen octahedra.

Figure 2(b) presents the behavior of the spontaneous polarization (displacement order parameter) P_s as a function of temperature over the same wide range as in Fig. 2(a). Data of the hysteresis loops for $T\geqslant T_{LH}$ in the region of T_{FP} are combined with pyroelectric charge measurements made around T_{LH}. They indicate, as expected, a small change in spontaneous polarization at the onset of the tilting transition. It may be noted that the numerical values for T_c and T_{FP} are not identical to those in Fig. 2(a) but appear slightly shifted towards lower temperatures. This may be due to the fact that in the measurements of the hysteresis loops, extra heating of the sample under the relatively high driving field is known to take place. The thermocouple, which is not in good thermal contact with the sample, may register a temperature closer to that within the furnace than that of the sample. The data, read automatically from the digital scope, include extra polarization just above the true transition temperature. As is well known, at first-order discontinuous transitions in ferroelectric perovskites, the single hysteresis loops evolve towards double loops which, when imperfectly compensated due to the high conductivity of the sample, may give rise to tails in the apparent spontaneous polarization above the transition temperature, whose precise value may become blurred. It should be pointed out that the estimated ratio between the true (single-crystal) polarization and the apparent (ceramic) polarization is about 1.15,[15] which has been used to correct the data shown in Fig. 2(b).

IV. THEORETICAL ANALYSIS

To analyze the phase transition sequence

$$F_{RL}(\eta_s>0;p_s>0;\Delta p_s>0)\to F_{RH}(\eta_s=0;p_s>0;\Delta p_s=0)$$

$$\to P_C(\eta_s=p_s=\Delta p_s=0),$$

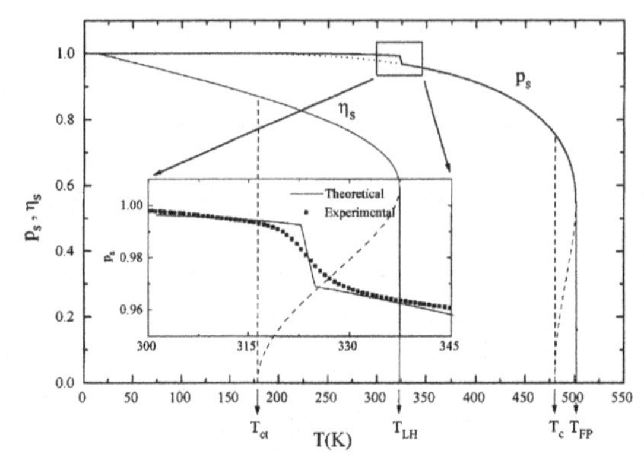

FIG. 3. Calculated temperature dependence of the tilt and polarization order parameters for PZT ($x=0.035$) indicating that the extra polarization associated with the tilt Δp_s is well described, below T_{LH} by $p_s=(1+p_s\eta_s)^{-1}(1-p_s^2)\eta_s$, Eq. (9). [See text for details on the fitting procedure for $\eta_s(T)$ and $p_s(T)$]. Dashed lines indicate calculated metastable region between low-temperature and high-temperature phases. The dotted line for $p_s(T)$ below T_{LH} is the calculated polarization in the absence of tilt.

we first examine separately the polarization order parameter (p_s) and the tilt order parameter (η_s) using a common order-disorder statistical approach, and then we investigate the weak coupling between both order parameters, which gives rise to the increase in the polarization order parameter (Δp_s) apparent at $T\leq T_{LH}$ [see Fig. 2(b)].

A. Polarization

If there are N interacting elementary dipoles per unit volume in the solid, of which N_2 are pointing one way and N_1 in the opposite way, $(N_1+N_2)=N$, in thermal equilibrium we have

$$N_2(k_B\Theta_D/h)e^{-\phi_d/k_BT}e^{-E_{\rm eff}\mu/k_BT}$$
$$=N_1(k_B\Theta_D/h)e^{-\phi_d/k_BT}e^{E_{\rm eff}\mu/k_BT}, \quad (1)$$

where $(k_B\Theta_D/h)e^{-\phi_d/k_BT}$ is the jump probability per unit time per unit dipole, μ, for an effective field $E_{\rm eff}=0$, and ϕ_d is the height of the energy barrier between the two potential minima corresponding to the two possible orientations ($E_{\rm eff}=0$). The net dipolar polarization (practically identical to the total polarization, P) would then be

$$P_d=(N_2-N_1)\mu=N\mu\tanh(E_{\rm eff}\mu/k_BT), \quad (2)$$

where μ, as before, is the elementary dipole moment per unit cell.

The effective field may be expanded in powers of the polarization, taking into account that, for an external field $E=0$, $E_{\rm eff}(P_d)=-E_{\rm eff}(-P_d)$. Thus,

$$E_{\rm eff}=E+\beta P_d+\gamma P_d^3+\delta P_d^5+\cdots, \quad (3)$$

where β,γ,δ are constant (i.e., temperature independent) coefficients.

Using the dimensionless variables, $e=E/\beta N\mu$, $p=P_d/N\mu=P/N\mu$, and substituting $T_c=\beta N\mu^2/k_B$, $g\equiv(\gamma/\beta)N^2\mu^2$, $h\equiv(\delta/\beta)N^4\mu^4$, we obtain the equation of state

$$e=(T/T_c)\tanh^{-1}p-p(1+gp^2+hp^4+\cdots), \quad (4)$$

which specifies completely the temperature dependence of the spontaneous ($e=0$) order parameter $p(0)\equiv p_s$. (This dimensionless field e should not be confused, obviously, with the structural parameter **e**, which describes the O_3 rotation.)

The temperature dependence of the inverse dielectric constant can also be obtained easily from the equation of state, Eq. (4), as $\epsilon^{-1}(T)=(T_c/C)(de/dp)$.

B. Tilt

Similarly, we can try to describe the temperature dependence of the tilt order parameter in the F_{RL} ferroelectric antiferrodistortive phase, in which the unit cell is doubled, using a statistical approach as follows. If there are N' interactive unit cells per unit volume, consisting of N_1' with the two oxygen octahedra within the unit cell tilted in the sequence $(+\omega,-\omega)$, and N_2' with the two oxygen octahedra tilted in the opposite sequence $(-\omega,+\omega)$, the net "staggered" tilt per unit volume θ_t of the pseudocubic three-dimensional arrangement of cells is

$$\theta_t=(N'_2-N'_1)2|\omega|=N'2|\omega|\tanh(|X_{\rm eff}|2|\omega|/k_BT), \quad (5)$$

where

$$|X_{\rm eff}|=X+\beta'\theta_t+\gamma'\theta_t^3+\delta'\theta_t^5+\cdots \quad (6)$$

is the generalized (torsional) field and $|\omega|$ is the absolute value of the rotation angle of a single oxygen octahedron in the unit cell (the other oxygen octahedron within the unit cell will have, automatically, a rotation angle with the same value in the opposite direction). Because the F_{RL} phase presents an antiferrodistortive deformation with respect to the higher temperature rhombohedral phase, F_{RH}, $X_{\rm eff}$ is an effective staggered field (torsional field in our case), conjugate with the staggered tilt strain.

The equation of state in dimensionless variables is

$$x=\left(\frac{T}{T_{ct}}\right)\tanh^{-1}\eta-\eta(1+g_t\eta^2+h_t\eta^4+\cdots), \quad (7)$$

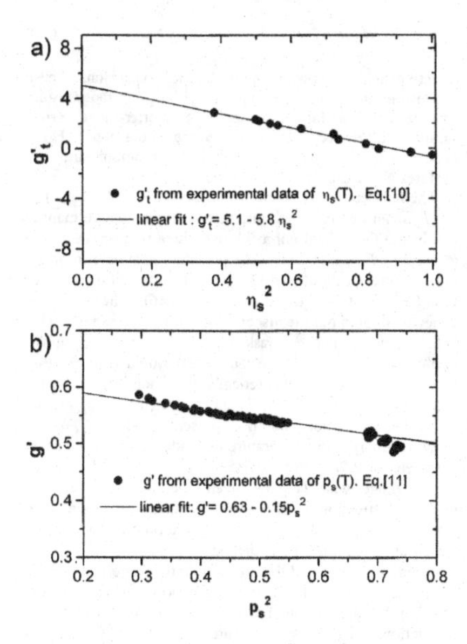

FIG. 4. (a) Plot of $(g'_t)_{\rm exp} = f_t[T/T_{\rm ct}, \eta_s^2]$ vs η_s^2, where $T_{\rm ct}$ and $I_{\rm RL}(0)$ are adjusted by a least-squares fit to get the best linear dependence of g'_t vs η_s^2. $I_{\rm RL}(T)$ are the actual measured values of temperature and integrated intensity. It may be noted that $[I_{\rm RL}(0)]^{1/2}$ comes out very close to $[I_{\rm RL}(20\ {\rm K})]^{1/2}$. (b) Plot of $(g')_{\rm exp} = f(T/T_c, p_s^2 = P_s/P_{so})$ vs p_s^2, where T_c and P_{so} are similarly adjusted. In this case P_{so} is still somewhat larger than $P_s(T \approx 450\ {\rm K})$ which is the maximum value actually measured. Points corresponding to $T > T_{\rm FP}$ (transition temperature) have been omitted.

where $x = X/\beta' N' 2|\omega|$; $\eta = \theta_t/N' 2|\omega|$, which can also be interpreted as $\langle \mathbf{e}(T)\rangle / \langle \mathbf{e}(0\ {\rm K})\rangle$, due to the fact that $\mathbf{e} = \tan \omega/4\sqrt{3} \approx \omega/4\sqrt{3}$; $T_{\rm ct} = \beta' N'(2|\omega|)^2/k_B$, $g_t \equiv (\gamma'/\beta')N'^2(2|\omega|)^2$ and $h_t \equiv (\delta'/\beta')N'^4(2|\omega|)^4$. Here $T_{\rm ct} < T_{\rm LH}$ is the effective Curie temperature for the transition involving O_3 tilting, which is different from $T_c < T_{\rm FP}$, the Curie temperature for the ferroparaelectric transition involving Pb and Zr/Ti displacement or order-disorder orientations.

The spontaneous ($x=0$) tilt order parameter $\eta(0) \equiv \eta_s \equiv \mathbf{e}_s/\mathbf{e}_{so}$ is therefore given by Eq. (7) with $x=0$. Note that $\mathbf{e}$ can be interpreted, in general, as a common tilt for all unit cells (displacive transition) at any given temperature, or as an average tilt $\langle \mathbf{e} \rangle$ for a statistical distribution of the tilts through the lattice (order-disorder transition). Very often, transitions have a mixed displacive/order-disorder character which should, to some extent, be taken into account by a generalized effective field which includes terms with higher order powers, as in Eq. (6).

C. Polarization-tilt coupling

Let us assume that there is a weak coupling between tilt and polarization, or, in other words, that a unit cell tilted in one specific direction favors atomic displacements in one of the two opposite directions perpendicular to the plane of the tilt. In this case, the total interaction energy to be substituted into Eq. (2) is $W_d + W_t = E_{\rm eff}\mu + X_{\rm eff} 2\omega$, instead of $W_d = E_{\rm eff}\mu$ only, and we therefore have

$$p_s + \Delta p_s = \tanh\left[\frac{W_d}{k_BT} + \frac{W_t}{k_BT}\right] = \frac{\tanh\left[\frac{W_d}{k_BT}\right] + \tanh\left[\frac{W_t}{k_BT}\right]}{1 + \tanh\left[\frac{W_d}{k_BT}\right]\tanh\left[\frac{W_t}{k_BT}\right]}. \tag{8}$$

Taking into account that for $E=0$, $\tanh[W_d/k_BT] = p_s$ and $\tanh[W_t/k_BT] = \eta_s$, we obtain directly from Eq. (8)

$$\Delta p_s = (1 + p_s\eta_s)^{-1}(1 - p_s^2)\eta_s, \tag{9}$$

which relates the increase in polarization to the tilt in a very simple way. The prefactor $(1 + p_s\eta_s)^{-1}$ varies smoothly from 0.5 at $T=0$ K to 1.0 at $T = T_{\rm LH}$. The second factor $(1 - p_s^2)$ is zero at $T=0$ K and is still much less than unity at $T = T_{\rm LH}$ if $T_{\rm LH}$ is substantially lower than $T_{\rm FP}$ as in the present case.

Figure 3 shows the temperature dependence of the spontaneous polarization and the tilt, in excellent agreement with the observed behavior.

V. DISCUSSION AND CONCLUSIONS

As mentioned above, the temperature dependence of the tilt order parameter, $\eta_s(T)$, and the polarization order parameter, $p_s(T)$, are well described by Eqs. (7) and (4) with $x=0$ and $e=0$, respectively. In Figs. 1, 2(b), and 3, which display Eqs. (7) and (4), full line indicates equilibrium states (heating) and dashed line corresponds to ideal metastable behavior (cooling). The fitting procedure was the following: the experimental values for $\eta_s(T) = [I_{\rm RL}(T)/I_{\rm RL}(0)]^{1/2}$ and $p_s(T) = P_s(T)/P_s(0)$, with initial values for $I_{\rm RL}(0)$, $T_{\rm ct}$ and $P_s(0)$, T_c chosen as discussed below, were substituted, respectively, into

$$g'_t \equiv \frac{\left(\frac{T}{T_{\rm ct}}\right)\tanh^{-1}\eta_s(T)/\eta_s(T) - 1}{\eta_s^2(T)} \approx g_t + h_t\eta_s^2 \tag{10}$$

and

$$g' \equiv \frac{\left(\frac{T}{T_c}\right)\tanh^{-1}p_s(T)/p_s(T) - 1}{p_s^2(T)} \approx g + hp_s^2. \tag{11}$$

Here the experimental values for g'_t, defined by the actual value of the tilt at a given temperature, are plotted vs η_s^2, using as normalized parameters $I_{\rm RL}(0)$ and $T_{\rm ct}$. If Eqs. (7) and (4) describe correctly the observed behavior, g'_t vs η_s^2 and g' vs p_s^2 should result in linear plots, giving automatically (g_t, h_t) and (g, h). The quality of these least-square-fitted linear plots was assessed through $R_t=|\Delta g_t'/g_t'|^2$ and $R=|\Delta g'/g'|^2$, respectively, summing up over all data points, where $|\Delta g_t'|$ and $|\Delta g'|$ are the differences between experimental and calculated values. The values of $I_{\mathrm{RL}}(0\ \mathrm{K})$, T_{ct} needed to obtain $\eta_s(T)$ and T/T_{ct}, and P_{so}, T_c to get $P_s(T)$ and T/T_c in Eqs. (10) and (11), are not known experimentally, so the fitting procedure is carried out by changing them to get the best linear fit. The initial values are chosen from the experimental data knowing that $I_{\mathrm{RL}}(0\ \mathrm{K})\approx I_{\mathrm{RL}}(20\ \mathrm{K})$, $T_{\mathrm{ct}}\leqslant T_{\mathrm{LH}}=323$ K, $P_{so}\geqslant P_s$ (430 K) and $T_c\leqslant T_{\mathrm{FP}}=501.0$ K.

Then, varying T_{ct} and $I_{\mathrm{RL}}(0\ \mathrm{K})$ [fixed in practice because the neutron data include points for $I_{\mathrm{RL}}(20\ \mathrm{K})\approx I_{\mathrm{RL}}(0\ \mathrm{K})$ with very small statistical error], we proceed to minimize the least-square error of the fit to get a final T_{ct} [see Fig. 4(a)]. Likewise, experimental values for g', defined by the actual value of the spontaneous polarization at a given temperature, are plotted vs p_s^2, using as normalized parameters initial estimates for P_{so} and T_c. Varying again (within narrow limits) T_c and P_{so} and optimizing the linear fit, we get final values for P_{so} and T_C [see Fig. 4(b)].

This procedure leads to $R\leqslant10^{-4}$ for $I_{\mathrm{RL}}(20\ \mathrm{K})/I_{\mathrm{RL}}(0\ \mathrm{K})=0.985$ and $T_{\mathrm{ct}}=178.2$ K (with estimated uncertainties of the order of 1%), and to $R\leqslant10^{-3}$ for $P_{so}=33.0\ \mu\mathrm{C/cm^2}$, $T_c=480.0$ K (with estimated uncertainties of the order of 5%). These fits resulted in linear plots giving $g_t=5.1\pm0.1$, $h_t=-5.8\pm0.2$ for the tilt order parameter, and $g=0.62\pm0.25$, $h=-(0.15^{+1.10}_{-0.15})$ for the polarization order parameter. It should be noted that the set of experimental data for $\eta_s(T)$ covered the range $0.06\leqslant T/T_{\mathrm{LH}}\leqslant1.00$, while the available set for $p_s(T)$ covered only $0.85\leqslant T/T_{\mathrm{FP}}\leqslant1.00$. This is the main reason for the larger uncertainties in the latter.

The occurrence of nonvanishing values $g_t>\frac{1}{3}$, $h_t<0$, and $g>\frac{1}{3}$, $h<0$ in the expressions for $\eta_s(T)$ and $p_s(T)$ implies a first-order (discontinuous) character for the transitions, in agreement with the observed behavior. Physical meaning can be attributed to the effective field coefficients β_t, $\gamma_t(g_t)$, $\delta_t(h_t)$ and β, $\gamma(g)$, $\delta(h)$ in Eqs. (6) and (3), considering the effective field expressions as multipolar expansions. These expansions include successive dipolelike (long-range), quadrupolelike (short-range), octupolelike (shorter-range) terms, summing up the contributions over the whole lattice. However, detailed calculations of this kind in rhombohedral perovskites are nontrivial.

The temperature dependence of the weak peak in $\epsilon(T)$ at $T\approx T_{\mathrm{LH}}$ can be described only in a semiquantitative manner within the theoretical approach used here to describe $p_s(T)$ at the F_{RL}-F_{RH} transition. The presently available information on the trend of $\Delta p_s(T_{\mathrm{LH}})$ with composition (x) for $x=0.035$,[9] $x=0.10$,[6] and $x=0.40$ (Ref. 16) indicates a tendency to smooth out the discontinuity with increasing x. This behavior might also be masked by the increasing compositional inhomogeneity of the samples. It would be interesting to be able to predict theoretically the composition dependence of $T_{\mathrm{LH}}(x)$ and $\Delta p_s(T_{\mathrm{LH}}(x))$ in terms of $g_t(x)$ and $h_t(x)$, but this is not possible at present.

In summary, the temperature dependence of the tilt order parameter of PZT with $x=0.035$, previously investigated in a narrow range near[9] T_{LH}, has been determined by means of neutron diffraction in the whole temperature range from $T=20$ K to $T_{\mathrm{LH}}=323$ K. The associated polarization change $P_s(T)$ at $T\approx T_{\mathrm{LH}}$ has been determined.

It may be concluded that the simple two-order-parameters statistical theory outlined in Sec. IV accounts well for the coupling between tilt and polarization determined by neutron diffraction and dielectric measurements, especially in view of the fact that the data were obtained not from single crystals, but from ceramic samples.

ACKNOWLEDGMENTS

Financial support from CICyT (Grant No. PB93-1253/94), Iberdrola (Grant No. INDES 94/95), and Comunidad de Madrid (Grant No. AE00138-94) is gratefully acknowledged. Work at Brookhaven is supported by the U.S. Department of Energy, Division of Materials Sciences, under Contract No. DE-AC02-76CH00016.

---

[1] See for instance, *Ferroelectric and Related Substances: Oxides*, Landolt-Börnstein, New Series, Group 3, Vol. 16, Pt. a (Springer-Verlag, Berlin, 1981).

[2] G. Shirane, K. Suzuki, and A. Takeda. J. Phys. Soc. Jpn. 7, 12 (1952).

[3] J. C. Burfoot and G. W. Taylor. *Polar Dielectrics and Their Applications* (McMillan, London, 1979).

[4] H. Barnett, J. Appl. Phys. **33**, 1606 (1962).

[5] C. Michel, J. M. Moreau, G. D. Achenbach, R. Gerson, and W. J. James, Solid State Commun. 7, 865 (1969).

[6] A. M. Glazer, S. A. Mabud, and R. Clarke, Acta Crystallogr. B **34**, 1060 (1978).

[7] A. M. Glazer, Acta Cryst. B **28**, 3384 (1972).

[8] H. D. Megaw and C. N. W. Darlington, Acta Cryst. A **31**, 161 (1975).

[9] B. Noheda, T. Iglesias, N. Cereceda, J. A. Gonzalo, H. T. Chen, Y. L. Wang, D. E. Cox, and G. Shirane, Ferroelectrics **184**, 251 (1996).

[10] B. Noheda, N. Cereceda, T. Iglesias, G. Lifante, J. A. Gonzalo, H. T. Chen, and Y. L. Wang, Phys Rev. B **51**, 16 388 (1995).

[11] X. Dai, J-F. Lie and D. Viehland, J. Appl. Phys. **77**, 3354 (1995).

[12] T. R. Halemane, M. J. Haun, L. E. Cross, and R. E. Newnham, Ferroelectrics **62**, 149 (1985).

[13] T. R. Halemane, M. J. Haun, L. E. Cross, and R. E. Newnham, Ferroelectrics **70**, 153 (1986).

[14] J. A. Gonzalo, *Effective Field Approach to Phase Transitions and Some Applications to Ferroelectrics* (World Scientific, Singapore, 1991).

[15] D. Berlincourt and H. A. Krueger, J. Appl. Phys. **30**, 1804 (1959).

[16] A. Amin, R. E. Newnham, L. E. Cross, and D. E. Cox, J. Solid State Chem. **37**, 248 (1971).

Diffuse Phase Transitions and Random-Field-Induced Domain States of the "Relaxor" Ferroelectric $PbMg_{1/3}Nb_{2/3}O_3$

V. Westphal and W. Kleemann
Angewandte Physik, Universität Duisburg, W-4100 Duisburg 1, Germany

M. D. Glinchuk
Institute of Solid State Physics, Ukrainian Academy of Sciences, SU-252180 Kiev, Ukraine
(Received 14 October 1991)

The diffuseness of the ferroelectric phase transition in $PbMg_{1/3}Nb_{2/3}O_3$ is proposed to be due to quenched random electric fields originating from charged compositional fluctuations. They are responsible for the extreme critical slowing down, the freezing into nanometric ferroelectric domains, and the slow relaxation of the polarization below $T_c \sim 212$ K. Barkhausen jumps during poling exclude glassiness, which was conjectured previously. At T_c a ferroelectric anomaly of the dielectric permittivity appears, if the random fields are overcome by an external electric field.

PACS numbers: 77.80.Bh, 64.70.Kb, 77.20.+y, 78.20.Fm

Diffuse phase transitions (PT) in "relaxor" ferroelectrics, extending over a finite range of temperatures, ΔT, have been a challenging subject to both experimentalists and theorists ever since their detection nearly forty years ago [1]. As a rule [2], they occur in disordered ionic structures, in particular, in solid solutions. Within the Curie range ΔT, the dielectric permittivity achieves very high values and displays a large dispersion, which is reminiscent of that found for orientational glasses [3]. This is, in fact, the model which has recently [4] been applied to the best-known relaxor system, $PbMg_{1/3}Nb_{2/3}O_3$ (PMN) [5], the structure of which is pseudocubic with vanishing spontaneous polarization P_s at all temperatures T [6]. It has to be stressed, however, that static freezing has not been evidenced yet, since the system continues to slow down beyond the laboratory time scale [7] in the vicinity of the decay temperature of the remanent polarization, $T_0 \sim 200$ K [4].

It has been argued [4] that the glassy behavior of PMN might be due to frustrated correlations between superparaelectric moments [8]. These are due to ferroelectric symmetry breaking on a nanometric scale, which was evidenced very recently by x-ray and neutron diffraction [9]. Profile analysis of the diffraction lines shows that correlated clusters with ⟨111⟩ distortions develop upon cooling from about 600 K to T_0. They are about 10 nm in diameter at low T. On the other hand, quenched compositional fluctuations extending over about 3 nm with one-by-one ordering of Mg^{2+} and Nb^{5+} cations on the perovskite B sites have been observed by transmission electron microscopical imaging [9]. These nanodomains carry negative charges and are thus intense sources of quenched random electric fields. These have to be taken into account together with compositional and, hence, space-charge fluctuations on smaller length scales in the interpretation of the observed diffuse PT.

In this Letter we propose that random fields (RF) are, indeed, at the origin of the observed tremendous slowing down of the dynamics of PMN and its eventual freezing into a domain state on a nanometric length scale. We refer to the original idea of Imry and Ma [10], pointing out the stability of domain states due to the local fluctuations of quenched microscopic fields in the case where the PT is driven by an order parameter with continuous symmetry. This situation seems approximately to apply to PMN despite its eightfold dipolar degeneracy, which reflects cubic anisotropy. It should be noted that only in the case of large cubic anisotropy is relaxation towards long-range order (LRO) expected [11], as in the case of Ising systems [10]. Very probably the PT of the low-anisotropy system PMN is destroyed by virtue of the RF. The observed "diffuse PT" merely signifies the rounding and the slow dynamics which is left. The ground state of PMN is, hence, disordered on macroscopic scales and therefore resembles a dipolar glass in many respects. There are, however, a number of arguments which are clearly in favor of the RF mechanism and the basic ferroelectric nature of PMN. In addition to the observation of ⟨111⟩ distorted nanodomains [9] we present new experimental data showing that (i) Barkhausen jumps of microdomains control the low-T poling process, and that (ii) a ferroelectric anomaly of the dielectric permittivity appears slightly above T_0 when applying a moderate poling field. The latter phenomenon, which was already observed previously [12,13], is explained in the spirit of Andelman and Joanny's idea [14] that the external field helps privileged domains to grow up to macroscopic size. The appearance of a sharp PT under similar conditions was recently evidenced on the ferroelectric RF Ising system $KTaO_3$:Li (KTL) with $x_{Li} = 0.063$ [15].

Optical experiments have been carried out on a single-crystalline platelet-shaped sample of PMN with polished (001) face, edges parallel to [110] and [$1\bar{1}0$], respectively, and size 2×1 mm^2. Two lateral faces of size 2×1 mm^2 were aluminized in order to apply electric fields E along [110]. Linear birefringence (LB) was measured at $\lambda = 589$ mm by use of standard methods [16] including a polarizing microscope in order to select strain-free sample sections. The field-induced LB is due to the polarization preferentially aligned along two of the eight dipolar ⟨111⟩

directions, [111] and [11$\bar{1}$]. At sufficiently high fields parallel to [110] only these two dipolar states are occupied. The polarization affects the LB directly via second-order electro-optics and indirectly via electrostrictive strain. The (001)-plane LB then reads $\Delta n = g_{44} n_0^3 P_s^2/2$ [17], where n_0 and g_{44} are the cubic refractive index and the total polarization optic coefficient, respectively. At lower fields and in the paraelectric regime, P_s^2 has to be replaced by $\delta P^2 \propto 3(P_\parallel^2 - P_\perp^2)/2$, where $P_\parallel$ and $P_\perp$ are the polarization components along [110] and [1$\bar{1}$0], respectively. The dielectric permittivity, ϵ' and ϵ'', was measured at a frequency $f = 1$ kHz along the [110] direction of another sample with thickness 0.35 mm under electric fields up to $E = 4$ kV/cm with an automatic bridge, HP 4192 A.

Figure 1 shows the intraplanar LB observed upon (i) field heating after zero-field cooling (ZFC/FH) with $E = 3.3$ kV/cm between 5 and 300 K, followed by (ii) field-cooling (FC) down to 5 K with $E = 3.3$ kV/cm, and (iii) subsequent ZFH again up to 300 K. Using equal heating and cooling rates, $|dT/dt| \sim 25$ mK/s, the limit of ergodicity of the system appears at $T_f \sim 234$ K, where FC and ZFC/FH curves split apart. Because of the slow dynamics of PMN, however, T_f only signifies a dynamic freezing temperature at frequencies in the hertz regime.

At $T \le 160$ K the system appears completely frozen. On the one hand, zero LB in the ZFC/FH curve indicates a random distribution of spontaneously polarized domains and their complete immobility. On the other hand, nearly perfect coincidence of the large induced LB upon FC *and* subsequent ZFH shows the thermal stability of the polarized state once frozen in. The wiggly shape of the LB curves is due to interference effects by multiple reflection at the sample surfaces [18]. Defreezing of the induced polarization starts above 180 K quite rapidly, such that $\Delta n \sim 0$ at $T_0 \sim 200$ K. Similar observations of the field-induced strain [6] hint at the existence of an

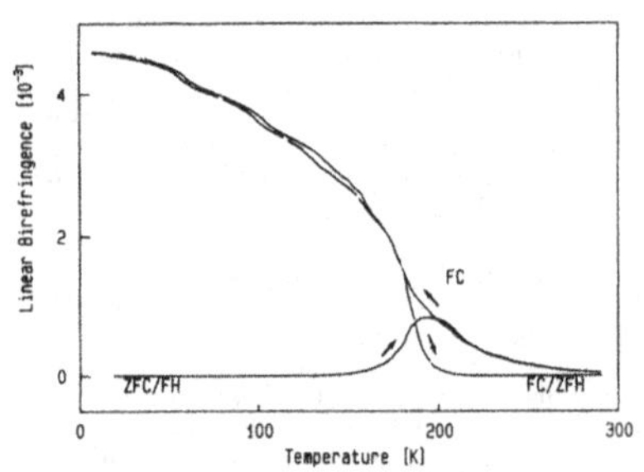

FIG. 1. Temperature dependence of the linear birefringence in the (001) plane of PMN induced by an electric field $E = 3.3$ kV/cm along [110].

equilibrium phase-transition temperature T_c in the absence of random fields, $T_0 < T_c < T_f$. Unlike KTL, $x_{Li} = 0.063$ [14], which is a random-field Ising system, the electret state of PMN does not persist up to T_c, owing to its lower anisotropy connected with eightfold orientational degeneracy.

Hysteretic effects are seen in field cycles of the induced LB as shown in Fig. 2 for three temperatures. In the paraelectric regime, at 285 K, the LB varies roughly as E^2, since $P \propto E$ in the low-field regime [Fig. 2(a)]. It is seen, however, that the cycles are not fully reversible despite their very low frequency, $f \sim 1$ cycle/h. At $T = 221$ K [Fig. 2(b)], under the same conditions, the hysteresis becomes substantially larger with a coercive field $E_c = 0.8$ kV/cm and a remanence of about $0.1\Delta n(E_{max})$, where $E_{max} = 2.7$ kV/cm. It should be noticed that Δn still increases after inverting the scanning direction at $E = \pm E_{max}$. This clearly indicates very sluggish relaxation in the submillihertz regime, which is even more pronounced in the ferroelectric temperature range at $T = 181$ K [Fig. 2(c)].

Time dependences of the field-induced LB as measured at $T = 221$ and 181 K are shown in Fig. 3. After ZFC, a field $E_0 = 3.3$ kV/cm was first applied in order to induce polar order and uniaxial strain. At both temperatures the LB rises sublinearly with time t, but apparently nonexponentially. This is seen from the finite slope, $|d\Delta n/dt|$, in the long-t limit. There is obviously *no* simple temporal dependence, e.g., according to a stretched exponential function. Occasionally, additional points of inflection do occur in both curves. Presumably they refer to subsequent different domain growth processes. Even steplike decreases of the LB are often observed. They can be traced back to the strong generation of intermediate "90°" domains (i.e., $P_s \parallel$ [1$\bar{1}$1],[1$\bar{1}\bar{1}$]), which cause stripe domain patterns as observed recently under a polarizing microscope [12].

Figure 3(a) shows, after intermediate relaxation in zero field (see below), the poling process in a reversed field, $E = -E_0$. Under this condition, after initial monotonic sublinear increase of the LB, starting at $\Delta n \sim 0$, a

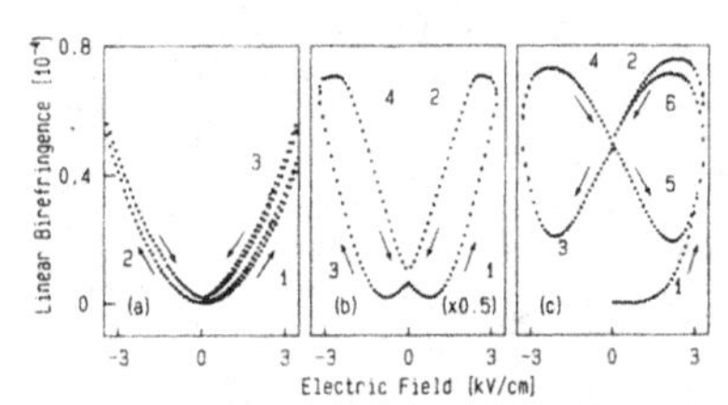

FIG. 2. Field dependences of the induced linear birefringence at (a) 285, (b) 221, and (c) 181 K, cycled as indicated by arrows and successive numbers.

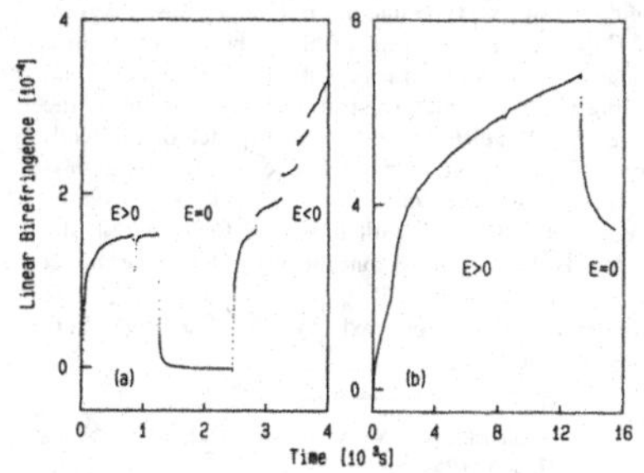

FIG. 3. Time dependence of the linear birefringence at (a) 221 and (b) 181 K measured successively under an applied ($E>0$), removed ($E=0$), and inverted electric field ($E<0$), where $|E|=3.3$ kV/cm for $E\neq 0$.

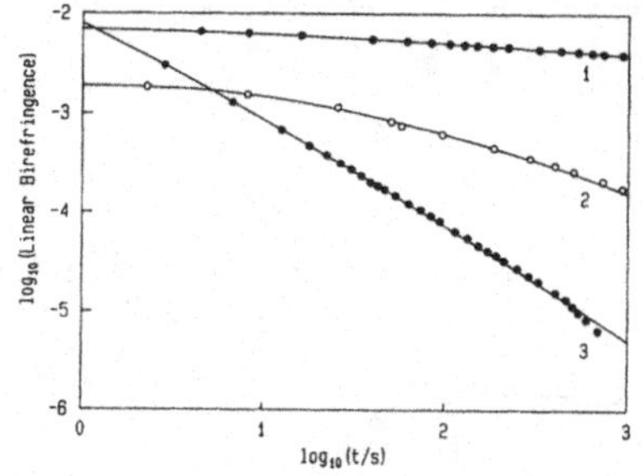

FIG. 4. Time dependence of the linear birefringence after isothermal poling with $E=3.3$ kV/cm at (1) 181, (2) 201, and (3) 221 K, respectively. Only part of the measured data (circles) is shown together with best-fit decay functions (see text).

number of discontinuities occur within a period of about 10^3 s. Obviously, avalanchelike depinning of microdomains takes place, thus giving rise to optically detected Barkhausen jumps. These are not compatible with glassy reorientation, which takes place on submicrometric length scales, hence, continuously and monotonically.

The decay curves of the polarization as measured after removing the poling field at 221 and 181 K (Fig. 3, curves "$E=0$") are monotonic and smooth. They are shown in a double logarithmic plot together with data taken in a similar way at $T=201$ K, in Fig. 4. It is seen that the decay slows down tremendously at decreasing temperature. In addition, the convexity of the functions increases significantly upon lowering T.

This impression is confirmed by testing the usual [19] set of decay functions applicable to random systems: logarithmic, stretched exponential, power, and generalized power laws. Best fits for the data at 181 and 201 K (solid lines in Fig. 4) are provided by the generalized power law

$$\Delta n(t)=\Delta n(\tau)\exp\{-b[\ln(t/\tau)]^{\beta}\},\quad t\geq\tau\,. \qquad (1)$$

Whereas the parameter $\Delta n(\tau)$ merely describes the starting conditions, the exponents $b=0.02$ and 0.11 and the characteristic relaxation times $\tau=2.2$ and 0.3 s, respectively, reflect the acceleration of the decay with increasing T. Remarkably, however, we find an exponent $\beta=1.73$ in both cases, which indicates a substantial deviation from a simple power law, $\beta=1$. On the other hand, the 221-K data fitted by Eq. (1) yield $\beta=1.14$, hence, nearly power-law behavior (solid line in Fig. 4). A fit of the 221-K data by a power law is only slightly worse despite the use of only two fitting parameters.

Equation (1) describes the fluctuations of the order-parameter autocorrelation function $C_l(t)$ of droplets in a homogeneous surrounding with an inverted order parameter under the constraint of quenched RF [19]. This model is probably valid also in the present case, where the order parameter **P** is homogeneous at the beginning, $\mathbf{P}_{\parallel}$, and droplets with differently orientated polarization, $\mathbf{P}_{\perp}$, are thermally excited. The LB measures the spatial and temporal average of the quadrupolar order parameter, $\langle P_{\parallel}^2\rangle-\langle P_{\perp}^2\rangle$. Its variation is, hence, directly proportional to $C_l(t)\equiv\langle P_{\perp}^2(t)\rangle$. Contrary to the Ising case [19], however, the droplets in PMN tend to grow instead of decaying because of the low anisotropy and the high directional degeneracy of the order parameter in PMN. The validity of Eq. (1) in the case of droplet formation with thick domain walls [10] remains, hence, to be shown.

The situation at higher temperatures, $T=221$ K (curve 3), is qualitatively different, since a pure power law appears adequate. In fact, very probably an essentially paraelectric behavior is probed, since the Curie temperature of the system is $T_c\sim 212$ K (see below). Tentatively, we suppose that the temporal behavior of the LB describes the growth of a percolating paraelectric cluster within the homogeneously polarized single domain, which is stochastically pinned to the RF distribution. Very probably the paraelectric cluster is fractal. Its growth is hence described by a power law, e.g., as observed on fractal antiferromagnetic domains in diluted Ising antiferromagnets in zero field [20]. As a result of the considerable smearing of the ferroelectric transition, however, a transition regime between low-T generalized and high-T simple power-law behavior will be encountered. This might explain the slight departure of β from unity when fitting the 221-K data by Eq. (1).

Probably the most convincing piece of evidence for the ferroelectric nature of PMN is the observation of an extra peak in the dielectric permeability when applying an ordering field to the domain state. This is shown in Fig. 5 by comparing the ZFC and FH curves, ϵ' and ϵ'' vs T, between 120 and 300 K. Very clearly, the field, $E=4$

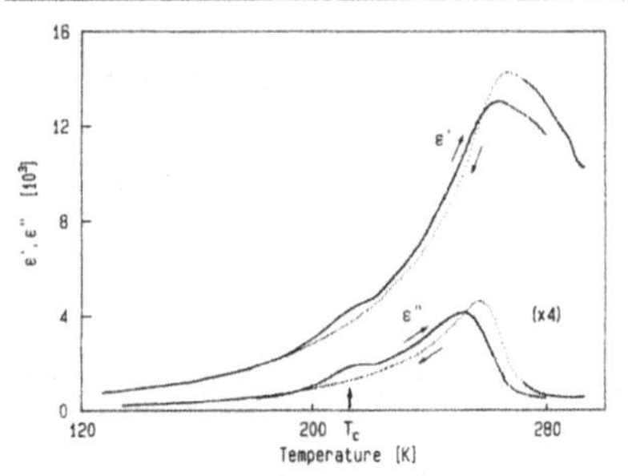

FIG. 5. Dielectric permittivity (ϵ', ϵ'') measured at 10^3 Hz upon zero-field cooling and successive field heating with $E = 4$ kV/cm as indicated by arrows.

kV/cm, induces net, albeit broadened, peaks at $T_c \sim 212$ K. They are superimposed on the usual broad background, which is slightly decreased and shifted towards lower T as observed previously [12,21]. This is a consequence of partial poling of the network of *nonpercolating* ferroelectric clusters, which start to grow below $T = 600$ K [9,22]. The extra peak at T_c, however, is attributed to the *percolating* cluster due to the Andelman-Joanny–type [14] suppression of the RF nanodomain structure. Sharper peaks of ϵ' vs T at T_c are induced by using higher fields, $E = 20$ kV/cm [12,13]. PMN, hence, behaves similarly to KTL with $x_{Li} = 0.063$, whose ferroelectric ϵ anomaly, however, is observable even in zero field after FC. This signifies a basic difference between both systems, KTL being much closer than PMN to the RF Ising universality class, which has a long-range ordered ground state in three dimensions [10].

Summarizing, the RF field interaction is very probably at the heart of the extremely smeared ferroelectric phase transition of PMN. Our arguments *against* an orientational glass interpretation are (i) remanence curves $\langle P^2 \rangle$ vs T, which mark lower limits of the transition temperature T_c; (ii) Barkhausen jumps indicating discontinuous domain rearrangements; (iii) field-induced peaks of ϵ' and ϵ'', which designate unsmearing by virtue of RF compensation; and (iv) generalized temporal power-law decay of the isothermal remanent polarization, which characterizes the reconstitution of a nanodomain state in zero field.

Clearly, the relaxational behavior of PMN and its connection with nonergodicity requires more attention, which has so far not been devoted to near-Heisenberg RF systems. Here we notice that the recently [4] reported de Almeida–Thouless-type line in the E-vs-T phase diagram of PMN does *not* prove dipolar glassy behavior. A similar phase line is also observed in the three-dimensional RF Ising model without frustration [20]. Moreover, the observed [4] Vogel-Fulcher-type divergence of the leading relaxation time at $T_f = 217$ K does not unambiguously signify glassy freezing. An identical expression, $\tau = \tau_0 \exp[a(T-T_c)^{-\theta\nu}]$ with $\theta\nu \sim 1$, is found within the activated dynamic scaling concept of the RF Ising model [19].

This work was supported by the Deutsche Forschungsgemeinschaft.

[1] G. A. Smolenskii and V. A. Isupov, Dokl. Akad. Nauk SSSR **97**, 653 (1954).
[2] M. E. Lines and A. M. Glass, *Principles and Applications of Ferroelectrics and Related Materials* (Clarendon, Oxford, 1979), p. 287.
[3] U. T. Höchli, K. Knorr, and A. Loidl, Adv. Phys. **39**, 405 (1990).
[4] D. Viehland, M. Wuttig, and L. E. Cross, Ferroelectrics **120**, 71 (1991).
[5] G. A. Smolenskii *et al.*, Fiz. Tverd. Tela (Leningrad) **2**, 2906 (1960) [Sov. Phys. Solid State **2**, 2584 (1960)].
[6] G. Schmidt *et al.*, Phys. Status Solidi (a) **63**, 501 (1981); G. Schmidt, Ferroelectrics **104**, 205 (1990).
[7] E. G. Nadolinskaya *et al.*, Fiz. Tverd. Tela (Leningrad) **29**, 3368 (1987) [Sov. Phys. Solid State **29**, 1932 (1987)].
[8] L. E. Cross, Ferroelectrics **76**, 241 (1987).
[9] E. Husson, M. Chubb, and A. Morell, Mater. Res. Bull. **23**, 357 (1988); N. de Mathan *et al.*, Ferroelectrics (to be published); N. de Mathan *et al.*, J. Phys. Condens. Matter (to be published).
[10] Y. Imry and S. K. Ma, Phys. Rev. Lett. **35**, 1399 (1975); A. Aharony, Solid State Commun. **28**, 607 (1978).
[11] T. Nattermann, Ferroelectrics **104**, 171 (1990).
[12] H. Arndt *et al.*, Ferroelectrics **79**, 145 (1988).
[13] R. Sommer, N. K. Yushin, and J. J. van der Klink, Ferroelectrics (to be published).
[14] D. Andelman and J. F. Joanny, Phys. Rev. B **32**, 4818 (1985).
[15] H. Schremmer, W. Kleemann, and D. Rytz, Phys. Rev. Lett. **62**, 1896 (1989).
[16] F. J. Schäfer and W. Kleemann, J. Appl. Phys. **57**, 2606 (1985).
[17] W. Kleemann, F. J. Schäfer, and D. Rytz, Phys. Rev. Lett. **54**, 2038 (1985).
[18] D. P. Belanger, Ph.D. thesis, University of California, Santa Barbara, 1981 (unpublished).
[19] D. A. Huse and D. S. Fisher, Phys. Rev. B **35**, 6841 (1987).
[20] U. Nowak and K. D. Usadel, Phys. Rev. B **44**, 7426 (1991).
[21] N. de Mathan *et al.*, Mater. Res. Bull. **25**, 427 (1990).
[22] P. Bonneau *et al.*, J. Solid State Chem. **91**, 350 (1991).

3

Ferroelectrics 2001–2021

(Sr, Mn)TiO_3: A Magnetoelectric Multiglass

V. V. Shvartsman, S. Bedanta, P. Borisov, and W. Kleemann*
Angewandte Physik, Universität Duisburg-Essen, Lotharstrasse 1, 47048 Duisburg, Germany

A. Tkach and P. M. Vilarinho
Department of Ceramics and Glass Engineering, CICECO, University of Aveiro, 3810-193, Aveiro, Portugal
(Received 8 February 2008; published 16 October 2008)

By close analogy with multiferroic materials with coexisting long-range electric and magnetic orders a "multiglass" scenario of two different glassy states is observed in $Sr_{0.98}Mn_{0.02}TiO_3$ ceramics. Sr-site substituted Mn^{2+} ions are at the origin of both a polar and a spin glass with glass temperatures $T_g \approx 38$ K and ≤ 34 K, respectively. The structural freezing triggers that of the spins, and both glassy systems show individual memory effects. Thanks to strong spin-phonon interaction within the incipient ferroelectric host crystal $SrTiO_3$, large higher order magnetoelectric coupling occurs between both glass systems.

DOI: 10.1103/PhysRevLett.101.165704 PACS numbers: 64.70.P−, 75.50.Lk, 77.22.−d, 77.84.Dy

Recent years have seen a growing research interest in materials exhibiting the linear magnetoelectric (ME) effect [1,2]. It describes the cross-linking dependence of the magnetization and polarization on applied electric and magnetic fields, respectively. Since it promises to reach largest values in materials with both high electric and magnetic susceptibility [3], a most favorable situation might be realized in multiferroic materials, where two different ferroic states, e.g., ferroelectric and ferromagnetic, coexist [4]. However, the occurrence of linear magnetoelectricity is very rare because of its high symmetry demands. For this reason it may even be absent in multiferroics as exemplified, e.g., by $BiMnO_3$ with $T_C^{FM} = 100$ K [5] and $T_C^{FE} = 440$ K [6], respectively. Only higher order magnetoelectricity is predicted similarly as for systems without well-defined symmetry [7].

In this situation it seems meaningful to investigate the converse situation, namely, the ME coupling of disordered systems, e.g., given by glassy materials which have by definition neither electric nor magnetic periodic long-range order. In this Letter we show that the class of ME materials may, indeed, be extended to those undergoing transitions into glassy (or frozen metastable) states. These are well known to occur as a result of competing interactions and topological frustration, where the glass transition temperature T_g separates the ergodic high temperature regime, $T > T_g$, from the nonergodic low-T one. At $T < T_g$ true thermodynamic equilibrium can be reached only asymptotically. Structural and spin glass states at low T are established by cooperative random freezing of the orientational [8] and of the spin degrees of freedom, respectively [9]. Here we report on the simultaneous occurrence of a polar cluster glass and a magnetic spin glass state and on their mutual higher order ME coupling in $(Sr_{1-x}Mn_x)TiO_3$ (SMnT, for short) with $x = 0.02$.

$SrTiO_3$ is an incipient ferroelectric, whose polar instability is suppressed by quantum fluctuations, such that the system remains in the paraelectric state down to 0 K [10]. Polar order in STO may be induced by ionic substitutions as in $(Sr_{1-x}Ca_x)TiO_3$ [11]. In the related solid solution SMnT with $x \leq 0.03$, slim polarization vs electric-field hysteresis loops and a broad strongly frequency dependent peak of the dielectric permittivity are found [12]. The polar state in these compounds is due to off-center shifts of the Mn^{2+} cations at the 12-fold coordinated A-cation Sr^{2+} positions within the perovskite structure as depicted in Fig. 1(a) and evidenced by energy dispersive X-ray spectra [12] and by ESR techniques [13]. The off-center Mn^{2+} cations are assumed to create dipoles, which induce polar clusters in the highly polarizable $SrTiO_3$ host lattice [14]. The situation resembles the related (nonmagnetic) system $(K_{1-x}Li_x)TaO_3$ (KLT) [8], which undergoes a transition into a generic glass state at $T_g < 40$ K for $x < 0.022$ [15]. In the case of SMnT [Fig. 1(a)] a six-state Potts glass referring to six discrete directions of the polar order parameter [16] is supposed to occur.

Ceramic samples SMnT with $x = 0.02$ were prepared by a mixed oxide technology described elsewhere [12].

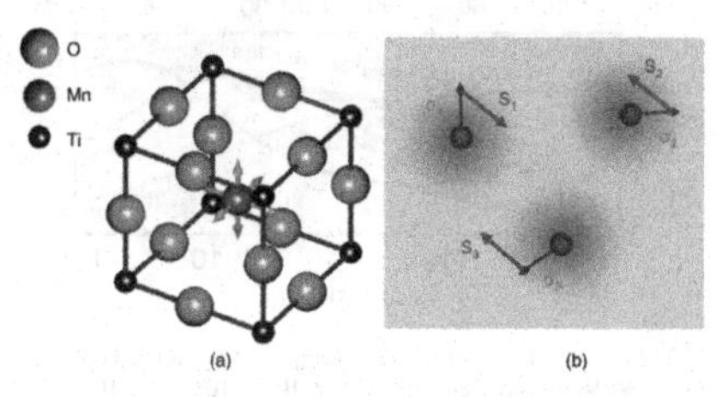

FIG. 1 (color online). (a) Displacement vectors of an off-center Mn^{2+} ion in its cage of 12 surrounding oxygen ions of A-site doped $SrTiO_3$. (b) Frustrated arrangement of three antiferromagnetically interacting Mn^{2+} spins, S_j, $j = 1$, 2 and 3, occupying frozen off-center positions σ_j of the structural glass backbone.

Figure 2(a) shows the temperature dependence of the real part of the dielectric permittivity $\varepsilon'(T)$ recorded at frequencies $10^{-1} \leq f < 10^6$ Hz. The position of the peak temperature T_m of $\varepsilon'(T)$ is well described by a power law of the respective frequency, $f(T_m) \propto (T_m/T_g^e - 1)^{z\nu}$ [Fig. 2(b)], which is a typical manifestation of glassy critical behavior [17,18]. Best fits of the data yield the electric glass temperature $T_g^e = 38.3 \pm 0.3$ K (arrow), and the dynamical critical exponent, $z\nu = 8.5 \pm 0.2$, which compares well with that of spin glass systems [17].

An essential indicator of the suspected structural glass state is the memory effect, which arises after isothermally annealing below T_g^e. Figure 2(b) shows the "hole" burnt into $\varepsilon'(T)$ after a wait time $t_w \approx 10.5$ h at $T_w = 32.5$ K (arrow). We find an indentation at the "annealing" temperature T_w, $\Delta\varepsilon' = \varepsilon'_{\rm wait}(T_w) - \varepsilon'_{\rm ref}(T_w) \approx -6$. It resembles that observed on the structural glass $KTa_{0.973}Nb_{0.027}O_3$ [19] and signifies the asymptotic approach to the glassy ground state at T_w, via a decrease of the susceptibility. Since the glassy ground state varies as a function of the temperature, the system is "rejuvenating" at T sufficiently far from T_w [17]. Hence, the burnt "hole" is localized around T_w, while a global decrease would be expected for an ordinarily relaxing metastable system. The small amplitude of the "hole" indicates that only a small fraction of dipoles is actually freezing. Indeed, it should be reminded that the structural glassy freezing of SMnT does not signify complete immobilization of all hopping Mn^{2+} ions [Fig. 1(a)]. Instead, just one percolating cluster freezes with $\tau \to \infty$, while ramifications and clusters of smaller size are still able to relax at $f > 0$. This is confirmed by the temperature dependence of the broad distribution of relaxation times as mimicked by ε'' vs $\log f$ [Fig. 2(c)]. Its low-f tail gradually lifts upon cooling and attains a finite amplitude at $f_{\rm min} \to 0$ as $T \to T_g^e$. Similar relaxation spectra of $K_{0.989}Li_{0.011}TaO_3$ [15] and of the spin glass manganese aluminosilicate [20] were taken as fingerprints of glass transitions.

Figure 3(a) shows the temperature dependences of the real and imaginary parts of the magnetic ac susceptibility, $\chi' - i\chi''$. Pronounced peaks with comparably weak polydispersivity [cf. Fig. 2(a)] are observed for frequencies $0.1 \leq f \leq 10$ Hz slightly below $T_g^e \approx 38$ K [inset to Fig. 3(a)]. Convergence of the χ' vs T peaks occurs at $T_g^m \leq 34$ K [see also Fig. 3(b)], which is about 1 order of magnitude larger than T_g^m observed on other manganese doped insulators like aluminosilicate [21]. Since the average distance between Mn^{2+} ions in SMnT ($x = 0.02$), $\langle d \rangle \approx 1.5$ nm, is too large as to warrant stability of the spin glass structure solely by frustrated superexchange, an amplification effect needs to be considered. By analogy with the antiferromagnetic quantum paraelectric $EuTiO_3$ [7] we assume the applicability of the Hamiltonian

$$H^{me} = -\delta_{me} \sum_{\langle i,j \rangle} \sum_{\langle k,l \rangle} \mathbf{S}_i \mathbf{S}_j \boldsymbol{\sigma}_k \boldsymbol{\sigma}_l, \qquad (1)$$

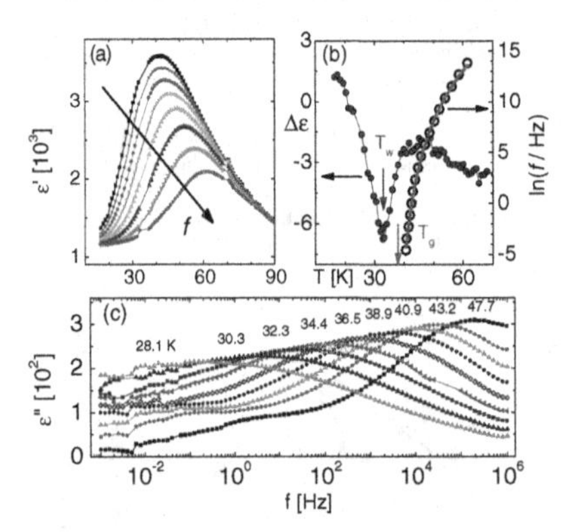

FIG. 2 (color online). (a) $\varepsilon'(T)$ of $Sr_{0.98}Mn_{0.02}TiO_3$ recorded at $E_{\rm ac} = 60$ V/m and frequencies $f = 10^{-1}$, 10^0, 10^1, 10^2, 10^3, 10^4, 10^5, and 0.4×10^6 Hz. (b) Frequency dependence of the peak temperature (T_m) of $\varepsilon'(T)$ taken from (a) together with a critical power law fit (solid line) and difference curve $\Delta\varepsilon' = \varepsilon'_{\rm wait} - \varepsilon'_{\rm ref}$ vs T obtained at $f = 10$ Hz upon heating after ZFC from 110 K with and without waiting for 10.5 h at $T_w = 32.5$ K, respectively. (c) $\varepsilon''(f)$ of $Sr_{0.98}Mn_{0.02}TiO_3$ recorded at frequencies $10^{-3} \leq f \leq 10^6$ Hz.

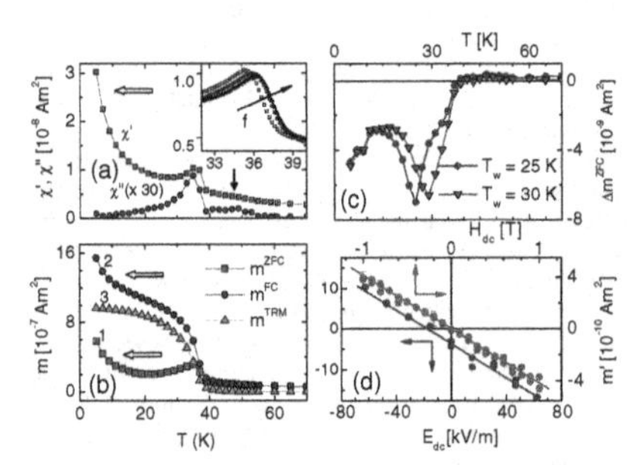

FIG. 3 (color online). (a) Temperature dependences of the real and imaginary parts, χ' and χ'', of the magnetic ac susceptibility of $Sr_{0.98}Mn_{0.02}TiO_3$ measured at $f = 1$ Hz with a field amplitude $\mu_0 H_{\rm ac} = 0.4$ mT (inset: expanded view of χ' around T_g for frequencies $10^{-1} \leq f \leq 10$ Hz). (b) Temperature dependences of $m^{\rm ZFC}$ (curve 1), $m^{\rm FC}$ (2), and $m^{\rm TRM}$ (3) measured in $\mu_0 H = 10$ mT. (c) Difference curves of $m^{\rm ZFC}$ after ZFC from 110 to 5 K with and without intermittent stop of $t_w = 10^4$ s at $T_w = 25$ and 30 K. (d) Real part m' of the ME ac susceptibility measured at $T = 10$ K and $f = 1$ Hz in an ac electric field with amplitude $E_{\rm ac} = 62.5$ kV/m vs $H_{\rm dc}$ ($-1 \leq \mu_0 H_{\rm dc} \leq 1$ T; upper line) and vs $E_{\rm dc}$ ($-62.5 \leq E_{\rm dc} \leq 62.5$ kV/m under a bias field $\mu_0 H_{\rm dc} = 1$ T; lower line).

which describes the biquadratic coupling of spins $\mathbf{S}_{i,j}$ and structural pseudospins $\boldsymbol{\sigma}_{i,j}$ represented by unit vectors along $\langle 100 \rangle$, and has explained both the observed large spin-phonon coupling [7,22] and the magnetocapacitive effect in $EuTiO_3$ [23]. The components of the pseudospins $\boldsymbol{\sigma}_{i,j}$ mimic the off-center displacements of the A site dopant ions (Mn^{2+}) being strongly coupled to the displacements of the B site (Ti^{4+}) ions, which participate in the optic soft mode. δ_{me} is an effective coupling constant. The displacement correlation functions couple to the $S = 5/2$ Heisenberg spins $\mathbf{S}_{i,j}$ of the $3d^5$ configuration of the Mn^{2+} ions. Large effects are expected around T_g, where the structural pair-correlation functions $\langle \boldsymbol{\sigma}_k \boldsymbol{\sigma}_l \rangle$ maximize, promote increased spin pair-correlation functions $\langle \mathbf{S}_i \mathbf{S}_j \rangle$, and give rise to weak anomalies of the magnetic susceptibility already above T_g [Fig. 3(a), vertical arrow]. Below $T_g^e \approx$ 38 K, the onset of a percolating network of frozen polar clusters probably triggers the freezing of the spin degrees of freedom. While above T_g^e freezing of the Mn^{2+} spins is suppressed by the hopping motion of the Mn^{2+} ions, those spins residing on the percolating polar glass cluster have a chance to achieve a well-defined collective state at $T < T_g^e$. Its glassy nature is provided by frustrated antiferromagnetic superexchange interaction via Mn-O-Mn chains similarly as in manganese aluminosilicate spin glasses [21]. Figure 1(b) shows a sketch of a possible local spin configuration at three sites of the frozen structural cluster.

Figure 3(b) shows the temperature dependence of the total magnetic moment m induced by $\mu_0 H = 10$ mT upon field heating after zero-field cooling (ZFC) to 5 K (m^{ZFC}, curve 1), subsequently upon field cooling (FC) again to $T = 5$ K (m^{FC}, curve 2), and finally as the thermoremanent magnetic moment upon heating in zero field (m^{TRM}, curve 3). Typical of spin glass behavior, irreversibility occurs below $T_g^m \approx 34$ K together with a peak in $m^{ZFC}(T)$. In addition, the concave shape of $m^{TRM}(T)$ around 40 K signifies the decay of metastable field-induced rather than of spontaneous magnetization.

It is noticed that Curie-type low-T paramagnetic components, $\Delta m \propto 1/T$, occur in both m^{ZFC} and m^{FC}, similarly as in the magnetic susceptibility, $\chi' \propto 1/T$ [Figs. 3(a) and 3(b), horizontal arrows]. This indicates that a sizable fraction of uncoupled spins remains paramagnetic down to lowest temperatures. Its fraction is estimated by approximating the low-field $m^{ZFC}(T)$ curve in Fig. 3(b) by Curie's law, $m^{ZFC} = (N m_0^2 / 3 k_B T) \mu_0 H$, both at high and low temperatures [N = number of paramagnetic ions, k_B = Boltzmann's constant, $m_0 \approx 5\mu_B \approx m(Mn^{2+})$ [24] calculated from the Curie constant at $T > 250$ K and the total number of Mn^{2+} ions, $N \approx 1.3 \times 10^{19}$]. A $1/m^{ZFC}$ plot (not shown) distinguishes a flat slope between 250 and 300 K from a steeper one below 30 K. This manifests the condensation process into the spin glass phase. From Curie's law we estimate that about 60%–70% of the Mn^{2+} ions remain paramagnetic at low T. This complies with the idea that only those spins are subject to glassy freezing, which reside on the rare percolating dipolar glass cluster, and underlines the close neighborhood of frozen dipoles and spins [Fig. 1(b)].

A crucial test of the spin glass phase is the memory effect similarly as verified in the polar glass [Fig. 2(b)]. Figure 3(c) shows the differences between m^{ZFC} data recorded with and without an intermittent stop, $\Delta m^{ZFC}(T) = m_{wait}^{ZFC}(T) - m_{ref}^{ZFC}(T)$, obtained after wait times $t_w = 10^4$ s at $T_w = 25$ and 30 K ($< T_g^m$). Sharply defined "hole burning" dips occur exactly at the respective T_w similarly as in Fig. 2(b). They clearly evidence memory and rejuvenation of the spin system like in atomic and superspin glasses [17].

The proposed biquadratic coupling, Eq. (1), enters the free energy density expansion [25] together with additional possible ME coupling terms, while spontaneous polarization and magnetization and bilinear ME coupling (due to the lack of an appropriate crystalline symmetry [26]) are absent:

$$F(\mathbf{E}, \mathbf{H}) = F_0 - \frac{1}{2}\varepsilon_0 \varepsilon_{ij} E_i E_j - \frac{1}{2}\mu_0 \mu_{ij} H_i H_j - \frac{\gamma_{ijk}}{2} H_i E_j E_k - \frac{\beta_{ijk}}{2} E_i H_j H_k - \frac{\delta_{ijkl}}{2} E_i E_j H_k H_l \quad (2)$$

under Einstein summation. The electric-field-induced magnetization components of

$$\mu_0 M_i = -\partial F / \partial H_i = \mu_0 \mu_{ij} H_j + \beta_{ijk} E_j H_k + \gamma_{ijk}/2 E_j E_k + \delta_{ijkl} H_j E_k E_l \quad (3)$$

are measured using ME SQUID susceptometry [27]. It involves ac and dc electric and magnetic external fields, $E = E_{ac} \cos\omega t + E_{dc}$ and H_{dc}, and records the first harmonic complex ac magnetic moment, $m'(t) = (m' - im'') \cos\omega t$, where

$$m' = (\beta E_{ac} H_{dc} + \gamma E_{ac} E_{dc} + 2\delta E_{ac} E_{dc} H_{dc})(V/\mu_0) \quad (4)$$

(V = sample volume) and $m'' \approx 0$ at the measurement frequency $f = \omega/2\pi = 1$ Hz. The orientation averaged coupling constants, β, γ, and δ were measured after cooling the sample in zero external fields to $T = 10$ K.

First, we established that $\mu_0 H_{dc} = 0$ at arbitrary values of E_{dc} yields $m' \equiv 0$ within errors. This means that both the bilinear and the quadratic ME responses are absent, $\alpha = 0 = \gamma$, as expected for systems without magnetic long-range order [28].

Second, we applied $E_{ac} = 62.5$ KV/m and $E_{dc} = 0$, and cycled the magnetic field, $|\mu_0 H_{dc}| \leq 1$ T [Fig. 3(d), upper line]. The slope of the emerging linear hysteresis-free cycle yields $\beta = \mu_0 \Delta m' / V E_{ac} \Delta H_{dc} \approx -3.0 \times 10^{-19}$ s/A. The corresponding electric-field dependent "*paramagnetoelec-*

tric" susceptibility, $\Delta m'/\Delta H_{dc} \propto E_{ac}$, is allowed whenever the paramagnetic ions are located at sites with broken inversion symmetry [29] and do not require magnetic long-range order [28]. However, since the linear dependence of $\Delta m'/\Delta H_{dc}$ on E_{ac} requires global inversion asymmetry, we assume the occurrence of some net polarization P_r, either by self-poling on zero-field cooling or during the first poling cycle at 10 K. Being metastable throughout the polar glass state, P_r controls the bilinear coupling, until $\beta \to 0$ as $T \to T_g$ on heating (not shown). This complies with the structure of Fig. 1(a), which becomes centrosymmetric above T_g. Details have yet to be explored.

Third, we applied $E_{ac} = 62.5$ kV/m and $\mu_0 H_{dc} = 1$ T, and cycled an electric field, $|E_{dc}| \leq 62.5$ kV/m [Fig. 3(d), lower line]. According to Eq. (4) it reveals βEH and δHE^2 from the intercept at $E_{dc} = 0$ and from the slope of m' vs E_{dc}, respectively. Consistently, the same β value emerges as from the slope of the upper curve, while the biquadratic coefficient—being quantified for the first time to the best of our knowledge—is $\delta = \mu_0 \Delta m'/(2VH_{dc}E_{ac}\Delta E_{dc}) \approx -9.0 \times 10^{-24}$ sm/VA. The negative sign of δ seems to corroborate double glassiness, whose free energy is enhanced in uniform external fields. The value of δ allows to predict an E^2H^2-based magnetocapacitive effect $\Delta\varepsilon/\varepsilon' \approx -5 \times 10^{-4}$, where $\Delta\varepsilon = \delta H^2/\varepsilon_0 \approx -0.65$ inserting $\mu_0 H = 1$ T and $\varepsilon'(T = 10$ K, $f = 1$ Hz$) \approx 1300$ [Fig. 2(a)]. Its magnitude is about 2.5% of that found for $EuTiO_3$ at $\mu_0 H = 1$ T and $T = 4$ K, $\Delta\varepsilon/\varepsilon' \approx 2 \times 10^{-2}$ [23], matching to the dilution in SMnT, $x_{Mn} = 0.02$.

In conclusion, when replacing Sr^{2+} ions in $SrTiO_3$ by a small amount of magnetic Mn^{2+} ions, two processes are activated at low temperatures. On one hand, the Mn^{2+} ions take the role of electric and elastic pseudospins and undergo a transition into a polar 6-state Potts glass [16]. On the other hand, the $S = 5/2$ spins, being attached to the Mn^{2+} ions, couple via frustrated antiferromagnetic superexchange, reinforced by ME two-spin-pseudospin interaction. They freeze into a spin glass state after the polar degrees of freedom come to rest below $T_g^e = 38$ K. Both glassy phases are independently evidenced by their specific memory effects. Dipolar and magnetic "holes" have never before been burned into one and the same system. Strong ME coupling via both the generic "*magnetocapacitive*" E^2H^2 and the probably spurious "*paramagnetoelectric*" EH^2 effects manifests the importance of quantum fluctuations in $SrTiO_3$. It should be noticed that these coupling schemes would also be valid, if SMnT were in a nonequilibrium [18] instead of a generic dipole glass state. Corroborating experiments on oriented $(Sr, Mn)TiO_3$ single crystals are desirable, but up to now mere A-site substitution has not yet been achieved [30].

Financial support by the European Community within STREP NMP3-CT-2006-032616 (MULTICERAL) and the Network of Excellence FAME (Contract No. FP6-500159-1), by Deutsche Forschungsgemeinschaft via KL306/38-3 and SFB 491, and by DAAD via Acções Integradas Luso-Alemãs is gratefully acknowledged.

*Author to whom correspondence should be addressed. wolfgang.kleemann@uni-due.de

[1] M. Fiebig, J. Phys. D **38**, R123 (2005).
[2] W. Eerenstein, N.D. Mathur, and J.F. Scott, Nature (London) **442**, 759 (2006).
[3] W.F. Brown, R.M. Hornreich, and S. Shtrikman, Phys. Rev. **168**, 574 (1968).
[4] N. Hur *et al.*, Nature (London) **429**, 392 (2004).
[5] C.H. Yang *et al.*, Europhys. Lett. **74**, 348 (2006).
[6] T. Kimura *et al.*, Phys. Rev. B **67**, 180401(R) (2003).
[7] C.J. Fennie and K.M. Rabe, Phys. Rev. Lett. **96**, 205505 (2006); **97**, 267602 (2006).
[8] U.T. Höchli, K. Knorr, and A. Loidl, Adv. Phys. **39**, 405 (1990).
[9] K. Binder and A.P. Young, Rev. Mod. Phys. **58**, 801 (1986).
[10] K.A. Müller and H. Burkard, Phys. Rev. B **19**, 3593 (1979).
[11] J.G. Bednorz and K.A. Müller, Phys. Rev. Lett. **52**, 2289 (1984).
[12] A. Tkach, P.M. Vilarinho, and A.L. Kholkin, Appl. Phys. Lett. **86**, 172902 (2005); Acta Mater. **53**, 5061 (2005); **54**, 5385 (2006).
[13] V.V. Laguta *et al.*, Phys. Rev. B **76**, 054104 (2007).
[14] A. Tkach, P.M. Vilarinho, and A.L. Kholkin, J. Appl. Phys. **101**, 084110 (2007).
[15] F. Wickenhöfer, W. Kleemann, and D. Rytz, Ferroelectrics **124**, 237 (1991).
[16] F.Y. Wu, Rev. Mod. Phys. **54**, 235 (1982).
[17] P.E. Jönsson, Adv. Chem. Phys. **128**, 191 (2004).
[18] Although a fit to the traditional Vogel-Fulcher "law" $f_m(T) = f_0 \exp(T_a/[T_{VF} - T])$ is only slightly worse than the dynamical critical slowing-down fit [Fig. 2(b)] with reasonable parameters $f_0 = 4.74 \times 10^9$ Hz and $T_a = 282$ K, it must be discarded because of its low $T_{VF} = 30$ K, which is in contradiction with hole burning observed at $T_w = 32.5$ K $> T_{VF}$ [Fig. 2(b)].
[19] P. Doussineau, T. de Lacerda-Aroso, and A. Levelut, Europhys. Lett. **46**, 401 (1999).
[20] L.E. Wenger, in *Lecture Notes in Physics* (Springer-Verlag, Berlin, 1983), Vol. 192, p. 60.
[21] F.S. Huang *et al.*, J. Phys. C **11**, L271 (1978).
[22] H. Wu, Q. Jiang, and W.Z. Shen, Phys. Rev. B **69**, 014104 (2004).
[23] T. Katsufuji and H. Takagi, Phys. Rev. B **64**, 054415 (2001).
[24] Ch. Kittel, *Introduction to Solid State Physics* (Wiley, New York, 1966), 3rd ed., p. 509.
[25] J.-P. Rivera, Ferroelectrics **161**, 165 (1994).
[26] L.D. Landau and E.M. Lifshitz, *Electrodynamics of Continuous Media* (Pergamon, Oxford, 1960), p. 119.
[27] P. Borisov, A. Hochstrat, V.V. Shvartsman, and W. Kleemann, Rev. Sci. Instrum. **78**, 106105 (2007).
[28] H. Schmid, Ferroelectrics **161**, 1 (1994).
[29] S.L. Hou and N. Bloembergen, Phys. Rev. **138**, A1218 (1965).
[30] A.G. Badalyan *et al.*, J. Phys. Conf. Ser. **93**, 012012 (2007).

ELSEVIER Physica A 308 (2002) 337–345

www.elsevier.com/locate/physa

Quantum tunneling versus zero-point energy in double-well potential model for ferrroelectric phase transitions

C.L. Wang[a,b,*], C. Arago[b], J. Garcia[b], J.A. Gonzalo[b]

[a]*Department of Physics, Shandong University, Jinan 250100, People's Republic of China*
[b]*Departamento de Fisica de Materiales. Facultad de Ciencias C-IV, Universidad Autonoma de Madrid, 28049 Madrid, Spain*

Received 11 December 2001

Abstract

The Ising model in a transverse field, which involves quantum tunneling in double potential wells, has been used for many years to describe ferroelectric phase transitions. An alternative model previously published takes into account the effect of a non-negligible zero-point energy in the double potential wells at the phase transition. A comparison of the respective Curie temperatures, spontaneous polarizations and susceptibilities shows that both models are nearly equivalent. Quantitative differences of the two methods are apparent at low temperature where the saturation polarization value with non-negligible zero-point energy is higher than the same quantity with the Ising model in a transverse field.

PACS: 77.80.Bh; 05.50.+q; 77.22.Ch

Keywords: Ising model; Phase transitions; Effective field theory; Zero point energy

1. Introduction

The Ising model in a transverse field was first employed by de Gennes [1] to describe order–disorder-type ferroelectrics. The model assumes ferroelectrics to consist of pseudo-spins with interactions in a transverse field. By including multi-spin interactions, this model can be used to describe first-order phase transitions in ferroelectrics [2]. The Ising model in a transverse field has applied also been successfully to study

* Corresponding author. Tel.: +86-531-8568569; fax: +86-531-8567031.
E-mail address: wangcl@sdu.edu.cn (C.L. Wang).

PII: S0378-4371(02)00606-4

surface and size effects in ferroelectrics. The influence of the surface on the Curie temperature and on the excitation spectrum has been obtained in semi-infinite systems [3,4]. The dependence of Curie temperature on film thickness as well as on surface variation of exchange constant and transverse field have been studied under mean-field theory [5,6] and effective field theory [7,8]. Later this model was extended to calculate polarization and susceptibility [9], as well as dynamical property [10] and other properties of ferroelectric films. In recent years, Ising model in a transverse field has been used to investigate the phase transition properties in ferroelectric superlattices [11], pyroelectric properties of ferroelectric multi-layers [12] and phase transition properties of ferroelectric particles [13].

On the other hand, the effective field approach, which is analogous to the Weiss' theory for phase transition in ferromagnetism, is another theoretical method to describe phase transition properties in ferroelectrics and other systems such as alloys and superconductors [14]. The inclusion of quadrupolar and higher-order terms into the effective field in second-order transitions in uniaxial ferroelectrics is shown to describe well the jump in specific heat at the Curie temperature of the triglycine sulfate family crystals [15]. The thermal hysteresis accompanying discontinuous ferroelectric transition can be estimated if one takes into account the quadrupolar contribution to the effective filed. The calculated values are in fair agreement with observed thermal hysteresis in several ferroelectrics belonging to different families [16]. A general equation of state using the effective field approach has been obtained for pressure and temperature-induced transitions in the ferroelectric triglycine selenate [17]. By including the zero-point energy, a simple quantum effective field approach has been developed to explain the Curie temperature in mixed ferroelectric system of rubidium/ammonium dihydrogen (RDP-ADP) [18] and tris–sarcosine calcium chloride/bromide (TSCC-TSCB) [19]. Recently, this quantum effective approach has successfully explained [20] the Curie temperature deviations from those to be expected from classical limit in triglycine sulfate/deuterated; triglycine sulfate (TGS-DTGS) and triglycine selenate/deuterated triglycine selenate (TGSe-DTGSe) mixed crystals [21,22].

2. Effective field approach

The effective field approach to ferroelectric phase transitions is certainly the simplest possible way to study phase transitions. In its simplest version ($E_{eff} = E + \beta P$), which takes into account only dipole interactions, it is capable of describing fairly well the main features of continuous ferroelectric phase transitions. A more general expression of effective field is

$$E_{eff} = E + \beta P + \gamma P^3 + \delta P^5 + \cdots , \tag{1}$$

given in terms of an external field E and odd powers of the polarization P, which correspond successively to dipolar, quadrupolar, octupolar contributions etc. The coefficients β, γ, δ may be expected to depend on the geometry of the lattice and on the spatial charge distribution within the unit cell. The equation of state is given by

Gonzalo [14]

$$P = N\mu \tanh\left(\frac{E_{eff}\,\mu}{k_B T}\right), \tag{2}$$

where N is the number of elementary dipoles per unit volume and μ is the electric dipole moment, k_B is the Boltzmann constant, and T is the absolute temperature. From this expression we can easily see that the low-temperature saturated polarization is $P_{S_0} = N\mu$. For simplicity, keeping terms only up to the dipole contribution in E_{eff} in Eq. (1), one can rewrite Eq. (2) as

$$P = N\mu \tanh\left(\frac{(E+\beta P)\mu}{k_B T}\right). \tag{3}$$

$P_{s_0} = N\mu$ is the spontaneous polarization at low temperature. Taking into account that as T approaches T_C from below at $E=0$, $P=P_s$ approaches zero, as a result the Curie temperature

$$T_C = \beta N\mu^2/k_B. \tag{4}$$

The equation of state given in Eq. (3) in implicit form can be made explicit as

$$E = \frac{k_B T}{\mu}\tanh^{-1}\left(\frac{P}{N\mu}\right) - \beta P. \tag{5}$$

The dielectric constant can be obtained easily from the above expression as follows:

$$\varepsilon^{-1} = \frac{1}{4\pi}\frac{\partial E}{\partial P}\bigg|_{E\to 0} = \frac{k_B T}{4\pi}\frac{N}{(N\mu)^2 - P_S^2} - \frac{\beta}{4\pi}. \tag{6}$$

For temperatures below the Curie temperature, i.e., at the ferroelectric phase, the above equation can be rewritten in the vicinity of Curie temperature as

$$\varepsilon^{-1} = -\frac{k_B}{2\pi N\mu^2}(T - T_C) = -\frac{2}{C}(T - T_C), \quad T \leqslant T_C. \tag{7}$$

Here, $C = 4 \times N\mu^2/k_B$ is the Curie–Weiss constant. When the temperature is higher than the Curie temperature, at the paraelectric phase ($P_s = 0$), we have

$$\varepsilon^{-1} = \frac{k_B T}{4\pi N\mu^2} - \frac{\beta}{4\pi} = \frac{1}{C}(T - T_C), \quad T \geqslant T_C. \tag{8}$$

Quantum mechanically, the zero-point energy should be included, so we have

$$\hbar\omega_0\left(\frac{1}{2} + \langle n\rangle_{T_C}\right) = k_B T_C^*, \tag{13}$$

where T_C^* refers to the classical Curie temperature in Ref. [19]

$$\langle n\rangle_{T_C} = [\exp(\hbar\omega_0/k_B T_C) - 1]^{-1} \tag{14}$$

is the thermal average of the number of energy quanta excited in each unit dipole at temperature $T = T_C$, above the zero-point energy level $E_0 = \frac{1}{2}\hbar\omega_0$, or Einstein energy [23]. From Eqs. (13) and (14), we can have

$$k_B T_C^* = \frac{\hbar\omega_0/2}{\tanh(\hbar\omega_0/2k_B T_C)} . \tag{15}$$

From the above we can see that the zero-point enrage rescales the real temperature to T^* or the quantum temperature scale [24,25]. Therefore, the quantum Curie temperature can be obtained from Eq. (3) as

$$k_B T_C = \frac{\hbar\omega_0/2}{\tanh^{-1}(\hbar\omega_0/2k_B T_C^*)} \tag{16}$$

or more explicitly in the form

$$k_B T_C = \frac{\hbar\omega_0/2}{\tanh^{-1}(\hbar\omega_0/2\beta N\mu^2)} \tag{16a}$$

Therefore, the dielectric constant at the paraelectric phase Eq. (8) can be rewritten as

$$\varepsilon^{-1} = \frac{\hbar\omega_0}{8\pi N\mu^2}\frac{1}{\tanh(\hbar\omega_0/2k_B T)} - \frac{\beta}{4\pi} . \tag{17}$$

At zero temperature,

$$\varepsilon^{-1}(T=0) = \frac{1}{4\pi N\mu^2}\left(\frac{1}{2}\hbar\omega_0 - \beta N\mu^2\right) , \tag{18}$$

which means that if the zero-point energy is large enough, there will be no phase transition to the ferroelectric phase. Eq. (17) is equivalent to the famous Barret formula [26], which is used to explain the low-temperature behavior of quantum ferroelectrics, such as $SrTiO_3$ and $CaTiO_3$ and their solid solutions.

3. Ising model in a transverse field

The starting point of the Ising model in a transverse field (IMTF) is the double-well potential in hydrogen bonds in ferroelectric such as the potassium dihydrogen phosphate. The spin-$\frac{1}{2}$ Hamiltonian IMTF in the general form [27] is

$$H = -\sum_i \Omega_i S_i^x - \frac{1}{2}\sum_{i,j} J_{ij} S_i^z S_j^z - 2\mu \sum_i E_i S_i^z , \tag{19}$$

where the first and third sums are over sites i and the second is over sites i and j. The first term involves the transverse field Ω_i, which is related to the tunneling frequency. S^x and S^z are spin-$\frac{1}{2}$ operators, S^x describes the tunneling effect., and S^z is related to the polarization. The third term accounts for the coupling to an external electric field E_i; μ is the dipole moment per Ising spin. With the usual notation, $\langle\cdots\rangle$ denotes a thermal average. The mean-field equation derived from Eq. (19) is

$$\langle \mathbf{S}_i \rangle = \frac{\mathbf{H}_i}{2|\mathbf{H}_i|}\tanh\left(\frac{|\mathbf{H}_i|}{2k_B T}\right) , \tag{20}$$

where

$$\mathbf{H}_i = \left(\Omega_i, 0, \sum_j J_{ij}\langle S_j^z\rangle + 2\mu E\right) \tag{21}$$

is the molecular field acting on spin at the ith site. The polarization P_i on site i is proportional to $\langle S_i^z \rangle$, namely $P_i = 2\mu \langle S_i^z \rangle$ and in the absence of an applied field, Eqs. (20) and (21) give for this average

$$\langle S_i^z \rangle = \frac{\sum_j J_{ij} \langle S_j^z \rangle + 2\mu E}{2[\Omega_i^2 + (J_{ij}\langle S_j^z \rangle + 2\mu E)^2]^{1/2}} \tanh \left\{ \frac{[\Omega_i^2 + (J_{ij}\langle S_j^z \rangle + 2\mu E)^2]^{1/2}}{2k_B T} \right\}, \quad (22)$$

In a uniform bulk material all Ω_i are equal and in the nearest-neighbor-exchange approximation, all exchange constants J_{ij} are equal to J. Therefore, the spontaneous polarization is related to $\langle S^z \rangle$ at zero external field in the form

$$1 = \frac{n_0 J}{2[\Omega^2 + (n_0 J \langle S^z \rangle)^2]^{1/2}} \tanh \left\{ \frac{[\Omega^2 + (n_0 J \langle S^z \rangle)^2]^{1/2}}{2k_B T} \right\}, \quad (23)$$

where n_0 is the number of nearest neighbors. At temperatures close to the Curie temperature, $\langle S^z \rangle$ approaches zero, and the mean-field value of the Curie temperature T_C is found by solving

$$k_B T_C = \frac{\Omega/2}{\tanh^{-1}(2\Omega/n_0 J)}. \quad (24)$$

The dielectric susceptibility can be found from Eqs. (22) and (23) as $\varepsilon = (1/4\pi)\partial P/\partial E$. For the ferroelectric phase

$$\varepsilon^{-1} = \frac{1}{16\pi N\mu^2} \frac{1}{f(\langle S^z \rangle)} - \frac{n_0 J}{16\pi N\mu^2}, \quad (25)$$

where the function $f(\langle S^z \rangle)$ is introduced to simplify the above equation, and has the form

$$f(\langle S^z \rangle) = \frac{1}{[\Omega^2 + (n_0 J \langle S^z \rangle)^2]} \left[\frac{\Omega^2}{n_0 J} + \frac{(1 - 4\langle S^z \rangle^2)(n_0 J)^2 - 4\Omega^2}{4k_B T} \langle S^z \rangle^2 \right] \quad (26)$$

and for paraelectric phase the dielectric constant becomes

$$\varepsilon^{-1} = \frac{2\Omega}{16\pi N\mu^2} \frac{1}{\tanh(\Omega/2k_B T)} - \frac{n_0 J}{16\pi N\mu^2}, \quad (27)$$

4. Discussion and comparison

First, we compare the expressions of the Curie temperatures obtained by both the methods. If we ignore the tunneling term (i.e., the transverse field) in the IMTF, the expression of Curie temperature Eq. (24) reduces to

$$k_B T_C = n_0 J/4, \quad (28)$$

By comparison with Eq. (4) obtained in the effective field approach, we can have the following relation for the parameters in the two models:

$$\beta N\mu^2 = n_0 J/4. \quad (29)$$

Therefore, β correspond the dipole–dipole interaction J in the IMTF. If the zero-point energy is included in the effective field approach, the expression to determine the Curie

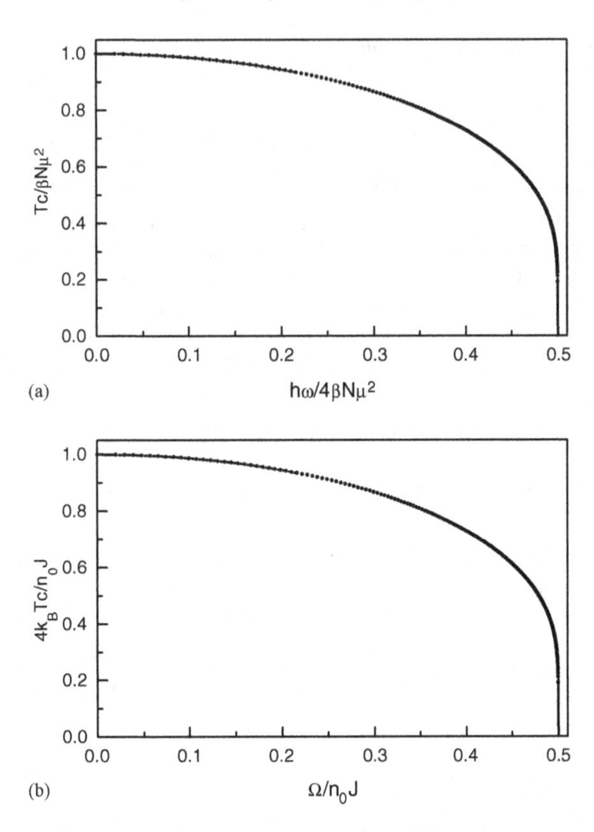

Fig. 1. Transition temperature as a function of zero-point energy (a) and as a function of tunneling frequency (b).

temperature is Eqs. (15) or (16). By comparison with Eq. (24) from IMTF, we can see that the tunneling frequency is equivalent to the zero-point energy in the effective field theory, i.e.,

$$\hbar\omega_0 = \Omega\,. \tag{30}$$

Eqs. (28) and (29) can be further confirmed by the dielectric constant at paraelectric phase (see Eq. (17) from effective field approach and Eq. (27) from IMTF.

In Fig. 1, we present the Curie temperature from the two models for comparison at different zero-point energies or tunneling frequencies. Fig. 1a is for effective field approach, and Fig. 1b for IMTF. From the figure we easily see that the Curie temperature are exactly the same, even the critical behavior near the critical value of zero-point energy or critical frequency, marked by arrows in the figures.

However, the two theoretical approaches do not always have exactly the same corresponding expressions. An example is the polarization at low temperature. From effective

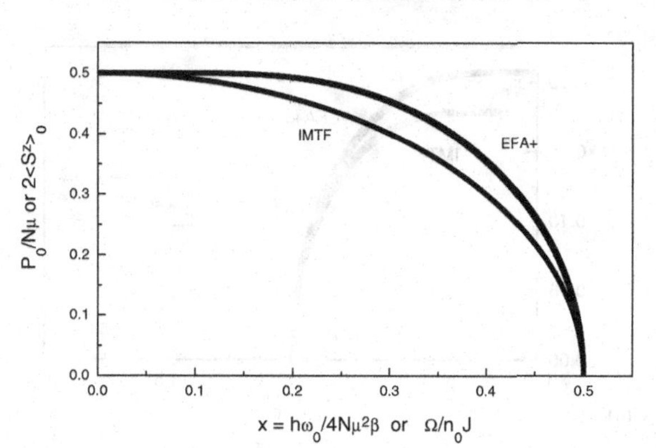

Fig. 2. The ground state polarization from Ising model in transverse field and effective field approach with zero-point energy.

field theory, as temperature approaches zero, $\langle n \rangle$ approaches zero; however, the energy does not tend to zero but to $\frac{1}{2}\eta\omega_0$, then the low-temperature spontaneous polarization is given by

$$P = N\mu \tanh\left(\frac{2\beta\mu P}{\hbar\omega}\right), \tag{31}$$

which is not the classical limit $N\mu$. It means that at zero temperature, there are still fluctuations, which prevent all dipole moments from aligning completely in the same direction. From the above we can get the critical value for the zero-point energy as

$$\hbar\omega_c = 2N\beta\mu^2 . \tag{32}$$

From Eq. (23) of IMTF at temperatures close to zero temperature we have a spin average given by

$$\langle S^z \rangle_0 = \tfrac{1}{2}\sqrt{1-(2\Omega/n_0 J)^2}\,, \tag{33}$$

which it is not equal to $\frac{1}{2}$ the classical limit. It means that not all spins are pointing in the same direction. Also, the critical value obtained from the above expression is

$$2\Omega_c = n_0 J . \tag{34}$$

From Eqs. (32) and (34), we can see that the critical values are the same if we recall expressions (28) and (29). The physics meaning behind the above two equations are the same. The zero-temperature polarization at different zero-point energies or tunneling is shown in Fig. 2 for comparison. The difference vanishes when there is no zero-point energy or tunneling frequency, and also near the critical value of the zero-point energy. The larger difference appears around the middle of the critical value.

In order to give a comparison in the whole temperature range, the temperature dependence of polarization and dielectric constant are shown in Fig. 2. To enhance the difference, we choose $\hbar\omega/4N\mu^{2\beta} = \Omega/n_0 J = 0.35$ in both diagrams. The difference in saturation polarization is largest at zero temperature (Fig. 3a) and as the temperature

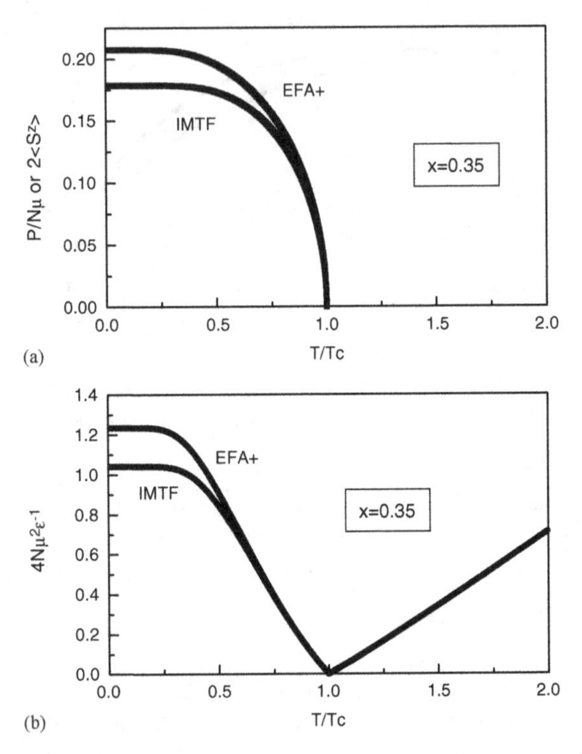

Fig. 3. The temperature dependence of polarization (a) and dielectric constant (b) at $\hbar\omega/4N\mu^2\beta=\Omega/n_0J=0.35$.

increases, the difference becomes smaller. The behavior of the polarization near the critical temperature is identical for both models. For the dielectric constants, as shown in Fig. 3b, the overall difference seems to be a little smaller than that of polarization in Fig. 3a, and there is no difference at the paraelectric phase.

5. Conclusion

From the comparison, we can see that when the zero-point energy is included in the effective field approach, the corresponding equations become the same or very similar to the equations from Ising model in a transverse field under mean-field approximation. The effective field approach can be used to described phase transitions in ferroelectrics with classical behavior as well as with quantum effects. The tunneling frequency in the Ising model in a transverse field is then analogous to the zero-point energy. The results from the two models can be said to be nearly equivalent, except the quantitative differences at low temperatures, when the zero-point energy or the tunneling frequency are away from the critical value.

Acknowledgements

Financial supports from DGICyT through Grant No. PB96-0037 is gratefully acknowledged. C.L. Wang is financially supported by China Scholarship Council to visit Universidad Autonoma de Madrid.

References

[1] P.G. Gennes, Solid State Commun. 1 (1963) 132.
[2] C.L. Wang, Z.K. Qin, D.L. Lin, Phys. Rev. B 40 (1989) 680.
[3] M.G. Cottam, D.R. Tilley, B. Zeks, J. Phys. C 17 (1984) 1793.
[4] I. Tamura, E.F. Sarmento, T. Kaneyoshi, J. Phys. C 17 (1984) 3207.
[5] C.L. Wang, W.L. Zhong, P.L. Zhang, J. Phys.: Condens. Matter 3 (1992) 4743.
[6] H.K. Sy, J. Phys.: Condens. Matter 5 (1993) 1213.
[7] X.Z. Wang, Y. Zhao, Physica A 193 (1993) 133.
[8] E.F. Sarmento, J.W. Tucker, J. Magn. Magn. Mater. 118 (1993) 133.
[9] C.L. Wang, S.R.P. Smith, D.R. Tilley, J. Phys.: Condens. Matter 6 (1994) 9633.
[10] C.L. Wang, S.R.P. Smith, J. Phys.: Condens. Matter 8 (1996) 3075.
[11] C.L. Wang, S.R.P. Smith, J. Kor. Phys. Soc. 32 (1998) S382.
[12] Y. Xin, C.L. Wang, W.L. Zhong, P.L. Zhang, Solid State Commun. 110 (1999) 265.
[13] C.L. Wang, Y. Xin, X.S. Wang, W.L. Zhong, Phys. Rev. B 62 (2000) 11423.
[14] J.A. Gonzalo, Effective Field Approach to Phase Transitions and Some Applications to Ferroelectrics, World Scientific Press, Singapore, 1991.
[15] B. Noheda, G. Lifante, J.A. Gonzalo, Ferroelectrics 15 (1993) 109.
[16] J.A. Gonzalo, R. Ramirez, G. Lifante, M. Koralewski, Ferroelectrics 15 (1993) 9.
[17] J.R. Fernandez del Castillo, J.A. Gonzalo, J. Przeslawski, Physica B 292 (2000) 23.
[18] J.A. Gonzalo, Phys. Rev. B 39 (1989) 12297.
[19] J.A. Gonzalo, Ferroelectrics 168 (1995) 1.
[20] C. Arago, M.I. Marques, J.A. Gonzalo, Phys. Rev. B 62 (2000) 8561.
[21] C. Arago, J.A. Gonzalo, J. Phys.: Condens. Matter 12 (2000) 3737.
[22] C. Arago, J.A. Gonzalo, Ferroelectrics Lett. 27 (2000) 83.
[23] E.K.H. Salje, B. Wruk, S. Marais, Ferroelectrics 124 (1991) 185.
[24] J. Dec, W. Kleemann, Solid State Commun. 106 (1998) 695.
[25] W. Kleemann, J. Dec, B. Westwanski, Phys. Rev. B 58 (1989) 8985.
[26] J.H. Barrett, Phys. Rev. 86 (1952) 118.
[27] R. Blinc, B. Zeks, Soft Modes in Ferroelectrics and Antiferroelectrics, North-Holland, Amsterdam, 1974.

Journal of Advanced Dielectrics
Vol. 2, No. 2 (2012) 1241007 (9 pages)

DOI: 10.1142/S2010135X1241007X

CHARACTERISTIC TEMPERATURES OF FIRST-ORDER FERROELECTRIC PHASE TRANSITION: EFFECTIVE FIELD APPROACH

C. L. WANG*
School of Physics, State Key Laboratory of Crystal Materials
Shandong University, Jinan 250100, P. R. China
wangcl@sdu.edu.cn

C. ARAGÓ[†] and M. I. MARQUÉS[‡]
Departamento de Física de Materiales
C-IV, Universidad Autónoma de Madrid
28049 Madrid, Spain
[†]*carmen.arago@uam.es*
[‡]*manuel.marques@uam.es*

Received 19 February 2012
Published 15 June 2012

The explicit expression of Helmholtz free energy has been obtained from the equation of state from effective field approach. From the Helmholtz free energy, four characteristic temperatures describing a first-order ferroelectric phase transitions have been determined. The physical meaning of coefficients in Landau-type free energy has been revealed by comparison with the expanding Helmholtz function. Temperature dependence of polarization under different bias, and hysteresis loops at different temperatures are presented and discussed. These results provide the basic understandings of the static properties of first-order ferroelectric phase transitions.

Keywords: Ferroelectricity; phase transition; effective field approach.

1. Introduction

The effective field approach for ferroelectricity, which is analog to the Weiss theory for ferromagnetism, is a useful procedure to describe phase transition properties in ferroelectrics and other systems such as alloys and superconductors.[1] The generation of the effective field expression to take into account in addition to dipolar, quadrupolar, octupolar, and higher order multipolar terms leads to a more general equation of state that describes better the dependence of polarization on temperature and external electric field. The cases of pure and deuterated members of the TGS family have been examined.[2] A changeover from continuous to discontinuous phase transition, and therefore a tricritical point has been obtained. Consideration of quadrupolar and octupolar contributions to the effective field improves the theoretical fit of polarization versus temperature for ferroelectric tris-sarcosine calcium chloride.[3] The thermal hysteresis accompanying discontinuous ferroelectric transition can be estimated if one takes into account

*Corresponding author.

quadrupolar contributions to the effective field. The calculated values are in fair agreement with the observed thermal hysteresis in several ferroelectrics belonging to different families.[4] The inclusion of quadrupolar and higher order multipolar terms into the effective field in uniaxial ferroelectrics has been shown to describe well the jump in specific heat at the Curie temperature of the triglycine sulfate family crystals.[5] A general equation of state using the effective field approach has been obtained to describe evolution of phase transition in ferroelectric triglycine selenate.[6] By including the zero point energy, a simple quantum effective field approach has been developed to describe the composition dependence of the Curie temperature in some mixed ferroelectrics systems.[7,8] Field induced phase transition properties have been studied using this quantum effective field approach,[9] a phase diagram has been obtained.

From thermodynamic theory, or Landau-type theory as is often called, we know that there exits four characteristic temperatures for a first-order ferroelectric phase transition.[10] They are the Curie–Weiss temperature T_0, which is the lowest temperature that the paraelectric phase can reach in cooling process. Ferroelectric limit temperature T_1 is the highest temperature that the ferroelectric phase can be maintained in heating process. Limit temperature of field induced phase transition T_2 is the highest temperature that ferroelectricity can be induced. The Curie temperature T_c is the balance temperature between the ferroelectric phase and the paraelectric phase, i.e. at Curie temperature T_c, free energy of ferroelectric phase is equal to that of paraelectric phase. When temperature is slightly higher than T_c, ferroelectric phase can still exist, but in a meta-stable state. When temperature is slightly lower than T_c, paraelectric phase can also exist in a meta-stable state.

In our last paper, characteristic temperatures T_0, T_1 and T_2, have been obtained from master equation of effective field approach.[9] However, the Curie temperature T_c, which is an important characteristic temperature from theoretical point of view, have not been determined as the free energy is needed. In this work, explicit expression of Helmholtz free energy from the master equation of effective field approach is obtained. A comparison between Landau-type free energy and the free energy of this work is made. Stability of ferroelectric phase is discussed and hysteresis loops at typical temperatures are presented.

2. Equation of State from Effective Field Approach

After the effective field approach (see Ref. 1) we consider that each individual dipole μ experiences a field that is due, not only to the applied external electric field E, but to all other dipoles of the sample. This mean field is expressed as a series of odd powers of polarization P that correspond successively to dipolar, quadrupolar, octupolar, etc. contributions.

$$E_{\text{eff}} = E + \beta P + \gamma P^3 + \delta P^5. \tag{1}$$

The coefficients β, γ and δ are expected to depend on the geometry of the lattice and on the spatial charge distribution within the unit cell. The energies associated with the two possible orientations of a given dipole are, therefore,

$$w_\pm = \pm E_{\text{eff}} \cdot \mu, \tag{2}$$

where μ is the elementary dipole moment. The partition function is the sum of the only two Boltzmann factors with w_+ or w_-, and the number of the dipoles pointing in the direction favored and opposed by the effective field is given respectively by

$$\begin{aligned} N_+ &= \frac{N}{Z}\exp\left(\frac{E_{\text{eff}}\cdot\mu}{k_BT}\right),\\ N_- &= \frac{N}{Z}\exp\left(-\frac{E_{\text{eff}}\cdot\mu}{k_BT}\right), \end{aligned} \tag{3}$$

where N is the total number of dipole per unit volume, and Z is the partition function.

$$N = N_+ + N_-, \tag{4}$$

$$Z = \exp\left(\frac{E_{\text{eff}}\cdot\mu}{k_BT}\right) + \exp\left(-\frac{E_{\text{eff}}\cdot\mu}{k_BT}\right). \tag{5}$$

The polarization is then given by

$$P = (N_+ - N_-)\mu \tag{6}$$

or

$$\begin{aligned} P &= N\mu\tanh\left(\frac{E_{\text{eff}}\mu}{k_BT}\right)\\ &= N\mu\tanh\left(\frac{E+\beta P+\gamma P^3+\delta P^5}{k_BT}\mu\right). \end{aligned} \tag{7}$$

Equation (7) can be rewritten in a more explicit form as

$$E = \frac{k_BT}{\mu}\tanh^{-1}\left(\frac{P}{N\mu}\right) - \beta P - \gamma P^3 - \delta P^5$$

or

$$E=\frac{k_BT}{2\mu}\ln\left(\frac{N\mu+P}{N\mu-P}\right)-\beta P-\gamma P^3-\delta P^5. \quad (8)$$

From above equation, static properties, such as the temperature dependence of polarization, hysteresis loop and susceptibilities, etc. can be obtained. As temperature T approaches transition temperature from below at $E=0$, $P=Ps$ approaches zero, one can easily get the transition temperature as

$$T_0=\frac{\beta N\mu^2}{k_B}. \quad (9)$$

Since the Curie temperature Tc is the same as Curie–Weiss temperature T_0 for a second-order phase transition, we denote this temperature as T_0 in Eq. (9). For a first-order phase transition, T_0 stands for the Curie–Weiss temperature in the Curie–Weiss law of dielectric susceptibility at paraelectric phase.

3. Helmholtz Free Energy

Helmholtz free energy can be obtained directly from the equation of state as the same way as in Ref. 11,

$$\begin{aligned}\Delta F=F-F_0&=\int_0^P E\cdot dP\\ &=\int_0^P\left\{\frac{k_BT}{2\mu}\cdot\ln\left(\frac{N\mu+P}{N\mu-P}\right)\right.\\ &\quad\left.-\beta P-\gamma P^3-\delta P^5\right\}\cdot dP \quad (10)\end{aligned}$$

or

$$\begin{aligned}\Delta F=&\frac{N\cdot k_BT}{2}\cdot\left\{\frac{P}{N\mu}\cdot\ln\left(\frac{N\mu+P}{N\mu-P}\right)\right.\\ &\left.+\ln\left(1-\left(\frac{P}{N\mu}\right)^2\right)\right\}-\frac{1}{2}\beta P^2\\ &-\frac{1}{4}\gamma P^4-\frac{1}{6}\delta P^6. \quad (11)\end{aligned}$$

Following the definition of the Helmholtz free energy

$$F=U-T\cdot S, \quad (12)$$

where U is the internal energy and S is the entropy,

$$S=k_B\ln\left(\frac{N!}{[N_+!N_-!]}\right). \quad (13)$$

The maximum entropy S_0 occurs at paraelectric phase, and since $N_+=N_-=N/2$,

$$S_0=k_B\ln\left(\frac{N!}{[(N/2)!]^2}\right). \quad (14a)$$

Defining the dimensionless maximum entropy per particle as,

$$s_0=\frac{S_0}{k_BN}=\frac{1}{N}\ln\left(\frac{N!}{[(N/2)!]^2}\right). \quad (14b)$$

From above equation, the maximum entropy s_0 versus different system size N is shown in Fig. 1. The entropy s_0 increases with systems size N, and approaches $\ln(2)$ as the system size approached infinite, as indicated by red arrow in Fig. 1. This is the same result that we find in Ref. 1 as, $S_0=k_B\ln(2^N)=Nk_B\ln(2)$ and it means that when the system size is finite, the entropy is lower than this maximum value.

Once we work out the maximum entropy, or the entropy at paraelectric phase, entropy difference between the ferroelectric phase and the paraelectric phase can be expressed as,

$$\Delta S=S-S_0=k_B\ln\left(\frac{[(N/2)!]^2}{N_+!N_-!}\right). \quad (15)$$

Using Stirling approximation, i.e.

$$\ln(N!)\approx N\ln(N)-N.$$

The entropy difference in Eq. (15) can be written as

$$\begin{aligned}\Delta S=&k_B[N\ln(N/2)-N-N_+\ln(N_+)\\ &+N_+-N_-\ln(N_-)+N_-]. \quad (16)\end{aligned}$$

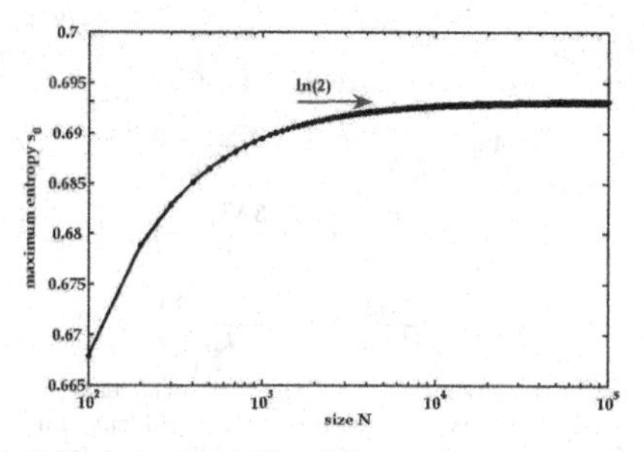

Fig. 1. (Color online) Dimensionless maximum entropy per site s_0 for different system sizes.

From Eqs. (4) and (6), we can have,

$$N_+ = \frac{N + P/\mu}{2}, \quad N_- = \frac{N - P/\mu}{2}. \tag{17}$$

Then Eq. (16) becomes

$$\Delta S = -\frac{Nk_B}{2}\left[\frac{P}{N\mu}\ln\left(\frac{N\mu + P}{N\mu - P}\right) + \ln\left(1 - \left(\frac{P}{N\mu}\right)^2\right)\right]. \tag{18}$$

The above expression is the first term of the free energy Eq. (11), that is, the entropy contribution to the free energy. From Eqs. (11) and (18), we can easily obtain the difference of internal energy at ferroelectric and paraelectric phase as,

$$\Delta U = -\frac{1}{2}\beta P^2 - \frac{1}{4}\gamma P^4 - \frac{1}{6}\delta P^6. \tag{19}$$

To see the temperature dependence of the coefficients in Landau free energy, we expand free energy Eq. (11) in power term of polarization as,

$$\begin{aligned}\Delta F = &\frac{Nk_B}{2}\left(T - \frac{N\mu^2\beta}{k_B}\right)\left(\frac{P}{N\mu}\right)^2 \\ &+ \frac{Nk_B}{12}\left(T - \frac{3N^3\mu^4\gamma}{k_B}\right)\left(\frac{P}{N\mu}\right)^4 \\ &+ \frac{Nk_B}{30}\left(T - \frac{5N^5\mu^6\delta}{k_B}\right)\left(\frac{P}{N\mu}\right)^6 \\ &+ \cdots .\end{aligned} \tag{20}$$

Usually Landau-type free energy is expressed as[12]

$$\Delta F = \frac{1}{2}A_{20}(T - T_0)P^2 + \frac{1}{4}A_4P^4 + \frac{1}{6}A_6P^6 + \cdots . \tag{21}$$

Hence we can have the coefficients of free energy as

$$\begin{aligned}A_{20} &= \frac{k_B}{N\mu^2}, T_0 = \frac{N\mu^2\beta}{k_B} \\ A_4 &= \frac{k_B}{3N^3\mu^4}\left(T - \frac{3N^3\mu^4\gamma}{k_B}\right) \\ A_6 &= \frac{k_B}{5N^5\mu^6}\left(T - \frac{5N^5\mu^6\delta}{k_B}\right)\end{aligned} \tag{22}$$

Here T_0 is the same as in Eq. (9). The coefficients A_2, A_4, A_6 are usually temperature dependent, but simple examples of first- and second-order ferroelectric phase transitions can be described assuming A_4 and A_6 independent of temperature in conventional Landau theory (see Ref. 10, pp. 71–72). From Eq. (22), we can see that these coefficients depend linearly on temperature dependent, and they correspond to quadrupolar and octupolar terms in the expression of effective field. As free energy in Landau theory is expanded around the transition temperature, we replace temperature T in A_4 and A_6 by Curie–Weiss temperature T_0, i.e.

$$\begin{aligned}A_4 &= \frac{1}{3(N\mu)^2}[\beta - 3(N\mu)^2\gamma] \\ A_6 &= \frac{1}{5(N\mu)^4}[\beta - 5(N\mu)^4\delta]\end{aligned} \tag{23}$$

As we know from Landau theory that to describe a first-order ferroelectric phase transition, A_4 has to be negative, i.e.

$$\beta - 3(N\mu)^2\gamma < 0 \tag{24}$$

or

$$\gamma > \frac{1}{3}\frac{\beta}{(N\mu)^2}. \tag{25}$$

This result agrees with previous works relating the tri-critical point behavior in phase transitions with quadrupole.[6]

4. Analysis of Helmholtz Free Energy

In order to simplify the algebra work, we define the following dimensionless parameters,

$$\begin{aligned}p &= \frac{P}{N\mu}, \quad \theta = \frac{T}{T_0} = \frac{k_BT}{N\mu^2\beta}, \\ g &= \frac{\gamma}{\beta}(N\mu)^2, \quad e = \frac{E}{N\mu\beta}\end{aligned} \tag{26}$$

Then the Helmholtz free energy Eq. (11) can be rewritten as, by omitting p^6 term for simplicity of numerical calculation in the following.

$$\Delta F = (N\mu)^2\beta\left\{\frac{\theta}{2}\cdot\left[p\cdot\ln\left(\frac{1+p}{1-p}\right) + \ln\left(1-p^2\right)\right] - \frac{1}{2}p^2 - \frac{1}{4}gp^4\right\} \tag{27}$$

and the dimensionless Helmholtz free energy can be defined as

$$\Delta f = \frac{\Delta F}{(N\mu)^2\beta}, \tag{28}$$

i.e.

$$f = \frac{\theta}{2} \cdot \left[p \cdot \ln\left(\frac{1+p}{1-p}\right) + \ln(1-p^2) \right] - \frac{1}{2}p^2 - \frac{1}{4}gp^4, \quad (29)$$

where Δ is neglected for simplicity in the above expression. The dimensionless internal energy and entropy are

$$u = -\frac{1}{2}p^2 - \frac{1}{4}gp^4, \quad (30a)$$

$$s = -\frac{1}{2} \cdot \left[p \cdot \ln\left(\frac{1+p}{1-p}\right) + \ln(1-p^2) \right]. \quad (30b)$$

Therefore the constant volume specific heat c_v is,

$$c_v = \frac{du}{d\theta} = -(p + gp^3)\frac{dp}{d\theta}$$
$$c_v = \theta\frac{ds}{d\theta} = -\frac{\theta}{2} \cdot \ln\left(\frac{1+p}{1-p}\right)\frac{dp}{d\theta}. \quad (31)$$

The dimensionless equation of state from Eq. (8) becomes

$$e = \theta \cdot \tanh^{-1}(p) - p - gp^3 \quad (32a)$$

or

$$e = \frac{\theta}{2} \cdot \ln\left(\frac{1+p}{1-p}\right) - p - gp^3. \quad (32b)$$

In the absence of external electric field, we can easily see that the two expressions of specific heat in Eq. (31) are equivalent. First equation in Eq. (31) has been used to explain the specific heat jump in TGSe.[5]

From Eq. (29), dimensionless Helmholtz free energy profile at different temperatures for $g = 2/3$ is shown in Fig. 2, and obviously it corresponds to a first-order phase transition, with $g > 1/3$. The dark solid lines are the free energy at the characteristic temperatures. The red dashed lines are the free energy at half way between the adjacent characteristic temperatures. We can see that at the Curie–Weiss temperature θ_0 there is only one minimum of free energy at finite polarization, which is the spontaneous polarization. This means that ferroelectric phase is stable below the Curie–Weiss temperature θ_0, since the free energy of ferroelectric phase is lower than that of paraelectric phase. The maximum of free energy is at $p = 0$, which means

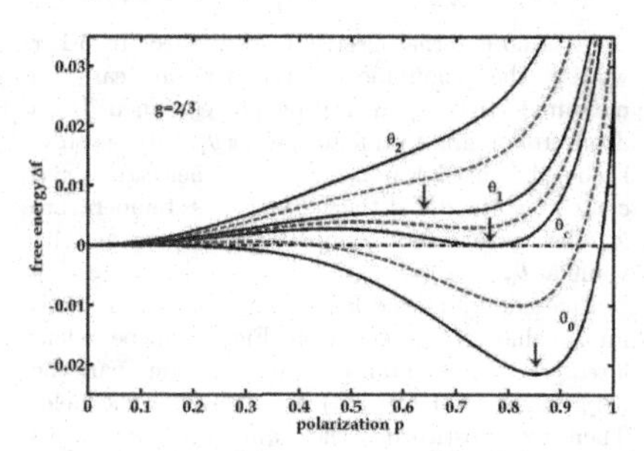

Fig. 2. (Color online) Dimensionless Helmholtz free energy profile at different temperatures.

paraelectric phase is unstable state. At Curie temperature θ_c, free energy of ferroelectric phase is the same as that of the paraelectric phase. This means that either ferroelectric or paraelectric phase can exist at Curie temperature. While the temperature is between the Curie–Weiss temperature θ_0 and Curie temperature θ_c, ferroelectric phase is still in stable state, but there appears another free energy minimum at $p = 0$, see the lowest red dashed curve. That means that the existence of paralectric phase is possible, but it is in a meta-stable state since the free energy of paraelectric phase is higher than that of ferroelectric phase. At temperature θ_1, free energy minimum appears only at $p = 0$, which suggests that paraelectric phase becomes a stable state. However, there is an inflexion point in the free energy as marked by the blue arrow. This means that no spontaneous polarization exists when temperature is higher than θ_1, i.e. this temperature is the up limit temperature for existence of ferroelectric phase. When temperature is between θ_c and θ_1, ferroelectric phase becomes a meta-stable state, while paraelectric phase is in stable state, since the free energy of ferroelectric phase is higher than that of paraelectric phase. The highest characteristic temperature θ_2 shown in Fig. 2 is called limit temperature for induced ferroelectric phase. That is because when the temperature is between θ_1 and θ_2, ferroelectric phase can be induced by applying an external electric field. However, this behavior cannot be understood from the free energy curve shown in Fig. 2, we will discuss this behavior in the following section. From Fig. 2, we can also see that once the system is in ferroelectric phase at low temperature, ferroelectric phase can be kept up to temperature θ_1 if there is no external electric field applied. In other words, the spontaneous polarization can be measured up to θ_1 in heating process, then drops down from a finite value to zero at θ_1. If the system is cooled from high temperature, paraelectric phase can be maintained at the Curie–Weiss temperature θ_0, then polarization jumps from zero to a finite value at θ_0.

Free energy at the ferroelectric phase, or minimum values of free energy in Fig. 2 can be calculated. First, spontaneous polarization can be determined from Eq. (32) with zero electric field. Then by substituting the value of spontaneous polarization in the free energy expression Eq. (29), we can have the free energy of ferroelectric state. The temperature dependent of Helmholtz free energy at ferroelectric state for different parameter g is shown in Fig. 3. The dashed lines are for second-order phase transitions, i.e. the cases of $g < 1/3$; and the solid lines are for first-order phase transitions, i.e. $g > 1/3$. It can be easily seen that there is only one transition temperature for second-order phase transition, i.e. Curie temperature or Curie–Weiss temperature denoted as θ_0 in Fig. 3. In this case, the free energy increases with increase of temperature, and reaches zero continuously at the transitions temperature θ_0. However, for first-order phase transition, the free energy curve split into two parts. The upper part corresponds to the maximum value of the free energy in Fig. 2. This part is not physically reasonable. The lower part curve is the solution we expected. We can see that the free energy increases with increase of temperature, and reaches zero at the Curie temperature θ_c. When the temperature is higher than the Curie temperature θ_c, free energy becomes positive, or the free energy of ferroelectric phase is higher than that of paraelectric phase. This means that ferroelectric phase becomes less stable state. Because it is still a minimum in free energy curve as can be seen in Fig. 2, ferroelectric phase is in a meta-stable state between temperature θ_c and θ_1. When temperature increases to θ_1, free energy terminates with a maximum value as shown in Fig. 3. No solution for free energy when temperature is higher than θ_1 means no ferroelectric phase can exist when temperature is higher than this temperature. We can also see from Fig. 3 that the thermal hysteresis i.e. temperature difference between θ_0 and θ_1 increases with increase of parameter g. This implies that the larger the quadrupolar moment or higher order multipolar moment, the stronger the first-order phase transition features.

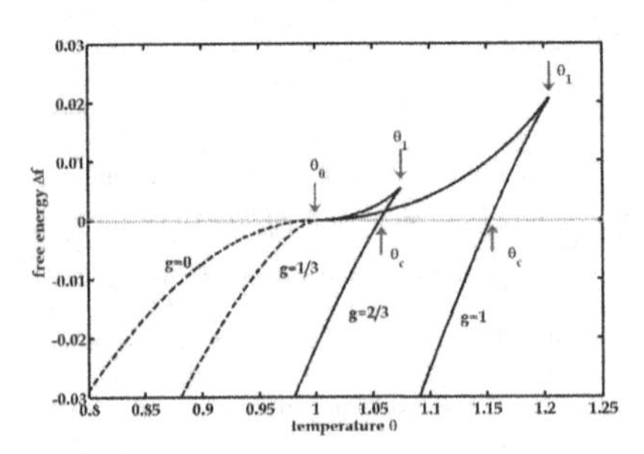

Fig. 3. Helmholtz free energy of ferroelectric phase for different parameter g.

5. Characteristic Temperatures and Phase Diagram

The four characteristic temperatures as defined above can be determined from Helmholtz free energy Eq. (29). The Curie temperatures θ_c is the temperature at which the free energy is the same both at ferroelectric and paraelectric phase. From Eqs. (29) and (30), the Curie temperature θ_c can be determined from,

$$f|_{\theta_c} = \frac{\theta_c}{2}\cdot\left[p_c\cdot\ln\left(\frac{1+p_c}{1-p_c}\right)+\ln(1-p_c^2)\right] - \frac{1}{2}p_c^2 - \frac{1}{4}gp_c^4 = 0,$$

$$\frac{d(f)}{dp}|_{\theta_c} = \theta_c\cdot\ln\left(\frac{1+p_c}{1-p_c}\right) - p_c - gp_c^3 = 0, \tag{33}$$

where p_c is the polarization at θ_c.

At the ferroelectric phase limit temperature θ_1, the free energy minimum becomes an inflexion point, so we can determine θ_1 from the following equations.

$$\left.\frac{d(f)}{dp}\right|_{\theta_1} = \theta_1\cdot\ln\left(\frac{1+p_1}{1-p_1}\right) - p_1 - gp_1^3 = 0,$$
$$\left.\frac{d^2(f)}{dp^2}\right|_{\theta_1} = \frac{\theta_1}{1-p_1^2} - 1 - 3gp_1^2 = 0, \tag{34}$$

where p_1 is the polarization at θ_1.

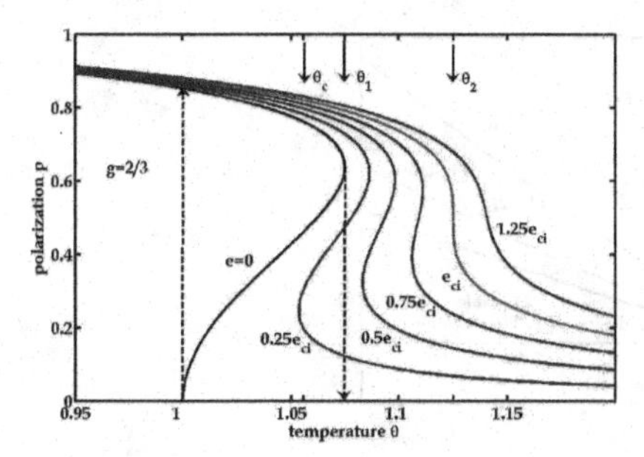

Fig. 4. (Color online) Temperature dependence of polarization under different bias.

In order to determine θ_2, we plot in Fig. 4 the temperature dependence of polarization under different bias fields obtained from Eq. (32), for $g = 2/3$. It can be seen that there is a critical field e_{ci} (see the red curve) under which thermal hysteresis disappears.

For a deeper understanding of the meaning of these parameters and the field induced ferroelectric behavior, hysteresis loops at temperatures θ_1 and θ_2, are shown in Fig. 5. The red arrows in Fig. 5 indicate the changing direction of the polarization with electric field, and the dashed-line parts are the nonphysical solutions. At θ_1, the hysteresis loop is series pinched, or nearly a double loop. Then a double hysteresis loop should to be expected at temperatures between θ_1 and θ_2. But at θ_2, the hysteresis loop disappears, in spite of $P-E$ relation still shows a strong nonlinearity.

From analysis of Figs. 4 and 5, we can see that the limit temperature θ_2 for induced field ferroelectric behavior, the critical field e_{ci} and the corresponding polarization p_2, can be determined from the equation system,

$$\left.\frac{d^2(f)}{dp^2}\right|_{\theta_2} = \frac{de}{dp} = \frac{\theta_2}{1-p_2^2} - 1 - 3gp_2^2 = 0$$
$$\left.\frac{d^3(f)}{dp^3}\right|_{\theta_2} = \frac{d^2e}{dp^2} = \frac{2p_2}{(1-p_2^2)^2}\theta_2 - 6gp_2 = 0 \tag{35}$$

whose solutions are

$$p_2^2 = \frac{3g-1}{6g}, \quad \theta_2 = \frac{(3g+1)^2}{12g} \tag{36}$$

and the critical field is

$$e_{ci} = \frac{\theta_2}{2} \cdot \ln\left(\frac{1+p_2}{1-p_2}\right) - p_2 - gp_2^3. \tag{37}$$

Once the four characteristic temperatures are determined, the phase diagram is straightforwardly obtained, as shown in Fig. 6. There are six different regions in this phase diagram. The Curie–Weiss temperature θ_0 is plotted as a horizontal line. When temperature is lower than the Curie temperature θ_c, only ferroelectric phase exists. Above θ_2, only paraelectric phase exists. When $g < 1/3$, second-order phase transition will occur. When $g > 1/3$, first-order phase transition is possible. For first-order phase transition and in the case of heating process, ferroelectric phase changes from a stable state below θ_c to meta-stable state between θ_c and θ_1, then into an inducible state between θ_1 and θ_2. In the case of cooling process, paraelectric phase first becomes a meta-stable at the Curie temperature θ_c, then completely disappears at the Curie–Weiss temperature θ_0.

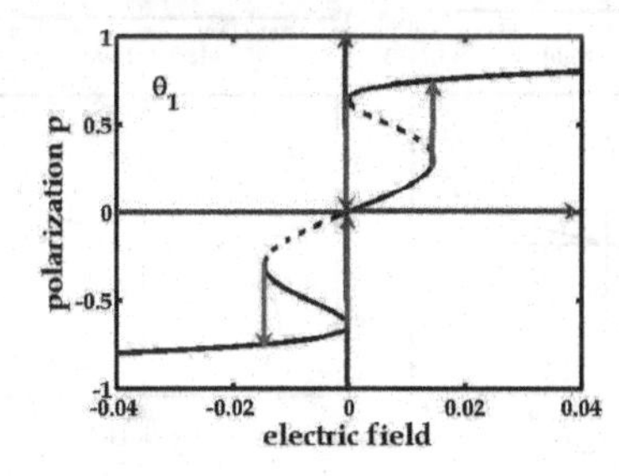

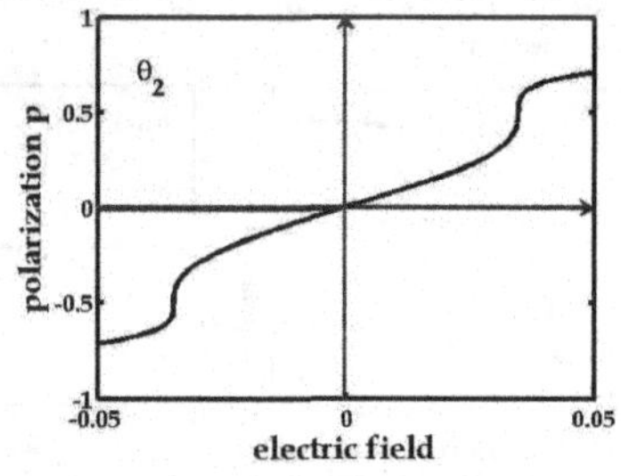

Fig. 5. (Color online) Hysteresis loops at temperature θ_1 and θ_2.

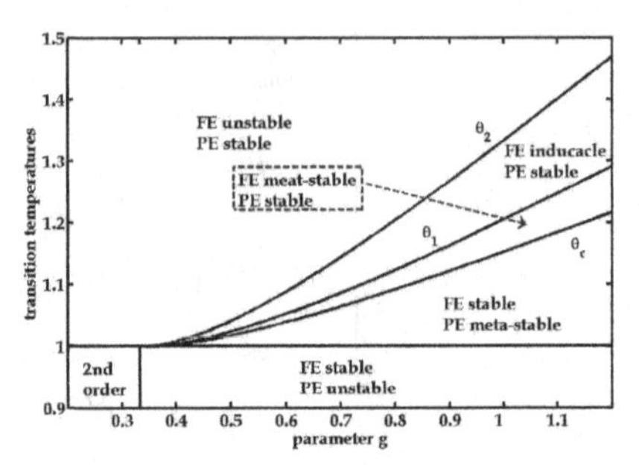

Fig. 6. Phase diagram.

6. Static Hysteresis Loops at Different Temperatures

The static hysteresis loops at different temperatures are shown in Fig. 7 for $g = 2/3$. The polarization changes with electric field following the red arrows. We can see that below Curie temperature θ_c, there are normal hysteresis loops, as shown in Figs. 7(a) and 7(b). When the temperature is between θ_c and θ_1, pinched loop can be observed, see Fig. 7(c). Between θ_1 and θ_2, a double hysteresis loop is found as expected. At θ_2 and higher temperatures no hysteresis loop appears but the strong nonlinear polarization–electric field relation stands. All these types of hysteresis loops have been observed in $BaTiO_3$ crystals.[12]

From Figs. 4 and 7, we can conclude that physical properties do not show any anomaly at the Curie temperature θ_c for first-order phase transitions. In other words, the Curie temperature is not accessible from experimental measurements. Curie–Weiss temperature can be determined experimentally from Curie–Weiss law of dielectric constant at paraelectric phase. The high limit temperature of

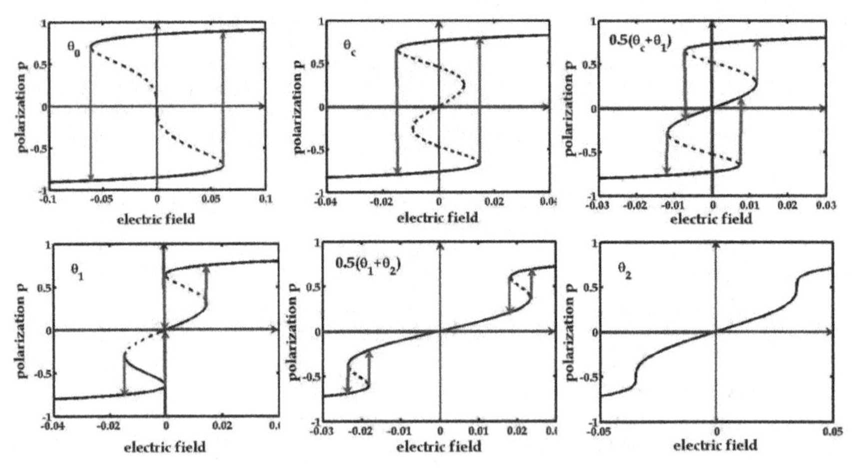

Fig. 7. (Color online) Hysteresis loop at different temperatures for $g = 2/3$.

ferroelectric phase θ_1 can be observed in anomalous of dielectric constant upon heating process. Another characteristic but often ignored temperature is that corresponding to the limit field induced ferroelectric phase, θ_2. Accurate determination of this temperature experimentally is not such straightforward.

7. Summary

Explicit expressions of Helmholtz free energy and entropy have been obtained from effective field approach. The characteristic temperatures describing phase transition are determined for both second- and first-order phase transitions. The physical meaning of the free energy expansion coefficients, and the stability of the ferroelectric phase around the transition temperature are discussed. A phase diagram, as well as the corresponding hysteresis loops for different critical temperatures, is displayed.

Acknowledgments

The authors gratefully acknowledge to MICINN of Spain for supporting the project FIS2008-00715 and National Natural Science Foundation of China under Grant Nos. 51172129, 51172128 and 51102153.

References

1. J. A. Gonzalo, *Effective Field Approach to Phase Transitions and Some Applications to Ferroelectrics* (World Scientific Press, Singapore, 2006).
2. R. Ramirez, M. F. Lapena and J. A. Gonzalo, *Ferroelectrics* **124**, 1 (1991).
3. G. Lifante, J. A. Gonzalo and W. Windsch, *Ferroelectrics* **135**, 277 (1992).
4. J. A. Gonzalo, R. Ramírez, G. Lifante and M. Koralewski, *Ferroelectrics* **15**, 9 (1993).
5. B. Noheda, G. Lifante and J. A. Gonzalo, *Ferroelectrics* **15**, 109 (1993).
6. C. Aragó and J. A. Gonzalo, *J. Phys.: Codens. Matter* **12**, 3737 (2000).
7. J. A. Gonzalo, *Phys. Rev.* **B39**, 12297 (1989).
8. C. Aragó, M. I. Marqués and J. A. Gonzalo, *Phys. Rev.* **B62**, 8561 (2000).
9. C. L. Wang, J. C. Li, M. L. Zhao, J. L. Zhang, W. L. Zhong, C. Arago, M. I. Marques and J. A. Gonzalo, *Physica A* **387**, 115 (2002).
10. M. E. Lines and A. M. Glass, *Principles and Applications of Ferroelectrics and Related Materials* (Clarendon Press, Oxford, 1977).
11. T. Mitsui, I. Tatszaki and E. Nakamura, *An Introduction to the Physics of Ferroelectrics*, Chapter 4 (Gordon and Breach Publisher, 1976).
12. W. J. Merz, *Phys. Rev.* **91**, 513 (1953).

Philosophical Magazine, 2015
Vol. 95, No. 7, 683–690, http://dx.doi.org/10.1080/14786435.2014.1000419 

Dielectric anomalous response of water at 60 °C

Juan C. del Valle[a], Enrique Camarillo[b], Laura Martinez Maestro[c]*, Julio A. Gonzalo[c,d], Carmen Aragó[c], Manuel Marqués[c], Daniel Jaque[c], Ginés Lifante[c], José García Solé[c], Karla Santacruz-Gómez[e], Roberto C. Carrillo-Torres[e] and Francisco Jaque[f]

[a]*Departamento de Química Física Aplicada, Universidad Autónoma de Madrid, Madrid, Spain;* [b]*Institut of Physics, UNAM, México DF, Mexico;* [c]*Departamento de Física de Materiales, Instituto Nicolás Cabrera, Madrid, Spain;* [d]*Poly-Technical School of the San Pablo-CEU University, Madrid, Spain;* [e]*Departamento de Física, Universidad de Sonora, Hermosillo, México;* [f]*Departamento de Biología, Universidad Autónoma de Madrid, Madrid, Spain*

(*Received 20 October 2014; accepted 10 December 2014*)

Recently, the paraelectric response of water was investigated in the range 0–100 °C. It showed an almost perfect Curie–Weiss behaviour up to 60 °C, but a slight change in slope of $1/\varepsilon_d$ versus T at 60 °C was overlooked. In this work, we report optical extinction measurements on metallic (gold and silver) nanoparticles dispersed in water, annealed at various temperatures in the range from 20 to 90 °C. An anomalous response at 60 °C is clearly detectable, which we associate to a subtle structural transformation in the water molecules at that temperature. This water anomaly is also manifested by means of a blue shift in the longitudinal surface plasmon resonance of the metallic nanoparticles for the solutions annealed at temperatures higher than about 60 °C. A reanalysis of $1/\varepsilon_d$ (T) for water in the whole temperature range leads us to conclude that the water molecule undergoes a subtle transformation from a low temperature (0–60 °C) configuration with a dipole moment $\mu_1 = 2.18$ D (close to the molecular dipole moment of ice) to a high temperature (60–100 °C) configuration with $\mu_2 = 1.87$ D (identical to the molecular dipole moment in water vapour).

Keywords: water; 60 °C

1. Introduction

Water is unquestionably the most abundant and the most important fluid on Earth. Water's unique physical and chemical properties make possible life on our planet. The possibility and actual blossoming of life to advanced forms of life in our planet, and in no other planet of our solar system, is a direct consequence of the abundance and unique properties of water. In fact, water plays a central role in determining the structure and functions of proteins and nucleic acids as well as those of membranes and ligand binding in plants and animals.

Water exists at atmospheric pressure in the range 0–100 °C. Its maximum density is only a few degrees (3.94 °C) above the freezing point, which results in top-down freezing. Water penetrates into rocks and, as a result of its anomalously high surface

*Corresponding author. Email: lm.maestro@uam.es

tension, water expands upon freezing and eventually turns rocks into fertile soil. As a solvent, water has also very special properties and plays an important role in a number of crucial biochemical processes.

On the other hand, water has also very special dielectric properties [1,2]. As a liquid, it exhibits a complex molecular configuration resembling first that of the solid [3–5] and at higher temperatures, as shown in this work, that of the vapour, with notable consequences. Recent works have shown that water has an almost perfect para-electric behaviour between 0 and 60 °C. However, a detailed review of previous works [6–10] shows that a clear change in the temperature dependence of the dielectric constant, ε_d, takes place at about 60 °C at ambient pressure (0.1 MPa). This anomaly, which has been repeatedly detected but surprising ignored for a long time, could play an important role in systems where liquid water is an essential component. One of these important systems consists in nanoparticles dispersed in water, which have important implications in nanomedicine. Indeed different types of nanoparticles are presently used for biomedical applications, and for this purpose, their hydrophilic character (dispersibility in water) becomes essential. Recently Fernández et al. [10] have studied the extinction spectra of gold nanorods (GNRs) of different dimensions synthesized in water at different temperatures, in the range 25–80 °C. These authors have reported an abrupt change in the wavelength peak position associated to the longitudinal surface plasmon resonance (LSPR), λ_{LSPR}, at 60–70 °C. The temperature coincidence between the change in the water dielectric constant anomaly and the anomalous shift in λ_{LSPR} as a function of synthesis temperature suggest a connection between both effects.

In order to get further experimental evidence on this connection, we have investigated the thermal behaviour of the optical extinction bands ascribed to LSPR of different metallic nanoparticles dispersed in water. Thus, we have systematically measured the optical extinction properties of both water-dispersed GNRs and silver nanoprisms (SNPs) after isochronal thermal treatments crossing the expected critical temperature, at around 60 °C. Since both the type metal of the nanoparticle employed as well and the composition of the surfactant coating were different, the presence of coincident isochronal annealing behaviour could be reasonably ascribed to the thermal change of water physical properties.

The experiments reported in this work confirm the existence of an anomaly in the water dielectric constant near 60 °C. This anomaly must be taken into account in all nanoparticles/water solutions, and so, it could have important implications in different biomedical applications that make use of these biocompatible solutions.

2. Results and discussion

2.1. *Dielectric properties of water*

Figure 1 presents the dependence of the water dielectric constant with temperature in the range 0–90 °C at atmospheric pressure. Here, it is important to mention that the data of this figure correspond to different original sources (and so reported at different times) and consequently have not been subject of any mathematical manipulation or smoothing [6–9]. Nevertheless, it is remarkable the excellent agreement amongst the different experimental sources. From these data, a bilinear relationship, not previously considered by any of the authors, is clearly observed; the crossing point of both linear fits being at

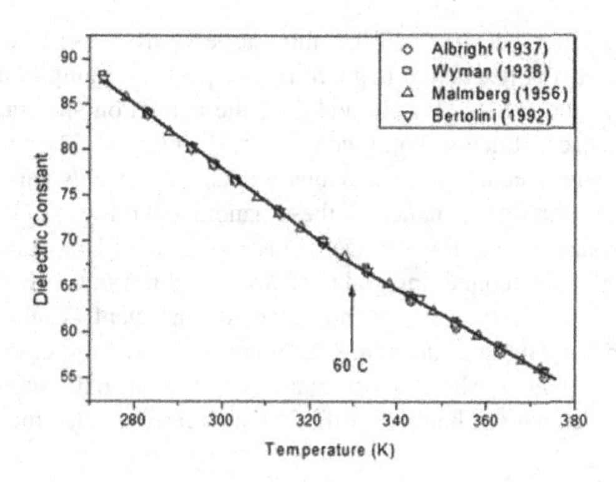

Figure 1. Temperature dependence of the water dielectric constant, ε_d, at 0.1 MPA in the temperature range 25–90 °C. Data obtained from Refs. [6–9].

≈60 °C. This bilinear trend of ε_d with T is also observed when the pressure increases, but the crossing point for 10 MPA appears around 105 °C [11].

Applying the well-known Curie–Weiss law, $\varepsilon_d^{-1} = (T - T_c)/C$, to the data of Figure 1 in the whole temperature interval from 0 to 100 °C, reasonable good statistics are found and the Curie constant, C, and the Curie temperature, T_c, could be estimated as $C = 1.2514 \times 10^4$ K and $T_c = 122.6$ K. On one hand, these values differ significantly from those obtained in Ref. [2] for the fit from 0 to 60 °C ($C = (1.5 \pm 0.3) \times 10^4$ K and $T_c = 90 \pm 14$ K). On the other hand, if we make a bilinear fit considering two separate temperature intervals, 0–60 and 60–100 °C, we get much better fits of the data with the Curie–Weiss law fitted by $T_{C1} = 100.1$ K and $C_1 = 1.432 \times 10^4$ K for the first interval and by $T_{C2} = 157.6$ K and $C_2 = 1.065 \times 10^4$ K for the second. This means that there is a subtle but significant change in the microscopic factors entering T_C and C as shown in Table 1, which gives the best bilinear fits to the paraelectric Curie–Weiss behaviour of liquid water for different relevant temperature intervals.

2.2. *Metallic nanoparticles: synthesis and optical properties*

The GNRs used in this work were supplied by Nanorods LLC. In the synthesis of these GNRs, Cetyltrimethylammonium bromide (CTAB) was used as surfactant to enable their dispersion in water. Two different colloidal solutions of GNRs were used, being differentiated by the GNR dimensions. GNR dimensions were, on average, 7 × 28 nm^2 (small GNRs, GNRs-s) and 16 × 77 nm^2 (large GNRs, GNRs-l). The corresponding concentrations in water were very similar: 1.3 × 10^{11} GNRs/cm^3 (small) and 1.0 × 10^{11} GNRs/cm^3 (large). The extinction spectra of both GNRs mainly consist in a broadband

Table 1. Values of T_C and C for the different temperature intervals.

Temperature interval (°C)	State	T_C (K)	C (K)
0–60 °C	Liquid (1)	100.1	1.432 × 10^4
60–100 °C	Liquid (2)	157.6	1.065 × 10^4

centred at 808 nm (small GNRs) or at 1030 nm (large GNRs), ascribed to the LSPR. The spectral position of this resonance is given by the particular longitudinal dimension and the aspect ratio of the GNRs [12]. In addition, the extinction spectra also display a narrower and less-intense extinction band, centred at 515 nm and 527 nm for small and large GNRs, respectively. According to previous works, these bands can be assigned to the transverse surface plasmon resonance of these nanometallic rods [12].

SNPs were synthesized at the Physics Department of Sonora University (Hermosillo, Mexico) in aqueous solution containing sodium citrate, sodium borohydride and polyvinyl pyrolidone. In this case, CTAD was not used as surfactant. The average size of SNPs was ≈20 nm before thermal annealing treatments. As for the case of GNRs, the extinction spectrum of SNPs shows two bands centred at 630 and 330 nm, also assigned to the optical excitation bands of different surface resonance modes.

2.3. *Thermal treatments*

The thermal stability of both GNRs and SNPs was studied in the 20–90 °C temperature range. The aqueous solutions containing the metallic nanoparticles were introduced into a 1 mm-thick quartz cell. Isochronal annealing treatments were then performed by introducing the quartz cells into a temperature-controlled bath. After each thermal treatment, the extinction spectrum was recorded at room temperature (23 °C). This implies that the observed effects must not be attributed to the measurement temperature but to morphological effects that the annealing treatments produce on the metallic nanoparticles. The extinction spectra were measured after each isochronal annealing temperature, and the spectral position of the extinction peak associated to the LSPR, λ_{LSPR}, was determined. For small GNRs, Figure 2(a), it is observed that the λ_{LSPR} value remains almost constant up to an annealing temperature of ≈67 °C. Treatments at higher temperatures lead to a remarkable blue shift, $\Delta\lambda$, of the extinction band, reaching a value of $\Delta\lambda \approx 40$ nm at 87 °C. TEM measurements revealed that the thermal annealing at this temperature modifies the rod shape, leading to a reduction in the GNRs aspect ratio close to ≈10%. As previously reported, [11] this change in the geometrical form of the GNRs is in accordance with the observed blue shift in λ_{LSPR}. For large GNRs, Figure 2(b), a slight and monotonous blue shift of the LSPR extinction band is observed for temperatures in the 20–70 °C range. As for small GNRs, annealing temperatures higher than 70 °C produce an abrupt shift towards shorter wavelengths, being the maximum induced shift as large as $\Delta\lambda \approx 25$ nm at 87 °C. Finally, Figure 2(c) shows the annealing behaviour for the water solution containing SNPs. For this solution, λ_{LSPR} remains constant in the annealing temperature range 23–60 °C, and above this temperature range, the extinction peak position becomes strongly dependent on the annealing temperature, reaching a wavelength shift value of $\Delta\lambda \approx 25$ nm at 87 °C. The presence of this critical temperature at around 60 °C was also found in the previous works of Zijlstra and co-workers, [13] who observed that the optical properties of GNRs were strongly modified when the synthesis temperature was increased over 60 °C. Figure 3(a) shows the variation of λ_{LSPR} as a function of synthesis temperature for these GNRs (values obtained from Ref. [13]). An abrupt blue shift of $\Delta\lambda \approx 100$ nm can be clearly observed in λ_{LSPR}, from 60 °C to about 80 °C annealing temperatures.

Since the singularity observed appears to be independent on the nature of the metal (silver or gold), the shape of the nanoparticles, as well as on the components used in

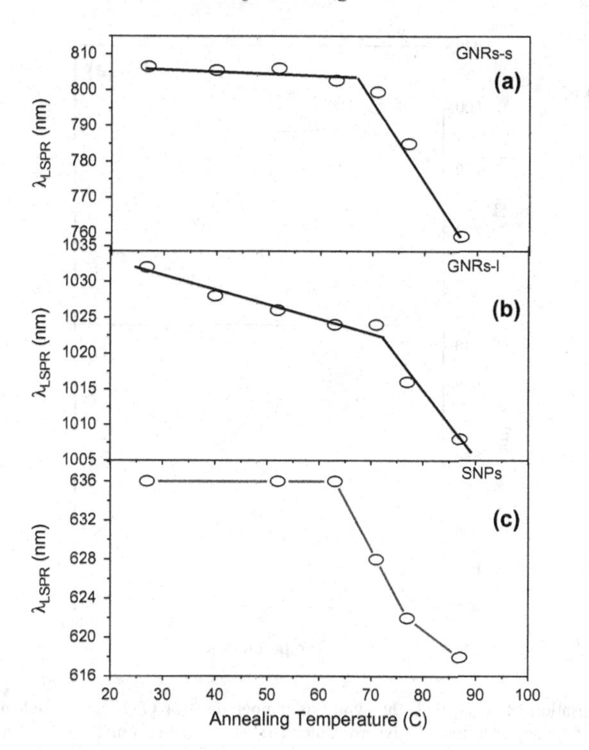

Figure 2. Variation of λ_{LSPR} with annealing temperature. (a) small GNRs; GNRs-s; (b) large GNRs; GNRs-L; (c) silver nanoprisms, SNPs. Lines are to guide the eyes.

the their synthesis (and so in their surface coatings), it is reasonable to think that this singularity is related to the anomaly found in the dielectric constant of water at around 60 °C. Indeed, the water dielectric constant is large and strongly temperature dependent, and therefore, it may cause inhomogeneity in size distribution of GNRs during nanoparticles synthesis. Thus, a reduction in the effective dielectric constant of the solvent (for instance, by adding glycerol in a glycerol/water ratio of 75% (v/v)) should lead to less temperature dependence of the extinction peak position. In this respect, Figure 3(b) shows the variation of λ_{LSPR} with temperature for GNRs dispersed in this mixed solvent, obtained from [14]. Once again, a blue shift of λ_{LSPR} takes place at around 60 °C. However in this case, due to the fact that the dielectric constant of the mixed solvent has been reduced, the shift, $\Delta\lambda \approx 14$ nm, is substantially reduced in respect to the case of sole water as solvent.

3. Discussion

In order to understand the anomaly displayed by the dielectric constant of water at about 60 °C, we have separately analysed its behaviour below and above this temperature.

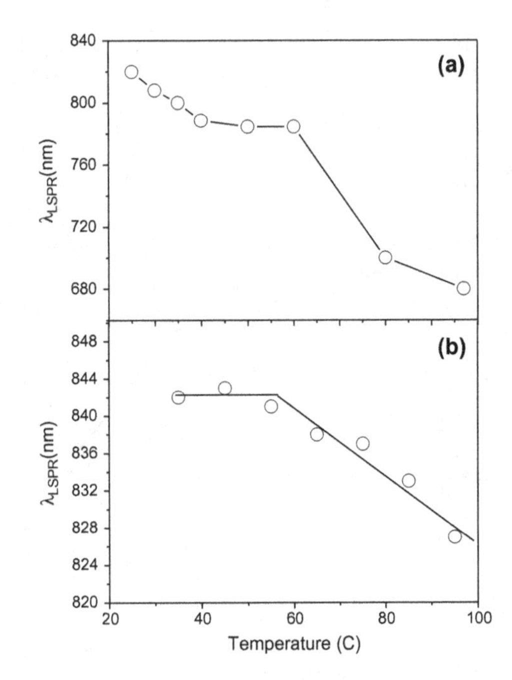

Figure 3. Variation of λ_{LSPR} with the synthesis temperature for GNRs dispersed in water (a) and dispersed in a binary solution of glycerol/water (b). Data taken from Refs. [13] and [14]. Lines are to guide the eyes.

In the interval 0–60 °C, the paraelectric behaviour of water is characterized by the Curie–Weis law, $(1/\varepsilon_d)=(1/C_1)(T-T_{C1})$, with $C_1 = 1.432 \times 10^4$ K and $T_{C1} = 100.1$ K. However, in the interval 60–100 °C, that behaviour is characterized by $(1/\varepsilon_d)=(1/C_2)(T-T_{C2})$, with $C_2 = 1.065 \times 10^4$ K and $T_2 = 157.6$ K.

The number of unit dipoles (water single molecules) per unit volume, N, changes very little in the whole interval 0–100 °C. Considering water density of 1 g/cm^3, this number is given by $N = 1/Mm_p = 3.32 \times 10^{22}$ dipoles/cm^3, where $M = 18$ is the molecular weight of water and $m_p = 1.67 \times 10^{-24}$ g is the proton mass. We can now estimate the dipolar moment in both temperature intervals:

$$\mu_1 = (k_B T_{C1}/\beta N)^{1/2} = \sim 2.17 \times 10^{18} \text{esu.cm} = 2.17\,\text{D} \quad (1)$$

and

$$\mu_2 = (k_B T_{C2}/\beta N)^{1/2} = 1.87 \times 10^{18} \text{esu.cm} = 1.87\,\text{D} \quad (2)$$

where k_B is the Boltzman constant and $\beta_1 = 4\pi T_{C1}/C_1 = 0.088$ and $\beta_2 = 4\pi T_{C2}/C_2 =$ 00.185 have been used [2]. The respective values for β, μ and $U = \beta N \mu^2$ (characteristic energy per dipole) are given in Table 2.

Table 2. Values of β, μ, and U in the different temperature intervals.

Temperature interval (°C)	State	β	μ (D)	U (eV)	Ref.
Below 0 °C	Ice	–	2.55	–	Johari [15]
0–60 °C	Liquid (1)	0.088	2.17	0.086	This work
60–100 °C	Liquid (2)	0.185	1.87	0.0155	This work
Above 100 °C	Vapour	–	1.86	–	Coulson [16]

Thus, we can see that in state (1) (below 60 °C), the water molecules have a relatively large dipole moment, $\mu_1 = 2.17$ D, relatively close to that in the ice state, while in state (2) (above 60 °C), the dipole moment is $\mu_2 = 1.87$ D, almost indistinguishable from that in the vapour state. The corresponding characteristic energy difference between states (1) and (2) is small.

$$U_2 - U_1 = k_B(T_{C2} - T_{c1}) = 0.0069 \text{ eV} \tag{3}$$

where $T_{C2} - T_{C1} = 57.1 \approx 60$ K.

4. Final conclusions and remarks

We have clearly demonstrated the existence of an anomaly in the dielectric constant of water at around 60 °C, which affects the optical extinction of metallic (gold and silver) nanoparticles dispersed in this solvent. The detected anomaly could be associated to a subtle change in the average effective dipole moment μ of water at about $T \approx 60$ °C, from $\mu_1 = 2.17$ D (which is close to the molecular dipole moment of ice) to $\mu_2 = 1.87$ D (almost identical to the molecular dipole moment of water vapour). This change in dipole moment must affect the effective charge density of the nanoparticle surface coating and therefore their thermal stability.

Further investigation on the water molecular dipole change at 60 °C, relative to room temperature, is under way. Nevertheless, it is important to remark that this water anomaly can have important implications in a number of properties of the nanoparticles used for biomedical applications, as the nanoparticle hydrophilic character can be strongly affected at this temperature.

Funding

This work has been supported by the Spanish Ministerio de Economía y Competitividad under the following projects, FIS-2012-36113 and MAT2013-47395-C4-1-R. Enrique Camarillo thanks to DGPA-UNAM for the financial support during his research stay at Madrid (Spain). Laura Martínez Maestro thanks the Secretaría de Estado de Investigación Desarrollo e Innovación of the Spanish Ministerio de Economía y Competitividad for a FPI grant.

References

[1] C. Reichardt and T. Welton, *Solvents and Solvent Effects in Organic Chemistry*, John Wiley & Sons, Weinheim, 2011.

[2] J. del Valle, C. Aragó, M. Marqués and J. Gonzalo, Ferroelectrics 466 (2014) p.166.

[3] F. Franks, *Water: A Matrix of Life*, Vol. 22, Royal Society of Chemistry, London, 2000.
[4] P. Ball, Chem. Rev. 108 (2008) p.74.
[5] G. Hura, J.M. Sorenson, R.M. Glaeser and T. Head-Gordon, J. Chem. Phys. 113 (2000) p.9140.
[6] P.S. Albright, J. Am. Chem. Soc. 59 (1937) p.2098.
[7] J. Wyman and E.N. Ingalls, J. Am. Chem. Soc. 60 (1938) p.1182.
[8] C. Malmberg and A. Maryott, J. Res. Nat. Bur. Stand. 56 (1956) p.2641.
[9] D. Bertolini, M. Cassettari and G. Salvetti, J. Chem. Phys. 76 (1982) p.3285.
[10] D.P. Fernández, Y. Mulev, A. Goodwin and J.L. Sengers, J. Phys. Chem. Ref. Data 24 (1995) p.33.
[11] M. Uematsu and E. Frank, J. Phys. Chem. Ref. Data 9 (1980) p.1291.
[12] P.K. Jain, K.S. Lee, I.H. El-Sayed and M.A. El-Sayed, J. Phys. Chem. B 110 (2006) p.7238.
[13] P. Zijlstra, C. Bullen, J.W.M. Chon and M. Gu, J. Phys. Chem. B 110 (2006) p.19315.
[14] A.S.A. Al-Sherbini, J. Nanomater. 2010 (2010) p.66.
[15] G. Johari, Philos. Mag. B 39 (1979) p.219.
[16] C. Coulson and D. Eisenberg, *Interaction of H_2O molecules in ice I. The dipole moment of an H_2O molecule in ice*, in *Proceedings of the Royal Society of London. Series A, Mathematical and Physical Sciences*, London, 1966, p.445.

PHYSICAL REVIEW B **74**, 184106 (2006)

Transition between the ferroelectric and relaxor states in $0.8Pb(Mg_{1/3}Nb_{2/3})O_3$-$0.2PbTiO_3$ ceramics

R. Jiménez,[1] B. Jiménez,[1] J. Carreaud,[2] J. M. Kiat,[2,3] B. Dkhil,[2] J. Holc,[4] M. Kosec,[4] and M. Algueró[1]
[1]*Instituto de Ciencia de Materiales de Madrid, CSIC, Cantoblanco, 28049 Madrid, Spain*
[2]*Laboratoire Structures, Propriétés et Modélisation des Solides, UMR8580-CNRS, École Centrale Paris, Grande Voie des Vignes, 92295 Châtenay-Malabry, France*
[3]*Laboratoire Léon Brillouin, CE Saclay, 91191 Gif-sur-Yvette Cedex, France*
[4]*Institute Jozef Stefan, Jamova 39, 1000 Ljubjlana, Slovenia*
(Received 11 July 2006; published 7 November 2006)

The phase transition between the ferroelectric and relaxor states for $0.8Pb(Mg_{1/3}Nb_{2/3})O_3$-$0.2PbTiO_3$ ceramics with high chemical homogeneity has been studied by measurements of the dielectric and elastic properties as a function of temperature and of thermal expansion. The room temperature ferroelectric phase structure has been studied in ceramics and powdered samples by Rietveld analysis of x-ray diffraction patterns and by ferroelectric hysteresis loops. Results indicate that the material has well defined, different transition and freezing temperatures, such as the transition is between a monoclinic Cm ferroelectric phase and the nonergodic glass state. The transition presents thermal hysteresis, not only in the transition temperature, but in the kinetics. This indicates that a quite sharp slowing down occurs in the temperature interval between 334 and 344 K: the transition temperatures on cooling and heating.

DOI: 10.1103/PhysRevB.74.184106 PACS number(s): 77.84.Dy, 68.35.Rh, 68.60.Bs

I. INTRODUCTION

A lot of attention is currently being paid to the $Pb(Mg_{1/3}Nb_{2/3})O_3$-$PbTiO_3$ (PMN-PT) relaxor-ferroelectric solid solution because of fundamental and applied issues. Phenomena such as the relaxor (R) state, the transformation between the relaxor and ferroelectric (F) states, and the morphotropic phase boundary (MPB) are not well understood yet. Selected compositions are already being used or are under consideration for a range of technologies, such as actuators and ultrasounds. PMN was one of the first relaxors to be reported,[1] and the material on which most of the work to understand the basis of the relaxor state has been done. PMN, as model relaxor, is characterized by a high, strongly dispersive electric polarizability. The dielectric constant shows a broad maximum with temperature of $\sim 20\,000\varepsilon_o$ at 265 K and 1 kHz, which position shifts toward higher temperatures with frequency within a range of 20 K (the so called Curie range) in the typical 100 Hz–1 MHz interval. Below the maximum, the permittivity also shows significant dispersion, permittivity decreasing when frequency increases.[2] No macroscopic phase transition occurs at the Curie range and the overall symmetry remains cubic down to 5 K, though diffuse scattering in x-ray and neutron diffraction has been associated with the presence of nanometer size polar regions within the cubic phase.[3] These polar nanoregions (PNRs) condense at the so called Burn's temperature, T_B, at $\sim$620 K, who first propose their existence from macroscopic measurements.[4] The temperature dependence of the diffuse scattering suggests that the volume fraction of PNRs increase as the temperature is decreased down to a temperature of $\sim$220 K, below which it maintains a constant value.[5] PNRs originate from site disorder and their dynamics are responsible for the relaxor characteristics, yet the actual mechanisms are still under debate.[6] Relaxors have been proposed to be dipolar glasslike systems.[7] In this model, the (interacting) PNRs present thermally activated polarization fluctuations above a freezing temperature, T_f, and evolve to a nonergodic glass state below T_f. Frustration of polar long range order results from competing interactions (random bonds). The freezing temperature is 217 K for PMN (Ref. 8) that is basically the temperature at which the volume fraction of PNRs, saturates. This model has been questioned, and random fields have been proposed to be the origin,[9] or to play a significant role,[10] in the relaxor state. Recently, neutron inelastic scattering experiments have shown the existence of a ferroelectric soft mode above T_B that becomes overdamped below this temperature, as a result of the condensation of PNRs.[11] This lowest energy transverse optic phonon recovers below 220 K, which seems to indicate that a well developed ferroelectric state, though short range, is established below that temperature.[12] The physical meaning of T_f, i.e., whether it is a freezing temperature or a transition temperature, is still unclear. A high enough electric field causes the development of a rhombohedral long range order below a temperature that is field dependent.[13] This ferroelectric phase undergoes a first order phase transition to the *R* state at 213 K during subsequent heating without field.[14] The addition of small amounts of PT shifts the Curie range so as it is at $\sim$313 K for 0.9 PMN–0.1PT.[15] On cooling, this compound presents a spontaneous *R*-*F* transition, also to a rhombohedral phase (F_R, $R3m$ space group), below room temperature (RT).[5] Further addition of PT causes the shift of the *R* state toward higher temperatures, and the F_R phase is stabilized at RT for 0.85 PMN–0.15 PT.[16] The F_R phase was thought to persist for 0.8 PMN–0.2 PT and 0.7 PMN–0.3 PT,[17] up to $\sim$0.65 PMN–0.35 PT, composition at which a morphotropic phase boundary (MPB) with a tetragonal phase (F_T, $4mm$) had been described earlier.[18] However, since the experimental discovery of a *Cm* monoclinic phase in $Pb(Zr,Ti)O_3$ (PZT) at the MPB,[19] several studies on the structure of MPB phases in relaxor-PT systems such as

PMN-PT, $Pb(Zn_{1/3}Nb_{2/3})O_3$-PT (PZN-PT) and $Pb(Sc_{1/2}Nb_{1/2})O_3$-PT (PSN-PT)[20–22] have reported the presence of monoclinic phases (with space groups *Cm* and *Pm*) around the MPB region. In PMN-PT, recent Rietveld analysis of powder XRD data have shown that two monoclinic phases (M_B and M_C with *Cm* and *Pm* space groups, respectively) exist between 0.73 PMN–0.27 PT and 0.65 PMN–0.35 PT, whereas a rhombohedral one was found for compositions with a PT content below 0.27 (Refs. 21 and 23). In contrast with these previous results, a very recent Rietveld study of powder neutron diffraction data for 0.75 PMN–0.25 PT has evidenced the growth of *Cm* monoclinic order (M_B type), from short range to long range, with decreasing temperature from 300 to 80 K.[24] This strong interest for PMN-PT with composition close to the MPB is motivated by the ultrahigh piezoelectricity and strain under the electric field of single crystals along the ⟨001⟩ pseudocubic direction and textured ceramics.[25,26] These materials are under consideration for the new generation of high sensitivity and high power piezoelectric devices.

Many aspects remain unclear in the structural evolution of these materials. A high temperature, ($T > 380$ K), *R* state has been shown to occur in 0.8 PMN–0.2 PT single crystals by neutron scattering experiments. The study also showed the persistence of PNRs up to 650 K, and the development of a rhombohedral distortion at 360 K,[27] in contradiction with more recent high *q*-resolution neutron scattering experiments on 0.8 PMN–0.2 PT single crystals that did not observe any rhombohedral distortion down to 50 K, though a significant broadening of the (220) Bragg peak was observed below 300 K.[28]

We have studied the temperature dependence of some macroscopic properties of this 0.8 PMN–0.2 PT composition on polycrystalline samples of high chemical homogeneity as an alternative means of studying the development of long range polar order. We reported preliminary results on the dielectric and elastic properties that clearly indicated the occurrence of a transformation between relaxor and ferroelectric states above room temperature.[29] This transition shows thermal hysteresis, not only in the temperature but also in the kinetics, which was discussed within the two stages model for the development of ferroelectric long range order in relaxor systems, recently proposed by Ye *et al.*[16] We present here a complete description of the structural, dielectric, and elastic properties and additional thermal expansion measurements that allow the phase transition to be described in detail and reveal aspects not reported before.

II. EXPERIMENTAL

0.8 PMN–0.2 PT ceramic samples were prepared from powders synthesized by mechanochemical activation of oxides, without any excess of PbO and MgO. This has been considered essential for properly controlling composition when addressing fundamental studies in the PMN-PT system.[21] Details can be found in Ref. 30. The technique provides nanometer-scale chemical homogeneity[31] and ceramics with high crystallographic quality.[30] Contamination from the WC:Co milling media was below 50 and 600 ppm of Co and W, respectively. Sintering was carried out at 1473 K for 1 h in a PbO atmosphere. A heating rate of 3 K min^{-1} was used. These conditions allowed ceramics with a grain size of ~4 μm and a low level of porosity (5–8%) to be obtained.[30] Room temperature phases in the ceramics were studied by x-ray diffraction and Rietveld analysis. High resolution x-ray diffraction measurements were carried out with a highly accurate two-axis diffractometer in a Bragg-Brentano geometry with Cu K_β wavelength issued from an 18 kW rotating anode generator. Structural refinements were accomplished with the XND program.[32] Structure of phases were also studied from powders obtained by gentle grinding of the ceramics with a pestle. A thermal treatment at 923 K for 1 h was carried out before structural characterization of both ceramics and powders for relaxing stresses.

Electrical characterization was carried out on ceramic discs on which Ag electrodes had been painted and sintered at 923 K. The dependence of the dielectric permittivity and losses on temperature was measured with a HP 4284A precision LCR meter. In addition to previously reported measurements that were dynamically accomplished at 1 K min^{-1}, static measurements were carried out. Temperature was varied in 2 K intervals between 298 and 475 K along a heating cooling cycle. Stabilization times longer than 30 min were used that provided a temperature stability better than 0.1 K. Forty-eight frequencies between 20 Hz and 1 MHz were scanned at each temperature. Ferroelectric hysteresis loops were also obtained by current integration. Voltage sine waves of 0.1 Hz frequency and amplitudes up to 1000 V were applied by the combination of a synthesizer/function generator (HP 3325B) and a bipolar operational power supply/amplifier (KEPCO BOP 1000 M). Both the current integrator and the software for loop acquisition and analysis were developed at CSIC.

Bending ceramic bars of $12 \times 2 \times 0.35$ mm^3 dimensions were machined for mechanical characterization. The low frequency Young's modulus and mechanical losses were measured as a function of temperature by dynamical mechanical analysis in a three point bending configuration. A stress sine wave of 12 MPa amplitude, superimposed on a static stress of 15 MPa, was applied to the bars. Unlike previously reported measurements that were dynamically accomplished at 3 K min^{-1} at a single frequency of 9 Hz, slower measurements at 0.1 K min^{-1} were carried out at frequencies of 4 and 30 Hz. This technique has been shown to be very suitable for studying phase transitions[33,34] and the dynamics of domain walls in ferroelectrics.[35] Finally, thermal expansion measurements were carried out on the same ceramic bars at 5 K min^{-1} with a constant force applied of 20 mN.

III. RESULTS

We have used the methodology currently used in the Rietveld analysis of MPB compounds. In particular, as in our previous works in PMN-PT (Ref. 20) and PSN-PT (Ref. 22), we have tested many structural models including phase mixing. Agreement factors for the ceramic and powder samples are given in Table I for the more relevant models. Regarding the ceramic, the best Rwp (agreement factor of the fitting on

TABLE I. Agreement factors of the Rietveld analysis of XRD data for 0.8 PMN–0.2 PT ceramics and powders (obtained from the ceramics, i.e., same size).

Sample	Symmetry	R_{wp}	GoF	R_{Bragg}
Ceramic	$R3m$	6.97	1.44	5.22
	Cm	6.71	1.39	2.73
Powder	$R3m$	6.99	1.62	3.67
	Cm	6.94	1.61	3.03
	Pm	8.26	1.92	3.54
	$Bmm2$	6.85	1.47	4.60
	0.18 $R3m$+ +0.82Cm	6.73	1.56	3.32

the profile pattern) and Gof (ratio ideally equal to one between Rwp and R exp, the latter roughly being a measurement of the data statistic) agreement factors were obtained for a pure monoclinic Cm phase or a pure rhombohedral $R3m$ phase. However, a significantly lower R_{Bragg} (agreement factor on the integrated intensity of the Bragg peaks) was obtained in the case of the Cm phase. $R3m$ and Cm phase mixing was also tested but gave poorer R_{Bragg} factor. For the powder sample, the lowest R_{Bragg} was obtained with the pure Cm phase whereas a mixing of monoclinic Cm (82%) and rhombohedral $R3m$ (18%) phases gave slightly better Rwp and GoF. Therefore the structure of both PMN–20 PT powders and ceramics can be considered as monoclinic Cm, yet the presence of a small amount of rhombohedral phase cannot be completely excluded. No differences in the monoclinic cells (a=5.699 Å), b=5.693 Å, c=4.030 Å, and β =89.88°) between both types of samples could be observed, and values are in good agreement with those reported earlier by Singh and Pandey.[21] It must be noted that whereas the determined monoclinic unit cell is very close to a rhombohedral one, the structure (symmetry) is definitively different. For instance, the polarization in the case of the monoclinic phase is not constrained to lie along a particular axis (as it is in the rhombohedral symmetry) but can rather be along any axis within the monoclinic plane of symmetry. It is also important to emphasize that even if the monoclinic phase gives the best agreement, the thermal factor refined for the Pb atoms remains very large (4.3 $Å^2$), which indicates (static and/or dynamic) local disordered displacements. This disorder of the Pb-atoms is a well-known feature of lead-based relaxors and is known to play a key role in these systems.[5]

Results of dielectric permittivity as a function of temperature are given in Figs. 1 and 2. Permittivity at 15 frequencies (selected out of the 48 for clarity) measured in static conditions during successive heating and cooling are shown in Fig. 1(a) and 1(b), respectively. On heating, the permittivity presents non-negligible dispersion in the range between RT and 340 K, permittivity decreasing when the frequency is increased. The permittivity then sharply increases between 340 and 350 K in a manner typical of a ferroelectric to paraelectric phase transition. And finally, above 350 K, a clear change to a relaxor-type behavior is observed: dispersion increases and the permittivity presents a broad maximum with temperature that shifts toward higher temperatures with frequency, from 364.5 K at 20 Hz to 373 K at 1 MHz. In contrast, permittivity measured on cooling presents a relaxor behavior down to RT, and a sharp decrease is not observed at any temperature. The thermal hysteresis is better illustrated in the inset of Fig. 1(b) that shows the permittivity at 10 kHz along a heating cooling cycle. The hysteresis is evident in the 320–350 K interval. The temperature dependence of the reciprocal permittivity at 10 kHz along the same cycle is shown in Fig. 2(a). Neither on heating nor on cooling is a Curie-Weiss behavior observed above the temperature of the maximum permittivity of 368 K. On the contrary, the reciprocal permittivity does present a linear behavior below a temperature of 341 and 330 K for heating and cooling, respectively. The same results were obtained when permittivity was dynamically measured at 1 K min^{-1}. The dependence of the temperature of the maximum dielectric permittivity on frequency along with the fit to a Vogel-Fulcher relationship is shown in Fig. 2(b). The freezing temperature, T_f, activation energy, E_g, and characteristic frequency, f_o, obtained from the fit are given in Table II. In this case, data correspond to dynamic measurements at 1 K min^{-1} on cooling, but very

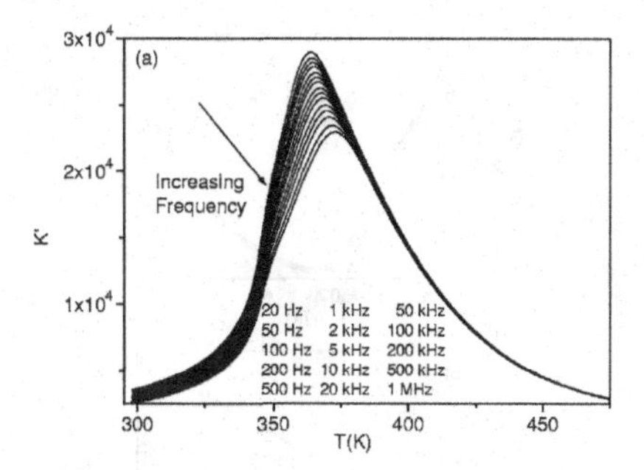

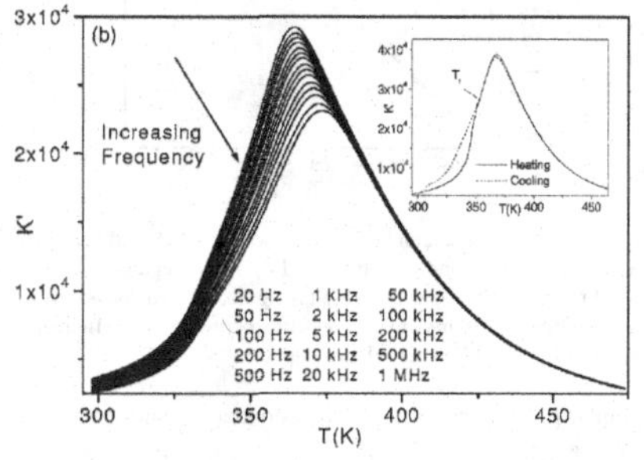

FIG. 1. Dielectric permittivity as a function of temperature at 15 frequencies measured in static conditions during successive (a) heating, and (b) cooling, for a 0.8 PMN–0.2 PT ceramic. The inset in (b) shows the permittivity at 10 kHz along a heating cooling cycle.

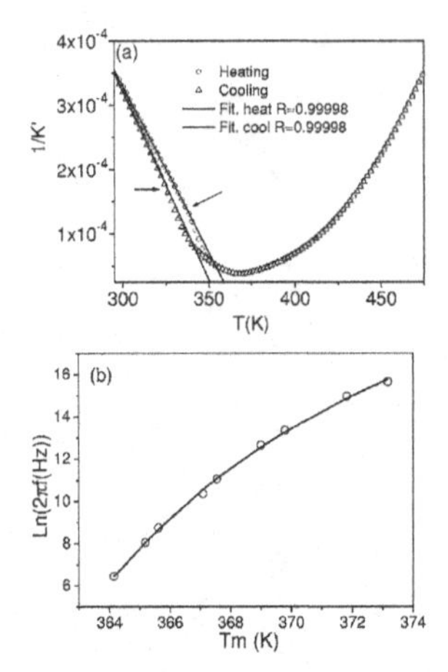

FIG. 2. (a) Temperature dependence of the reciprocal permittivity at 10 kHz along the same cycle of Fig. 1. The regression factors, R, of the linear fits are depicted in the figure, (b) dependence of the temperature of the maximum dielectric permittivity on frequency and fit to a Vogel-Fulcher relationship.

similar parameters were obtained on heating and from static measurements.

Results of the Young's modulus and mechanical losses as a function of temperature are given in Figs. 3 and 4. Differences between the two frequencies measured were not found. Young's modulus decreases with temperature between RT and 344 K, at which it presents a minimum, and then increases. The behavior between RT and 348 K is typical of a ferroelectric to paraelectric phase transition. 348 K is an inflection point, at which the derivative sharply decreases. There is a second inflection point at 358 K, at which the derivative increases again. This second feature was not observed in our previous measurements at 3 K min^{-1}, though it coincides with the temperature at which an amplitude dependence of the Young's modulus vanished.[29] The sharp minimum and inflection points at 344, 348, and 358 K observed during heating do not occur during cooling. Instead, a broad and asymmetric minimum is observed at 334 K. Mechanical

TABLE II. Parameters of the Vogel-Fulcher behavior for the relaxor state in the PMN-PT system.

	T_f (K)	E_a (meV)	F_o (Hz)
PMN[a]	217	79	10^{12}
0.9 PMN–0.1 PT[b]	291	41	1.03×10^{12}
0.8 PMN–0.2 PT[c]	350±2	30	2.5×10^{12}

[a]From Ref. 8.
[b]From Ref. 7.
[c]This work.

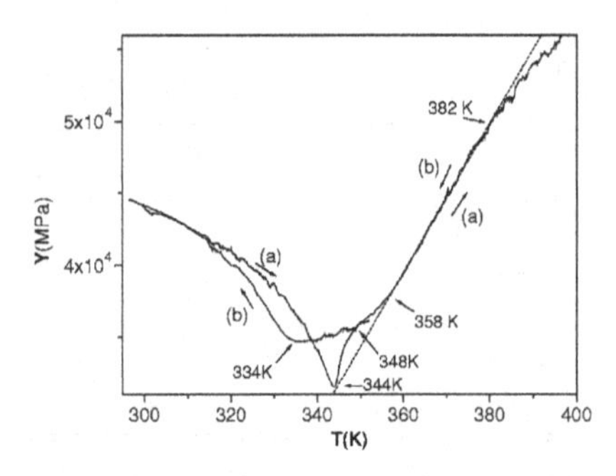

FIG. 3. Young's modulus as a fuction of temperature during successive (a) heating, and (b) cooling, for a 0.8 PMN–0.2 PT ceramic. Minima and inflection points are marked with arrows.

losses during heating present a well defined peak at 344 K, which position does not change with the heating rate [see Fig. 2(b)]. The sharp peak is not observed on cooling, but a broad maximum is found at 334 K.

Results of thermal expansion by dilatometry experiments during a heating cooling cycle are shown in Fig. 5. There is a small but sharp contraction at 345 K on heating. Above this temperature, the ceramic maintains a constant dimension until 358 K, from which it starts expanding. There is a clear increase of the derivative at 382 K. Above this point, the derivative keeps slowly increasing and a constant value is reached at 430 K. On cooling, the ceramic contracts with a decreasing derivative until 334 K, where a slight expansion occurs. No inflection points are observed above this temperature.

Finally, a typical room temperature ferroelectric hysteresis loop for these ceramics is shown in Fig. 6. A loop for a 0.7 PMN–0.3 PT ceramic with an analogous microstructure

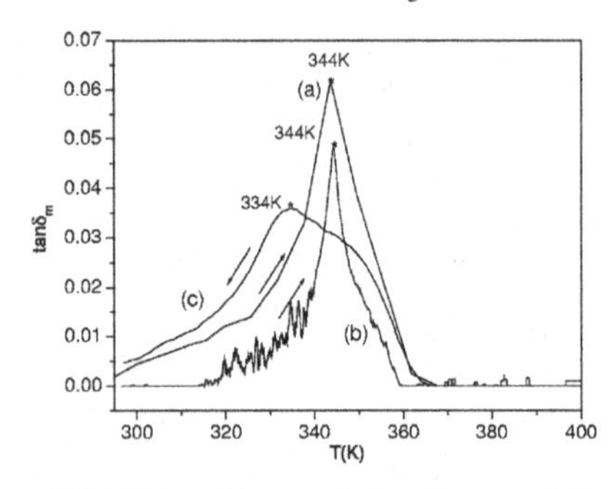

FIG. 4. Mechanical losses as a function of temperature during heating with (a) 2 K min^{-1}, (b) 0.1 K min^{-1} heating rates, and (c) cooling with a −0.1 K min^{-1} rate, for a 0.8 PMN–0.2 PT ceramic. Maxima are marked with stars.

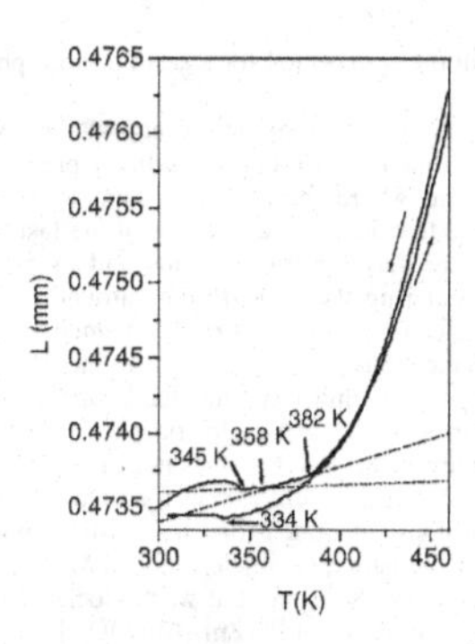

FIG. 5. Dilatometry results during successive heating and cooling for a 0.8 PMN–0.2 PT ceramic. Sharp changes in size and inflection points are marked with arrows.

(grain size and porosity), also processed from powders synthesized by mechanochemical activation, is included for comparison.

It is worth summing up the results on macroscopic properties before the discussion. On heating, the sharp increase of the dielectric permittivity occurs in the temperature range at which the Young's modulus has the minimum and where there is a maximum of mechanical losses (344 K). A contraction occurs at this temperature. The onset of the relaxor type behavior at 350 K occurs at the same temperature than the first inflection point of the Young's modulus. This temperature also corresponds to the freezing temperature T_f obtained from the Vogel-Fulcher relationship. The second inflection point of the Young's modulus at 358 K corresponds to the temperature at which the materials starts expanding after having maintained a constant size from 344 K. There is still a fourth feature at 380 K that only involves the thermal expansion, namely an increase of the derivative. On cooling, in contrast, the broad minimum of the Young's modulus at 334 K occurs at the same temperature at which the material slightly expands.

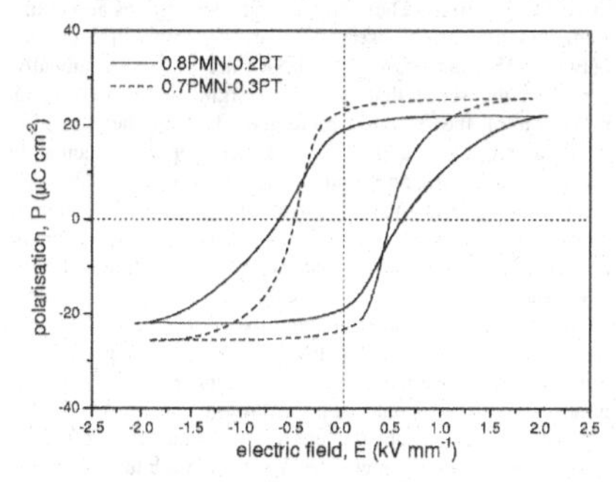

FIG. 6. Ferroelectric hysteresis loops for comparable 0.8 PMN–0.2 PT and 0.7 PMN–0.3 PT ceramics.

TABLE III. Ferroelectric shifts of lead and Mg/Nb/Ti cations in relation to the oxygens barycenter, and polarization (magnitude and direction) calculated with the apparent charges from Hewat (Ref. 36) for the ceramic and powder (obtained from the ceramics, i.e., same size) 0.8 PMN–0.2 PT samples with *Cm* space group.

	Ceramic	Powder
δ Pb-O (Å)	0.30	0.28
δ Ti/Mg/Nb-O (Å)	0.30	0.17
Magnitude (μC/cm^2)	58	48
Direction	[7 7 1]	[5 5 3]

IV. DISCUSSION

Our structural characterization indicates that 0.8 PMN–0.2 PT in the form of both ceramic and powder is monoclinic *Cm*, with the possible presence of a small amount of rhombohedral *R3m* phase in the powder. The monoclinic *Cm* structure of morphotropic lead-based compounds can be of two types: M_A type with $P_X=P_Y<P_Z$ components of polarization in the pseudocubic cell, and M_B type with $P_X=P_Y>P_Z$. In order to establish which one is the case for 0.8 PMN–0.2 PT, the magnitude and direction of the macroscopic polarization were calculated as it is currently done in the structural studies of ferroelectric compounds from our structural information by using the formula $P=\frac{e}{V_{cell}}\Sigma_N z''_i \delta i_i$ where e is the charge of the electron, V_{cell} the volume of the unit cell, N the number of ions in the cell, z''_i the apparent charge from Hewat[36] taking into account of the ionic polarization and δi the ionic relative displacement given by structural refinement (see Table III). As Mg, Nb, and Ti cations are, as usual, taken all on the same crystallographic site, an average displacement for the B cation is considered. As expected for this composition range of PMN-PT, *Cm* phases in both ceramic and powder are of the M_B type.[21]

These results are consistent with the recent evidence of short-range *Cm* monoclinic order (M_B type) in 0.75 PMN–0.25 PT powder at 300 K, which transforms into a long range M_B phase at 80 K.[24] As a matter of fact, the use of a M_B phase also gave very good agreement factors at 300 K in this latest work, in spite of previous reports of this composition being rhombohedral *R3m*.[21] Our results unambiguously show that the monoclinic local order observed in the 0.9 PMN–0.1 PT compound, evidenced by the observation of short range shifts of the Pb^{2+} cations along the ⟨110⟩ directions in addition to the ⟨111⟩ long range rhombohedral shift,[5] has transformed into a long range ordered *Cm* phase for 0.8 PMN–0.2 PT. This transformation of a rhombohedral phase with short ranged monoclinic order toward a long range monoclinic phase was reported for the first time in PMN-PT by Singh and Pandey,[21] though at higher PT contents, and was later also observed in PSN-PT.[22]

The short range monoclinic distortions within a rhombohedral phase were proposed to be the origin of the relaxor-

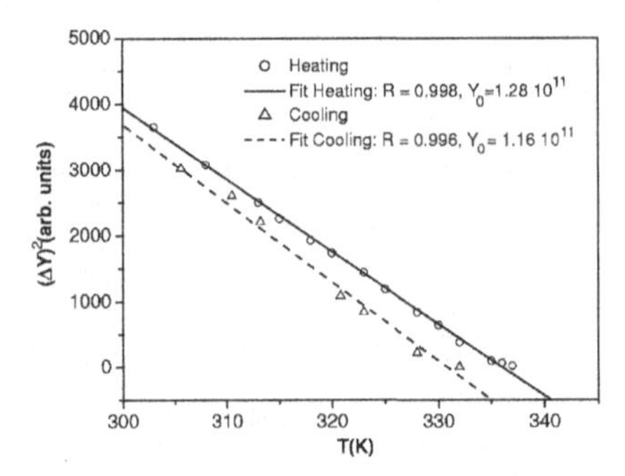

FIG. 7. Temperature dependence of ΔY^2, where $\Delta Y = Y - Y_o$: Y Young's modulus, Y_o the value of the minimum at the transition temperature, in the ferroelectric phase, measured during successive heating and cooling.

type dispersion of the permittivity still present in the ferroelectric phase of 0.9 PMN–0.1 PT.[5] Similar local cationic displacements along the ⟨110⟩ directions exist in rhombohedral $PbZr_xTi_{1-x}O_3$ (Ref. 37), for $0.53 < x < 0.62$ (Ref. 38), and in $Pb(Zn_{1/3}Nb_{2/3})O_3$.[39] These structural characteristics do not only cause permittivity dispersion, but distinctive ferroelectric hysteresis loops. This is also the case for the 0.8 PMN–0.2 PT ceramics reported here. Their ferroelectric loops present smaller polarization and are leant as compared with those for 0.7 PMN–0.3 PT ceramics with the same grain size and porosity, and well developed monoclinic *Cm* long range order.[21] Similar loops have also been observed for 0.75 PMN–0.25 PT single crystals along the ⟨110⟩ direction, and associated with a speckled domain configuration with sizes ranging from 8 μm^2 to less than 100 nm^2.[40]

On heating, the linear dependence of the reciprocal permittivity on temperature suggests that a ferroelectric to paraelectric, second order phase transition is approached. This is the same behavior observed for 0.75 PMN–0.25 PT ceramics.[41] The phase transition can also be studied with the Young's modulus. The mechanical coefficient, Y, is coupled with the electric polarization, P, because of the bilinear interaction mechanoelectric coupling term, ξsP, in the free energy expression, for which ξ is the coupling coefficient and s is the strain. For an isotropic system like a nonpoled ceramic (Ref. 42):

$$Y^2 = \left(\frac{\xi}{s}\right)^2 P^2. \quad (1)$$

Variations of strain in the ferroelectric phase are not large.[16,17] Therefore, the temperature dependence of the Young's modulus is basically due to the dependence of polarization. The temperature dependence of Y^2 on approaching the phase transition from RT is shown in Fig. 7. The increase of modulus, $\Delta Y = Y - Y_o$, where Y_o stands for the value of the minimum at 344 K must be used. A clear linear behavior is found, which indicates that $P^2 \propto (T - T_C)$ on approaching the phase transition as expected for a second order phase transition.

The transition on heating occurs at 344–345 K, the temperature at which the Young's modulus presents a sharp minimum, and where there is a maximum of mechanical losses. These features appear because of the fast decrease of the size of the ferroelectric domains and increase of their number on entering the critical range around the transition, where ferroelectric fluctuations occur. A small but sharp contraction of the material is observed at the transition. The behavior of the permittivity and the Young's modulus are that of a ferroelectric to paraelectric phase transition up to 350 K. However, the material does not linearly expand from 344 to 350 K as one would expect for such a transition, but maintains a constant dimension. This is the same behavior observed for the lattice parameter of 0.8 PMN–0.2 PT single crystals below 380 K,[28] and that was associated with a new X phase, also discussed in $Pb(Zn_{1/3}Nb_{2/3})O_3$-$PbTiO_3$, with an average cubic lattice but ferroelectric polarization.[43,44] At 350 K relaxor-type behavior starts, observed not only in the permittivity, but also in the Young's modulus, for which the derivative sharply decreases as the PNR's start contributing to the mechanical response.[45] The temperature 350 K also corresponds to the freezing temperature. As a matter of fact, the thermal expansion behavior above 344 K is similar to that of a relaxor.[5] All these observations suggest that the material is in a nonergodic glass state in the 344–350 K interval. The transition at 344 K is thus, very likely between a ferroelectric phase and the nonergodic glass state of the relaxor.

The figures for the freezing temperature, activation energy, and characteristic frequency of the relaxor state can be compared with those reported for PMN and 0.9 PMN–0.1 PT that have been included in Table II. Unlike the freezing temperature that increases with the amount of PT in a solid solution, the activation energy decreases when the amount of PT is increased. This activation energy is believed to be the product of an anisotropy energy and the PNR volume.[7] PNR volume has been reported to increase with PT (Ref. 46), so the anisotropy energy must decrease with the addition of PT. Characteristic frequencies of $\sim 10^{12}$ Hz are obtained for the three compositions. There are at least two regimes above the freezing temperature observed on heating. In the first regime, between 350 and 358 K, the PNRs are very mechanically active in the sense that provide a significant softening (a reduction of the Young's modulus). This is the range in which an amplitude dependence has been reported, such as Y decreases when the amplitude of stress increases.[29] This dependence suggests the movement of the PNRs boundaries across the material in this regime.[47] Thermal expansion is negligible, which indicates that the volume fraction of PNRs is constant. In the second regime above 358 K, the softening disappears, and the material starts expanding. These suggest that the volume fraction of PNRs start decreasing and that their dynamics change, so as their boundaries stop moving under stress. There could be a third regime above 380 K, the temperature at which the thermal expansion coefficient sharply increases, though the Young's modulus does not show any inflection point at this temperature.

On cooling, inflection points as those observed on heating at 380, 358, and 350 K are not found. Neither it is observed a sharp minimum of the Young's modulus nor a sudden decrease of permittivity that shows a relaxor type behavior down to room temperature. Instead, the Young's modulus presents a broad, asymmetric minimum at 334 K, a temperature at which a slight expansion occurs. These suggest that a phase transition to a ferroelectric phase occurs at this temperature. This is further supported by the linear behavior of the reciprocal permittivity and of the squared Young's modulus on cooling [see Figs. 2(a) and 7]. The results indicate that there is a strong thermal hysteresis, not only in the temperature of the transition that is decreased from 345 K to ~334 K, but also in the characteristic time scale of the transition that seems to increase, i.e., in the kinetics that slow down. This hysteretic behavior can be interpreted within the two stages model for the development of ferroelectric long range order in relaxor systems, recently proposed by Ye *et al.*[16] In this model, in a first stage at high temperature, PNRs start condensing at T_d (670 K for 0.8 PMN–0.2 PT).[27] Their number and size increases as the temperature is decreased until approaching the temperature of the phase transition. Then, the second stage begins that is characterized by the onset of ferroelectric fluctuations. As a matter of fact, this picture is confirmed by our results on heating, in which the ferroelectric fluctuations and the relaxor state are successively observed. The model proposes that the kinetics of the transition is controlled by the number of PNRs at the onset of the ferroelectric fluctuations, which depends on temperature. Therefore, the kinetics is slower the lower is the temperature of the transition, and controls the final states. The thermal hysteresis in the characteristic time scale of the transition for 0.8 PMN–0.2 PT would then be a consequence of the hysteresis in the temperature of the transition, and of the transition being slowed down in this interval. The slowing down has been experimentally observed here, and shown to be a quite sharp process that occurs between 334 and 345 K. 0.7 PMN–0.3 PT does not present such hysteresis in the kinetics,[29] for both transition temperatures (402 and 408 K) are above the range of temperatures at which the slowing down occurs, and therefore the macroscopic properties present well defined sharp features at the transition both on heating and cooling. On the other hand, transition temperatures for 0.9 PMN–0.1 PT are below this range, and the kinetics is always slow. As a consequence, the macroscopic properties do not reflect the transition either on heating or on cooling. For PMN, the transition would be extremely slow,[16] and whether a ferroelectric state is established at the end is under debate.[12]

It is worth commenting on the relation between the transition temperature and the freezing temperature. 0.8 PMN–0.2 PT has well defined, independent freezing T_f and transition T_c temperatures, the first being at a high temperature. This means that the ferroelectric phase transforms in a nonergodic glass state on heating, and that ferroelectric long range order does not develop from the relaxor state until the system has frozen. Another issue worth commenting on is the order of the phase transition. We discussed that the temperature dependence of the permittivity and Young's modulus on approaching the transition from the ferroelectric phase indicates second order character. However, the size of the material presents a discontinuity at the transition, and thermal hysteresis is evident, which both rather suggest a first order transition. This apparent contradiction could be related to the nature of the transition that is not either an order disorder or a displacive standard transition, but a transformation between short range and long range polar order. This may be a kind of percolation process of small (short-range order) monoclinic clusters, the PNRs. It must be noted that the R-F transition for 0.7 PMN–0.3 PT is between the relaxor state and the ferroelectric tetragonal phase, which then transforms into the thombohedral one at a lower temperature.[29] This suggest that PNRs are not monoclinic, but tetragonal for this latter composition. We have not discussed the origin of the sharp increase of the thermal expansion coefficient of 0.8 PMN–0.2 PT at 380 K, observed on heating. It is tempting to suggest that it is reflecting the transformation of the monoclinic PNRs into tetragonal ones.

V. CONCLUSIONS

A *Cm* monoclinic phase (of M_B type) has been evidenced for both ceramics and powder of 0.8 PMN–0.2 PT, in agreement with the recent structural characterization of Singh *et al.* for 0.75 PMN–0.25 PT.[24] This long range order monoclinic phase at room temperature may arise from a kind of percolation of small (short-range order) monoclinic clusters occurring at ~334 K. The ceramics present distinctive electrical properties that are most probably associated with the size of the monoclinic domains, which would be then smaller than for comparable 0.7 PMN–0.3 PT ceramics. The ferroelectric phase transforms into a relaxor state on heating. The system has well defined and different transition and freezing temperatures, the latter being the highest, so the transition is always between the ferroelectric phase and the nonergodic glass state of the relaxor. The transition presents thermal hysteresis, not only in the transition temperature, but in the kinetics. This behavior seems to indicate that a quite sharp slowing down occurs in the temperature interval between the transition temperatures on heating and cooling, i.e., between 334 and 344 K.

ACKNOWLEDGMENTS

This research has been funded by MEC through the Project No. MAT2005-01304, and by the EC through the Network of Excellence MIND (Multifunctional and Integrated Piezoelectric devices, Ref. No. E 515757-2).

[1] G. A. Smolenskii and A. I. Agranovskaya, Sov. Phys. Tech. Phys. **3**, 1380 (1958).
[2] L. E. Cross, Ferroelectrics **151**, 305 (1994).
[3] N. de Mathan, E. Husson, G. Calvarin, J. R. Gavarri, A. W. Hewat, and A. W. Morrell, J. Phys.: Condens. Matter **3**, 8159 (1991).
[4] G. Burns and F. H. Dacol, Phys. Rev. B **28**, 2527 (1983).
[5] B. Dkhil, J. M. Kiat, G. Calvarin, G. Baldinozzi, S. B. Vakhrushev, and E. Suard, Phys. Rev. B **65**, 024104 (2002).
[6] W. Kleemann, J. Dec, S. Miga, and R. Pankrath, Ferroelectrics **302**, 493 (2004).
[7] D. Viehland, S. J. Jang, L. E. Cross, and M. Wutting, J. Appl. Phys. **68**, 2916 (1990).
[8] D. Viehland, M. Wutting, and L. E. Cross, Ferroelectrics **120**, 71 (1991).
[9] V. Westphal, W. Kleemann, and M. D. Glinchuk, Phys. Rev. Lett. **68**, 847 (1992).
[10] R. Pirc and R. Blinc, Phys. Rev. B **60**, 13470 (1999).
[11] P. M. Gehring, S. Wakimoto, Z. G. Ye, and G. Shirane, Phys. Rev. Lett. **87**, 277601 (2001).
[12] S. Wakimoto, C. Stock, R. J. Birgeneau, Z. G. Ye, W. Chen, W. J. L. Buyers, P. M. Gehring, and G. Shirane, Phys. Rev. B **65**, 172105 (2002).
[13] H. Arndt, F. Sauerbier, G. Schmidt, and L. A. Shebanov, Ferroelectrics **79**, 145 (1988).
[14] G. Calvarin, E. Husson, and Z. G. Ye, Ferroelectrics **165**, 349 (1995).
[15] S. L. Swartz, T. R. Shrout, W. A. Schulze, and L. E. Cross, J. Am. Ceram. Soc. **67**, 311 (1984).
[16] Z. G. Ye, Y. Bing, J. Gao, A. A. Bokov, P. Stephens, B. Noheda, and G. Shirane, Phys. Rev. B **67**, 104104 (2003).
[17] O. Noblanc, P. Gaucher, and G. Calvarin, J. Appl. Phys. **79**, 4291 (1996).
[18] S. W. Choi, T. R. Shrout, S. J. Jang, and A. S. Bhalla, Ferroelectrics **100**, 29 (1989).
[19] B. Noheda, J. A. Gonzalo, L. E. Cross, R. Guo, S.-E. Park, D. E. Cox, and G. Shirane, Phys. Rev. B **61**, 8687 (2000).
[20] J. M. Kiat, Y. Uesu, B. Dkhil, M. Matsuda, C. Malibert, and G. Calvarin, Phys. Rev. B **65**, 064106 (2002).
[21] A. K. Singh and D. Pandey, Phys. Rev. B **67**, 064102 (2003).
[22] R. Haumont, B. Dkhil, J. M. Kiat, A. Al-Barakaty, H. Dammak, and L. Bellaïche, Phys. Rev. B **68**, 014114 (2003).
[23] A. K. Singh, D. Pandey, and O. Zaharko, Phys. Rev. B **68**, 172103 (2003).
[24] A. K. Singh and D. Pandey, J. Appl. Phys. **99**, 076105 (2006).
[25] S. E. Park and T. R. Shrout, J. Appl. Phys. **82**, 1804 (1997).
[26] E. M. Sabolsky, A. R. James, S. Kwon, S. Trolier-McKinstry, and G. L. Messing, Appl. Phys. Lett. **78**, 2551 (2001).
[27] T. Y. Koo, P. M. Gehring, G. Shirane, V. Kiryukhin, S. G. Lee, and S. W. Cheong, Phys. Rev. B **65**, 144113 (2003).
[28] G. Xu, D. Viehland, J. F. Li, P. M. Gehring, and G. Shirane, Phys. Rev. B **68**, 212410 (2003).
[29] M. Algueró, B. Jiménez, and L. Pardo, Appl. Phys. Lett. **87**, 082910 (2005).
[30] M. Algueró, A. Moure, L. Pardo, J. Holc, and M. Kosec,Acta Mater. **54**, 501 (2006).
[31] M. Algueró, J. Ricote, and A. Castro, J. Am. Ceram. Soc. **87**, 772 (2004).
[32] J. F. Bérar, IUCr. Sat. Meeting on Powder Diffractometry, Toulouse, 1990.
[33] B. Jiménez, A. Castro, and L. Pardo, Appl. Phys. Lett. **82**, 3940 (2003).
[34] R. Jiménez, A. Castro, and B. Jiménez, Appl. Phys. Lett. **83**, 3350 (2003).
[35] M. Algueró, B. Jiménez, and L. Pardo, Appl. Phys. Lett. **83**, 2641 (2003).
[36] A. W. Hewat, Ferroelectrics **6**, 215 (1974).
[37] D. L. Corker, A. M. Glazer, R. W. Whatmore, A. Stallard, and F. Fauth, J. Phys.: Condens. Matter **10**, 6251 (1998).
[38] Ragini, R. Ranjan, S. K. Mishra, and D. Pandey, J. Appl. Phys. **92**, 3266 (2002).
[39] G. Xu, Z. Zhong, Y. Bing, Z. G. Ye, and G. Shirane, Nat. Mater. **5**, 134 (2006).
[40] X. Zhao, J. Y. Dai, J. Wang, H. L. W. Chan, C. L. Choy, X. M. Wan, and H. S. Luo, Phys. Rev. B **72**, 064114 (2005).
[41] A. A. Bokov and Z. G. Ye, Appl. Phys. Lett. **77**, 1888 (2000).
[42] E. K. H. Salje, *Phase Transitions in Ferroelastic and Coelastic Crystals* (Cambridge University Press, Cambridge, 1993).
[43] K. Ohwada, K. Hirota, P. W. Rehrig, Y. Fujii, and G. Shirane, Phys. Rev. B **67**, 094111 (2003).
[44] G. Xu, Z. Zhong, Y. Bing, Z. G. Ye, C. Stock, and G. Shirane, Phys. Rev. B **67**, 104102 (2003).
[45] B. Jiménez and R. Jiménez, Phys. Rev. B **66**, 014104 (2002).
[46] A. D. Hilton, A. Randall, D. J. Barber, and T. R. Shrout, Ferroelectrics **93**, 379 (1989).
[47] A. E. Glazounov, A. K. Tagantasev, and A. J. Bell, Phys. Rev. B **53**, 11281 (1996).

Review

$SrTiO_3$—Glimpses of an Inexhaustible Source of Novel Solid State Phenomena

Wolfgang Kleemann [1,*], Jan Dec [2], Alexander Tkach [3] and Paula M. Vilarinho [3]

1 Applied Physics, University Duisburg-Essen, D-47048 Duisburg, Germany
2 Institute of Physics, University of Silesia, PL-40-007 Katowice, Poland; jan.dec@us.edu.pl
3 Department of Materials and Ceramic Engineering, CICECO—Aveiro Institute of Materials, University of Aveiro, P-3810-193 Aveiro, Portugal; atkach@ua.pt (A.T.); paula.vilarinho@ua.pt (P.M.V.)
* Correspondence: wolfgang.kleemann@uni-due.de; Tel.: +49-1575-226-3908

Received: 6 August 2020; Accepted: 30 September 2020; Published: 4 October 2020

Abstract: The purpose of this selective review is primarily to demonstrate the large versatility of the insulating quantum paraelectric perovskite $SrTiO_3$ explained in "Introduction" part, and "Routes of $SrTiO_3$ toward ferroelectricity and other collective states" part. Apart from ferroelectricity under various boundary conditions, it exhibits regular electronic and superconductivity via doping or external fields and is capable of displaying diverse coupled states. "Magnetoelectric multiglass (Sr,Mn)TiO_3" part, deals with mesoscopic physics of the solid solution $SrTiO_3$:Mn^{2+}. It is at the origin of both polar and spin cluster glass forming and is altogether a novel multiferroic system. Independent transitions at different glass temperatures, power law dynamic criticality, divergent third-order susceptibilities, and higher order magneto-electric interactions are convincing fingerprints.

Keywords: strontium titanate; quantum paraelectricity; quantum fluctuations; ferroelectricity; isotope exchange; external stress; polar metal; superconductivity; phase coexistence; magnetoelectric multiglass

1. Introduction

In this review, we focus onto two research lines of strontium titanate, $SrTiO_3$ (STO):

(1) the low-temperature phases around the quantum critical point of pure STO, and (2) the disordered electric and magnetic dipolar glassy phases in the solid solution STO: Mn. It is not intended to describe the full extent of all phenomena observed and to detail all of their properties from abundantly issued publications. We merely try to give an impression of some actual fields, which are partially related to our earlier cooperation with K. A. Müller and J. G. Bednorz.

STO is probably the most versatile perovskite-type oxide and one of the richest materials in terms of functionalities. In 1979, Müller and Burkard [1] reported that the planar permittivity of STO strongly increased upon cooling from ≈ 300 at room temperature and saturated with $\varepsilon'_{\langle 110\rangle} \approx 2.5{\cdot}10^4$ as $T \to 0$ (comparable to $\varepsilon'_{\langle 100\rangle}$ vs. T, cf. Figure 1, curve 1 [2]). They conjectured a "quantum paraelectric" ground state in close proximity to a ferroelectric (FE) one, where the centrosymmetric tetragonal lattice structure of STO becomes stabilized by quantum fluctuations of the nearly softening in-plane F_{1u} lattice mode.

Quantum corrections to the temperature were proposed to describe the critical behavior of STO from the beginning [1]. In order to account for the obvious deviations from the mean-field Curie–Weiss behavior, some of us proposed a generalized modified "quantum Curie–Weiss law" [3].

$$\varepsilon' = C/\left(T^Q - T_0^Q\right)^{\gamma} \quad (1)$$

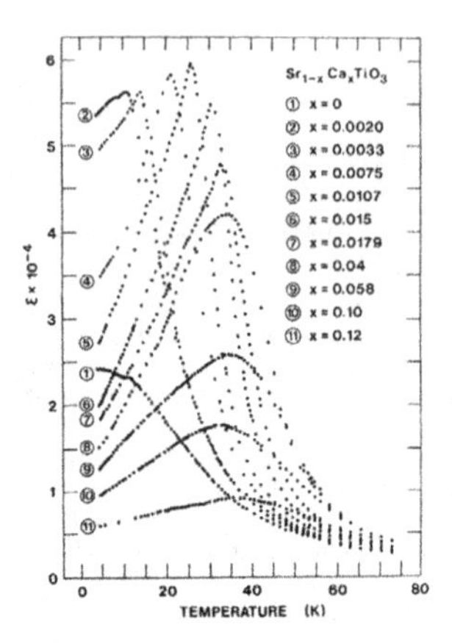

Figure 1. Temperature dependence of the dielectric permittivity $\varepsilon'_{[100]}$ of $Sr_{1-x}Ca_xTiO_3$ crystals with $0 \leq x \leq 0.12$ [2].

where C stands for the Curie constant, γ for the critical exponent, and $T_0^Q = T_S \coth(T_S/T_0)$ for the quantum critical temperature with classic critical temperature T_0 and saturation temperature T_S being related to the ground state energy of the quantum oscillator, $E_0 = k_B T_S$. Crucial novel ingredients are the free parameter γ and the quantum temperature scale, $T^Q = T_S \coth(T_S/T)$, which replaces T. The best fit to the STO data [3] within $2 \leq T \leq 110$ K yields C = (3 ± 2) 10^6, $T_S = (17 \pm 1)$ K, $T_0^Q \approx 0$, and the highly non-classic exponent $\gamma = 1.7 \pm 0.2$. Interestingly, a very similar value, $\gamma \approx 2$, was recently obtained on pure STO from a conventional power-law fit for $4 \leq T \leq 50$ K [4], where saturation effects deviated below ≈ 4 K. While our approach matches with this extreme quantum regime without any extra conditions, the Φ^4 model of [4] requires corrections due to long-range dipolar interactions and coupling of the electric polarization field to acoustic phonons.

2. Routes of $SrTiO_3$ toward Ferroelectricity and Other Collective States

Overcoming quantum paraelectricity and reaching stable long-range ordered states by proper treatments has remained a major challenge for ongoing research on STO.

(i) In 1976, Uwe and Sakudo [5] succeeded in stabilizing uniaxial ferroelectricity in STO at liquid He temperature by intraplanar symmetry breaking with uniaxial stress perpendicularly to the c axis, σ_{110}. This was the first experience to overcome quantum paraelectricity of STO by an external perturbation.

(ii) About 20 years later, in 1999, Itoh et al. [6] discovered a very efficient internal perturbation via isotopic exchange of ^{16}O by ^{18}O in order to establish the FE state at $T_0 \approx$ 25–50 K.

(iii) Again, 20 years later, in 2019, Nova et al. [7] realized transient FE in STO being metastable up to T > 290 K under optical strain due to intense femtosecond laser pulses, while Li et al. [8] recorded similar events after photoexcitation of quantum paraelectric STO with THz laser pulses in resonance with the FE soft mode at $T < 36$ K.

(iv) Metallic behavior of n-type STO has been achieved by substituting transition-metal dopants, e.g., La^{3+} for Sr^{2+} [9], Nb^{5+} for Ti^{4+}, or by reducing pure STO into $SrTiO_{3-\delta}$, $0 < \delta < 1$, where each oxygen vacancy generates two "doped" electrons [10]. The insulator-to-metal transformation occurs at a relatively low critical electron density $n \approx 10^{18}$ cm^{-3} [11], i.e., two orders of magnitude less than in the analogous case of barium titanate [12]. Here the word "metal" is not used in its common material meaning but merely stands for featuring metallic electronic conduction.

A combined method of substitution and reduction was utilized in the case of Cr-doped STO [11]. A significant spatial correlation between oxygen vacancies and Cr^{3+} ions in bulk was established in thermally reduced Cr-doped STO. In the presence of electron donors, the Cr atoms change their valence from 4+ to 3+. Consequently, this reduction drives a symmetry change of the crystal field experienced by the Cr ions from cubic to axial [11], which may be controlled by means of thermal annealing and/or doping with electron donors. This capability of controlling oxygen vacancies or transition-metal dopants has been essential for the development of semiconducting electronic devices [13].

(v) Theoretical predictions of superconductivity in degenerate semiconductors motivated research on reduced n-type STO, which revealed the critical temperature $T_c \approx 0.28$ K as early as 1964 [14]. However, 32 years later, perovskite-like cuprates were to open the door to modern high-T_c superconductivity with $T_c \approx 30$ K [15] and to the physics Nobel Prize [16]. On the other hand, gated n-type STO has reached at most only $T_c \approx 0.6$ K [17].

Meanwhile numerous other processes have made STO a nearly inexhaustible source of activating novel solid state phenomena that suggest future applications. This has remained an attractive research goal even more than 60 years after the pioneering experiments. Extending the initial idea of breaking the local symmetry by stress [5], Bednorz and Müller [2] introduced an A-site doping route by random replacement of Sr^{2+} ions with smaller Ca^{2+} ions in single crystals of $Sr_{1-x}Ca_xTiO_3$ (SCT). The local decrease of volume creates random strain ("negative stress field"), which has an enormous effect on the dielectric response for doping levels $0.002 \leq x \leq 0.12$, as shown in Figure 1 (curves 2–11). Sharp peaks occur at finite temperatures, $10 < T_m < 40$ K, which clearly hint at polar phase transitions (PTs). Their easy axes are actually lying along [110] and [1$\bar{1}$0] within the basal xy-plane and yield, e.g., $\varepsilon_{max}^{\langle 110 \rangle} = 1.1 \cdot 10^5$ for $x = 0.0107$ [2]. Discussion within a random-field concept of PTs reveals xy-type quantum ferroelectricity above $x_c = 0.0018$ along the a axes of the paraelectric parent phase and a PT into a random phase above $x_r \approx 0.016$ (Figure 1). Quantum corrections to the temperature (see Equation (1)) are essential to describe the critical behavior.

In order to understand more details, some of us measured the optical linear birefringence (LB), $\Delta n_{ac} = n_c - n_a$, where n_c and n_a are the principal refractive indices at light wavelength $\lambda = 589.3$ nm, being linearly polarized along the c and a axes of the SCT crystal, respectively, as functions of temperature, T [18]. It is well-known that the LB is sensitive to both the axial rotation of the TiO_6 octahedra, $<\Delta\Phi^2>$, below the antiferrodistortive phase transition temperature $T_a = 105$ K (for $x = 0$), and to the FE short-range order parameter, $<P_x^2>$, where $x \parallel <110>_c$ (Figure 1). Indeed, non-zero LB arises in pure STO at the transition temperature $T_a = 105$ K and at 115, 140, and 255 K (arrows) for $x = 0.002$, 0.0107, and 0.058, respectively, as shown in Figure 2. Additional FE anomalies, $\delta(\Delta n_{ac})$, are superposed at low T. Being non-morphic, they start smoothly with fluctuation tails and bend over into steeply rising long-range order parts below inflection points $T_1 \approx 15$, 28, and 50 K, respectively (arrows). These temperatures systematically exceed the ε' vs. T peak temperatures, $T_m = 14$, 26, and 35 K, respectively (Figure 1), where discontinuities of $d(\Delta n_{ac})/dT$ would be expected in case of PTs into long-range order. Absence of anomalies of this type and increasing differences, T_1–T_m, at increasing x hint at continuously growing smearing of the PTs. Simultaneously, as $T \to 0$, the polarization was calculated by use of the ordinary refractive index $n_o = 2.41$ and the electro-optic coefficient difference g_{11}-$g_{31} = 0.14$ m^4/C^2 as $< P_x^2 >^{1/2} = \{2\delta(\Delta n_{ac})/[n_o^3(g_{11}-g_{31})]\} = 9.8$, 29.4, and 42.6 mC/m^2 for $x = 0.002$, 0.0107, and 0.058, respectively [18]. Since $< P_x^2 >^{1/2}$ varies less than proportionally with x, comparatively incomplete FE order is observed. Further, the low-T polarization saturates, albeit slowly, at increasing electric field, E. This strongly hints at random-field induced nanodomains, whose average size increases with an applied ordering field. The increase of the average order parameter gives credit for disappearing domain walls as known from the domain-state FE $K_{0.974}Li_{0.026}TaO_3$ [19].

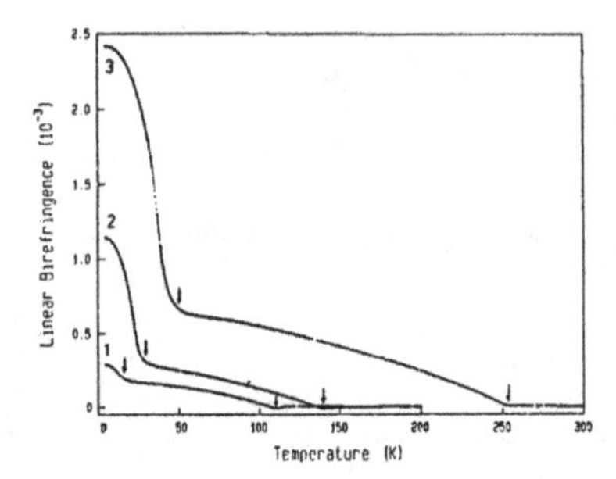

Figure 2. Linear birefringence Δn_{ac} vs. T measured at λ = 589.3 nm on crystallographic single domains of $Sr_{1-x}Ca_xTiO_3$ with x = 0.002 (1), 0.0107 (2), and 0.058 (3), respectively [18].

Further insight into FE SCT is gained from its relaxational behavior. Figure 3 shows the temperature dependence of the real and imaginary parts of the dielectric permittivity of SCT (x = 0.002), ε', and ε'' vs. T, at frequencies $10^3 \leq f \leq 10^4$ Hz [20]. In view of the rounded peaks of $\varepsilon'(T)$, a polydomain state of this FE is conjectured. This has first been interpreted within the concept of "dynamical heterogeneity" [21], which assumes a manifold of mesoscopic "dynamically correlated domains", relaxing exponentially with uniform single relaxation times. Their superposition defines the observed polydispersivity of the sample. It represents aggregates of polar clusters surrounding the quenched off-center Ca^{2+} dopant dipoles.

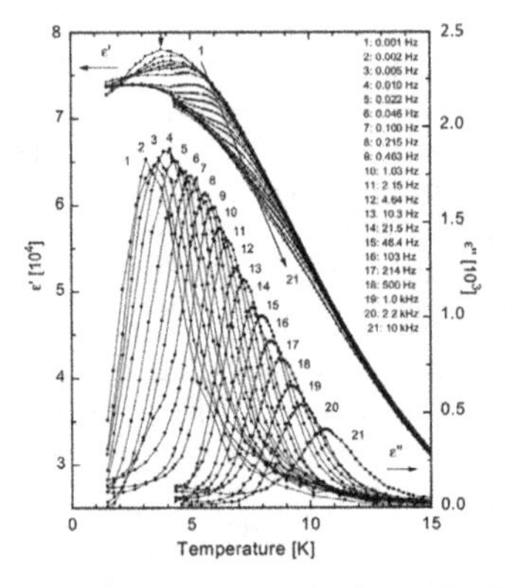

Figure 3. Real and imaginary parts of the permittivity, ε' and ε'' vs. T, of $Sr_{0.998}Ca_{0.002}TiO_3$ measured within $1.5 \leq T \leq 15$ K at frequencies $10^{-3} \leq f \leq 10^4$ Hz [20]. $T_g \approx 3.8$ K is indicated by an arrow.

It is noticed that ε' vs. T peaks at a "glass temperature", $T_g \approx 3.8$ K (Figure 3, arrow), in the quasi-static limit, f = 1 mHz, although at first glance, no glassy criticality as in spin glass is expected. However, in view of recently ascertained magnetic superspin glasses (SSG) of dipolarly coupled magnetic nanoparticles at low concentration [22], a related electric superdipolar glass (SDG) has become envisaged. It should behave like a relaxor ferroelectric [23] in terms of a superglassy critical power law behavior of the $\varepsilon''(f)$ vs. T peak position T_m.

$$f(T_m) \propto \left(T_m - T_g^e\right)^{zv}. \quad (2)$$

Evaluation over the whole range of frequencies, $10^{-3} \leq f \leq 10^{4}$ Hz, yields the expected dynamic critical exponent $z\nu \approx 10$ at $f > 1$Hz, while systematic deviations occur at lower f due to the well-known additional tunneling dynamics. Tests on the expected non-ergodicity of the SDG phase at $T < T_g$ upon zero-field- and field-cooled temperature cycles, respectively (cf. Section 3) are in preparation.

At higher concentration of Ca^{2+}, the polar nanoregions (PNRs) percolate into an FE ground state, as proven by first-order Raman scattering at the softening F_{1u} phonon mode in SCT ($x = 0.007$) at $T < T_0 = 18$ K [24]. SCT thus succeeds in demonstrating stable ferroelectricity. However, systematic research at increasing Ca content showed that T_0 is limited to $\approx$35 K, where the dielectric anomaly becomes increasingly smeared [25]. Better success was achieved by the classic method of stress-induced ferroelectricity in pure STO [5]. To this end Haeni et al. [26] utilized 50 nm thick films of STO, which were epitaxially grown with approximately +1.5% biaxial tensile strain on a (110) $DyScO_3$ substrate, while −0.9% uniform compression due to a $(LaAlO_3)_{0.29}(SrAl_{0.5}Ta_{0.5}O_3)_{0.71}$ (LSAT) substrate was barely active in this respect (Figure 4). The high permittivity in the films on $DyScO_3$, ε' up to 7000 at 10 GHz and room temperature, as well as its sharp dependence on an electric field is promising for device applications [4,26]. The observation of stress-induced ferroelectricity in STO films has confirmed theoretical predictions of Pertsev et al. [27]. While substrate induced tensile strain in epitaxial STO films via lattice parameter mismatch favors in-plane FE, compressive strain provides out-of-plane directed ferroelectricity [27]. This was observed by Fuchs et al. [28] at an STO film epitaxially grown on an STO substrate coated by compressive $YBa_2Cu_3O_7$.

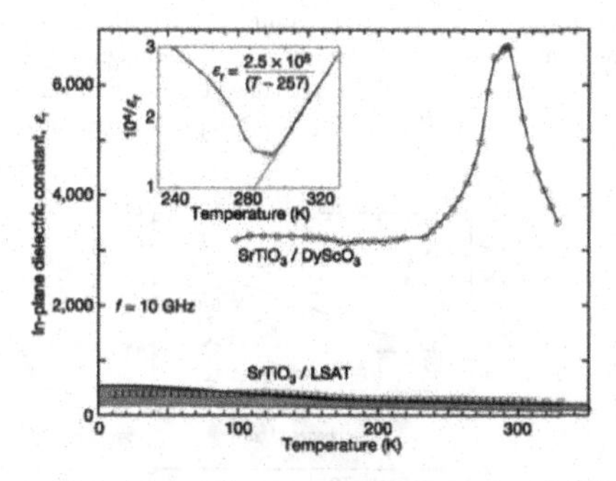

Figure 4. In-plane permittivity ε' vs. T of a strained 50 nm epitaxial STO/(110)$DyScO_3$ film at $f = 10$ GHz as compared to a compressed STO/LSAT film. The inset shows a Curie–Weiss fit to $(\varepsilon_r)^{-1}$ with $T_0 \approx 260$ K [26].

Another realization of room temperature ferroelectricity in STO confirms theoretical predictions of proximity effects at interfaces of metals to oxides containing PNRs such as, for example, STO [29]. Lee et al. [30] reported emergence of room temperature ferroelectricity at reduced dimensions, thus refuting a long-standing contradicting notion. Piezoelectric force microscopy (PFM) was able to evidence room-temperature ferroelectricity in strain-free epitaxial films with 24 unit-cell-thickness of otherwise non-ferroelectric STO (Figure 5). Following arguments from defect engineering in SCT, the authors claimed that electrically induced alignments of PNRs at Sr deficiency related defects are responsible for the appearance of a stable net of ferroelectric polarization in these films. This insight might be useful for the development of low-D materials of emerging nanoelectronic devices.

To systematically control ferroelectricity in thin films of STO at room temperature, Kang et al. [31] selectively engineered elemental vacancies by pulsed laser epitaxy (PLE). Sr^{2+} vacancies play an essential role in inducing the cubic-to-tetragonal transition, since they break the inversion symmetry, which is necessary for switchable electric polarization. The tetragonality turns out to increase with increasing vacancy density, thus strengthening the ferroelectricity, as shown in Figure 6a. This research has optimized tetragonality-induced ferroelectricity in STO with reliable growth control of the behavior. PFM yields stable hysteresis loops at room temperature, as shown in Figure 6b, where low and high laser fluences during PLE clearly demonstrate their key role in creating FE polarization. Similar propositions were made by the Barthélémy–Bibes group, which invoked both an electric field-switchable two-dimensional electron gas emerging in ferroelectric SCT films [32] and the non-volatile electric control of spin–charge conversion in an STO Rashba system [33].

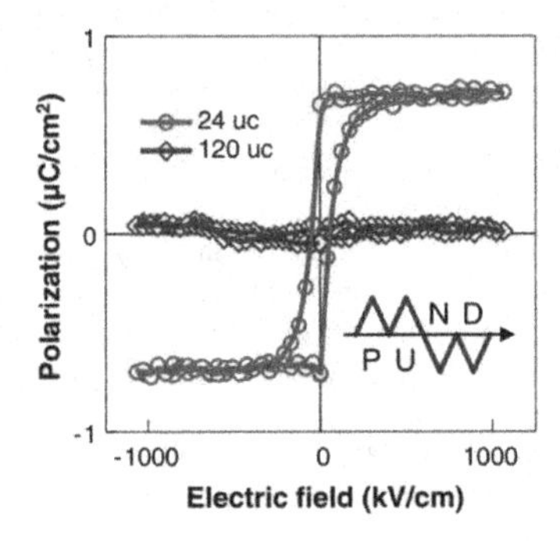

Figure 5. Polarization hysteresis of 24 and 120 unit-cell-thick STO films at room temperature, measured by using the double-wave PUND technique with a triangular *ac* electric field of 10 kHz (see schematic inset). The hysteresis component is obtained by subtracting the non-hysteretic (up (U) and down (D)) from the total (positive (P) and negative (N)) polarization runs [30].

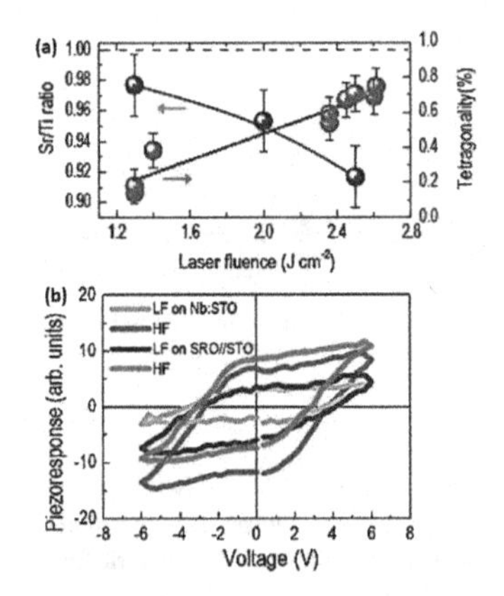

Figure 6. (**a**) Sr/Ti elemental concentration ratio (blue circles) and tetragonality measured at room temperature (red circles) plotted as functions of the laser fluence during pulsed laser epitaxy (PLE). (**b**) Ferroelectric hysteresis loops recorded by piezoelectric force microscopy (PFM) at 5-nm-thick $SrTiO_3$ films grown with low and high laser fluences (LF and HF, respectively; see (**a**)) on different bottom electrodes (STO:Nb and $SrRuO_3$/STO) [33].

Only recently has another insight into the ferroelectric state of compressively strained STO become available from high-angle annular dark-field imaging in scanning transmission electron microscopy. Salmani–Rezaie et al. [34] observed local polar regions in the room-temperature paraelectric phase of (001)-strained STO films, which were grown on (001) faces of LSAT and underwent an FE transition at low T. This unexpected feature was explained by a locally dipolar-ordered, but globally random phase of displaced Ti^{4+} columns, which underwent a disorder–order transition on cooling.

This Section started with different methods to establish long-range order, such as ferroelectricity (FE), metallicity (MT), or superconductivity (SC) in suitably modified STO [2,9,14]. Lately, more demanding procedures have become successful to stabilize the co-existence of apparently contradictory properties, e.g., FE-SC and FE-MT, which appear self-excluding at first glance. In this context, obscure terms such as, for example, "polar metal" and "metallic ferroelectric" or "ferroelectric metal" have been used interchangeably by the research community. Only recently have subtle distinctions of these variants with respect to their electric field switchability been clarified [12], although this topic still remains under debate.

Rischau et al. [35] showed that SC can coexist with an FE-like instability in oxygen-reduced ("*n*-doped") $Sr_{1-x}Ca_xTiO_{3-\delta}$ ($0.002 < x < 0.009, 0 < \delta < 0.001$), where both long-range orders are intimately linked. The FE transition of insulating SCT was found to survive in this reduced modification. Owing to its metallic conductivity, the latter does not show a bulk reversible electric polarization and hence cannot be a true ferroelectric. However, it shows anomalies in various physical properties at the Curie temperature of the insulator, e.g., Raman scattering evidences that the hardening of the FE soft mode in the dilute metal is identical with what is seen in the insulator. The anomaly in resistivity was found to terminate at a threshold carrier density (n^*), near to which the SC transition temperature is enhanced [35]. This evidences the link between SC pairing and FE dipolar ordering, a subject of current attention [36].

Moreover, it is widely accepted that the low-*T* phase of STO lies in the vicinity of a quantum critical point, where different phases (i.e., paraelectric, antiferrodistortive, FE, MT, and SC) with similar energies compete, while weak residual interactions may stabilize one or several of these states [37–39]. The coexistence of MT and FE states in STO has been addressed under the keyword charge transport in a polar metal by Wang et al. [40], who studied the low-*T* electrical resistivity in several $Sr_{1-x}Ca_xTiO_{3-\delta}$ single-crystals at $\delta > 0$ within $0.002 < x < 0.01$ (Figure 7). Since both MT and FE are dilute, the distance between mobile MT electrons and fixed FE dipoles can be separately tuned but kept much longer than the interatomic distance. This opens the chance of activating a Ruderman–Kittel–Kasuya–Yosida-like interaction [41] of carriers with local electric moments, which was originally proposed by Glinchuk and Kondakova [42]. They introduced this indirect interaction of FE off-center ions with conduction electrons in order to explain high FE transition temperatures in certain narrow-gap semiconductors with high conductivity, such as $Pb_{1-x}Ge_xTe$. In agreement with this theory, it is expected that the threshold concentration of carriers, n^*, is proportional to x, which indicates that it occurs at a fixed ratio between inter-carrier and inter-dipole distances.

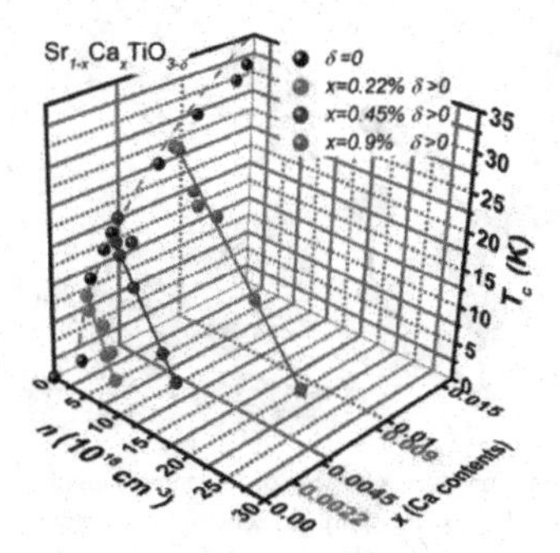

Figure 7. Ferroelectric (FE) phase transition temperatures T_c in insulating $Sr_{1-x}Ca_xTiO_3$ as functions of x (black balls [2]) and in metallic $Sr_{1-x}Ca_xTiO_{3-\delta}$ ($\delta > 0$) as functions of charge carrier density n and $x = 0.0022$, 0.0045 and 0.009 (green, blue, and red balls, respectively [40]).

Tomioka et al. [17] finally demonstrated the simultaneous occurrence of three states, FE, MT, and SC, by independently controlling two concentrations of electron-doped $Sr_{1-x}La_xTi(^{16}O_{1-z}{}^{18}O_z)_3$ single crystals. They precisely controlled the "dome-like" SC characteristic by n doping via the La^{3+} content, while independently enhancing T_c by substitution of $^{18}O^{2-}$ ions for $^{16}O^{2-}$. At an electron concentration of $n \approx 5 \times 10^{19}$ cm^{-3}, they found the apex of the SC dome at $T_c \approx 0.44$ K, where they subsequently shifted its height to a record-high $T_c \approx 0.6$ K by adjusting z(^{18}O).

Being arbitrarily close to the quantum critical point of non-centrosymmetric SC, experiments have thus come into reach to probe mixed-parity pairing mechanisms with topological aspects to their SC states, such as extremely large and highly anisotropic upper critical fields and topologically protected spin currents. A decisive step toward this aim was done by Schumann et al. [43] using La^{3+} or Sm^{3+} n-doped STO films on (001)-strained LSAT substrates. Being in their polar phase, they reveal enhanced superconducting T_c, while some of them show signatures of an unusual SC state, where the in-plane critical field is higher than both the paramagnetic and orbital pair breaking limits. Moreover, nonreciprocal transport is observed, which reflects the ratio of odd versus even pairing interactions. A similar highlight was observed in a gate-induced 2D SC of interfacial STO [44]. Due to its Rashba-type spin orbit interaction, it reveals nonreciprocal transport, where the inequivalent rightward and leftward currents reflect simultaneous spatial inversion and time-reversal symmetry breaking—an exciting prospect of forthcoming research on STO.

3. The Magnetoelectric Multiglass (Sr,Mn)TiO_3

The nature of glassy states in disordered materials has long been controversially discussed. In the magnetic community, generic spin glasses have long been accepted to undergo phase transitions at a static glass temperature T_g, where they exhibit criticality and originate well-defined order parameters [45]. In addition, disordered polar systems are expected to transit into generic "dipolar" or "orientational glass" states [46], which fulfil similar criteria as spin glasses. Hence, it appears quite natural to introduce the term "multiglass" for a new kind of multiferroic material revealing both polar and spin glass properties, which were discovered by some of us in the ceramic solid solution $Sr_{0.98}Mn_{0.02}TiO_3$ [47]. By various experimental methods [48–50] it has been ascertained that the Mn^{2+} ions are randomly substituting Sr^{2+} ions on A-sites in quantum paraelectric STO (Figure 8a), where they become off-centered due to their small ionic size and undergo covalent bonding with one of the twelve nearest neighboring O^{2-} ions. These elementary dipoles readily form polar nanoclusters with frustrated dipolar interactions, as illustrated in Figure 8b. It depicts the local cluster formation of Mn^{2+} ions with antiparallel electric dipole moments and antiferromagnetically correlated spins.

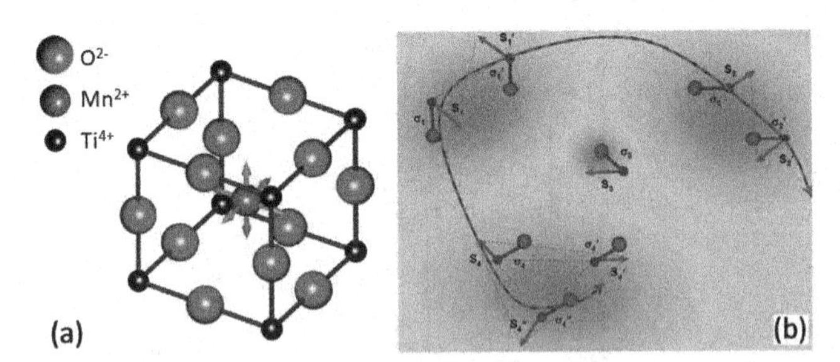

Figure 8. (**a**) *A* site substituted Mn^{2+} ion in its cage of 12 nearest neighboring oxygen ions in the ABO_3 lattice of STO going off-center along <100> [46]. (**b**) Schematic structure of $SrTiO_3$: Mn^{2+} highlighting a percolating multiglass path of randomly distributed Mn^{2+} ions (red–blue broken line) carrying dipole moments σ_j (blue lines) and spins S_j (red arrows) with electric dipolar and antiferromagnetic correlations, respectively, within polar STO clusters (red "clouds") [51].

The dipolar glass formation can easily be judged from the asymptotic shift of the dynamic dielectric susceptibility peak, $T_m(f)$, at frequencies within the range $10^{-1} \leq f \leq 10^6$ Hz in Figure 9a. It obeys glassy critical behavior according to Equation (2), where $z\nu = 8.5$ is the dynamic critical exponent and $T_g^e \approx 38$ K the electric glass temperature [51]. On the other hand, frustrated and random Mn^{2+}–O^{2-}–Mn^{2+} superexchange is at the origin of spin glass formation below the magnetic glass temperature $T_g^m \approx 34$ K. This temperature marks the confluence of three characteristic magnetization curves recorded in $\mu_0 H = 10$ mT after zero-field cooling (ZFC) to $T = 5$ K upon field heating (m^{ZFC}), upon subsequent field cooling (m^{FC}), and thereafter the thermoremanence (m^{TRM}) upon zero-field heating (ZFH) as shown in Figure 9c. It should be noticed that both glassy states have unanimously been confirmed by clear-cut individual aging, rejuvenation, and memory effects in their respective *dc* susceptibilities [51]. "Holes" burnt into the electric and magnetic susceptibilities by waiting in zero external field for 10.5 h at 32.8 K and for 2.8 h at 33 K, respectively, and subsequent heating with weak electric or magnetic probing fields are shown in Figure 9b,d, respectively. They corroborate the glassy ground states of both polar and magnetic subsystems and their compatibility with spin glass theory [45]. Observation of the biquadratic ME interaction in the free energy [47],

$$F(E,H) = F_0 - (\delta/2)E_i E_j H_k H_l (i,j,k,l = 1,2,3), \tag{3}$$

is compatible with the low symmetry of the compound and is thought to crucially reinforce the spin glass ordering, as schematically depicted in Figure 8b [51]. Similarly to the dielectric anomaly [52], the magnetic anomaly has been found to depend not only on the frequency, but crucially also on the Mn content, confirming its intrinsic origin [53]. Furthermore, apart from ceramics, both glassy states have also been detected in equivalent thin films [54].

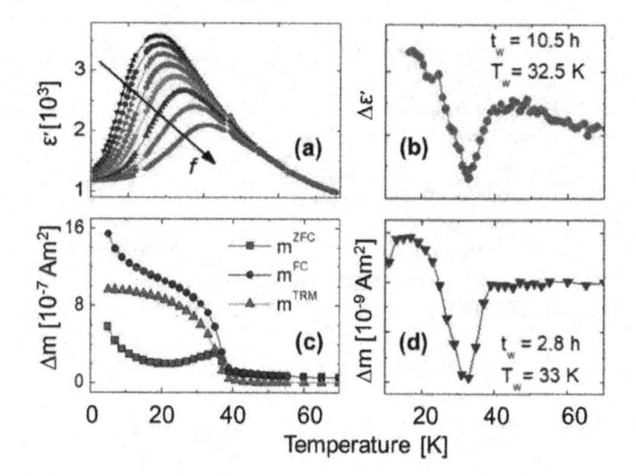

Figure 9. (**a**) Dielectric susceptibility $\varepsilon'(T)$ of $Sr_{0.98}Mn_{0.02}TiO_3$ ceramics recorded at frequencies $10^{-1} \leq f \leq 10^6$ Hz and (**c**) magnetization measured in $B = 10$ mT on field heating after ZFC (m^{ZFC}), on FC (m^{FC}), and on ZFH after FC (m^{TRM}). Holes $\Delta\varepsilon(T)$ and $\Delta m(T)$ burnt in zero fields at $T_{wait} = 32.5$ K for 10.5 h (**b**) and $T_{wait} = 33$ K for 2.8 h (**d**) confirm memory and rejuvenation of both electric and magnetic glassy subsystems [47].

Starting from a mean-field ansatz within the framework of a transverse Ising model [51], the complete theory of the ME multiglass is still under debate. In particular, the final steps for establishing the spin glass are missing. It is thought to emerge from multipolar interaction of spin clusters (Figure 8b) and probably comes close to the formation of a superspin glass as in systems of magnetic nanoparticles [22]. Since these probably consist of antiferromagnetic $MnTiO_3$ and carry merely surface magnetization [55], special care has to be taken.

In search of other ME multiglasses, we successfully examined also Mn^{2+} doped $KTaO_3$, which in the undoped case is a quantum paraelectric like STO, but nevertheless has slightly different properties on doping [56]. Other research groups have made similar experiments, and all of them reported considerable complexity [57–60]. Moreover, various other ME multiglasses have also been observed in disordered solid solutions such as $CuFe_{0.5}V_{0.5}O_2$ [61], La_2NiMnO_6 [62], Fe_2TiO_5 [63], and $(Ba_3NbFe_3Si_2O_{14})$:Sr [64].

4. Conclusions

STO still enjoys vivid interest in research and technological development. Having overcome the low-T bottleneck by advanced nanotechnologies, STO belongs to the most promising nanoelectronic materials. Unusual properties around the quantum critical point such as the co-existence of regular and superconductivity with ferroelectricity are still the focus of attention. On the other hand, the novel disordered phases of a superglass in $Sr_{0.998}Ca_{0.002}TiO_3$ and a multiglass in $Sr_{0.98}Mn_{0.02}TiO_3$ also still require dedicated activity.

Author Contributions: Conceptualization, W.K.; writing—original draft preparation, W.K.; writing—review and editing, W.K., J.D., A.T. and P.M.V. All authors have read and agreed to the published version of the manuscript.

Funding: This research received no external funding.

Acknowledgments: We are grateful to K.A. Müller and J.G. Bednorz for providing their outstanding single crystal samples of SCT and cooperating within common publications. In addition, we acknowledge valuable cooperation with A. Albertini, S. Bedanta, U. Bianchi, P. Borisov, A. Hochstrat, S. Miga, F.J. Schäfer, and V.V. Shvartsman.

Conflicts of Interest: The authors declare no conflict of interest.

Glossary

FC	field cooling
FE	ferroelectric or ferroelectricity
FH	field heating
LB	linear birefringence
LSAT	$(LaAlO_3)_{0.29}(SrAl_{0.5}Ta_{0.5}O_3)_{0.71}$
MT	metallic or metallicity
PLE	pulsed laser epitaxy
PNR	polar nanoregion
PT	phase transition
SC	superconductive or superconductivity
SCT	$Sr_{1-x}Ca_xTiO_3$
STO	$SrTiO_3$
TRM	thermoremanence
ZFC	zero-field cooling
ZFH	zero-field heating

References

1. Müller, K.A.; Burkard, H. $SrTiO_3$: An intrinsic quantum paraelectric below 4 K. *Phys. Rev. B* **1979**, *19*, 3593–3602. [CrossRef]
2. Bednorz, J.G.; Müller, K.A. $Sr_{1-x}Ca_xTiO_3$: An XY quantum ferroelectric with transition to randomness. *Phys. Rev. Lett.* **1984**, *52*, 2289–2293. [CrossRef]
3. Dec, J.; Kleemann, W. From Barrett to generalized quantum Curie-Weiss law. *Solid State Commun.* **1998**, *106*, 695–699. [CrossRef]
4. Rowley, S.E.; Spalek, L.J.; Smith, R.P.; Dean, M.P.M.; Itoh, M.; Scott, J.F.; Lonzarich, G.G.; Saxena, S.S. Ferroelectric quantum criticality. *Nat. Phys.* **2014**, *10*, 367–372. [CrossRef]
5. Uwe, H.; Sakudo, T. Stress-induced Ferroelectricity and soft phonon modes in $SrTiO_3$. *Phys. Rev. B* **1976**, *13*, 271–286. [CrossRef]

6. Itoh, M.; Wang, R.; Inaguma, Y.; Yamaguchi, T.; Shan, Y.-J.; Nakamura, T. Ferroelectricity induced by oxygen isotope exchange in strontium titanate perovskite. *Phys. Rev. Lett.* **1999**, *82*, 3540–3543. [CrossRef]
7. Nova, T.F.; Disa, A.S.; Fechner, M.; Cavalleri, A. Metastable ferroelectricity in optically strained $SrTiO_3$. *Science* **2019**, *364*, 1075–1079. [CrossRef]
8. Li, X.; Qiu, T.; Zhang, J.; Baldini, E.; Lu, J.; Rappe, A.M.; Nelson, K.A. Terahertz field–induced ferroelec-tricity in quantum paraelectric $SrTiO_3$. *Science* **2019**, *364*, 1079–1082. [CrossRef]
9. Tokura, Y.; Taguchi, Y.; Okada, Y.; Fujishima, Y.; Arima, T.; Kumagai, K.; Iye, Y. Filling dependence of electronic properties on the verge of metal-Mott-insulator. *Phys. Rev. Lett.* **1993**, *70*, 2126–2129. [CrossRef]
10. Spinelli, A.; Torija, M.A.; Liu, C.; Jan, C.; Leighton, C. Electronic transport in doped $SrTiO_3$: Conduction and potential applications. *Phys. Rev. B* **2010**, *81*, 155110. [CrossRef]
11. La Mattina, F.; Bednorz, J.G.; Alvarado, S.F.; Shengelaya, A.; Müller, K.A.; Keller, H. Controlled oxygen vacancies and space correlation with Cr^{3+} in $SrTiO_3$. *Phys. Rev. B* **2009**, *80*, 075122. [CrossRef]
12. Zhou, W.X.; Ariando, A. Review on Ferroelectric/polar metals. *Jpn. J. Appl. Phys.* **2020**, *59*, S10802. [CrossRef]
13. Alvarado, S.P.; La Mattina, P.; Bednorz, J.G. Electroluminescence in $SrTiO_3$: Cr single-crystal nonvolatile memory cells. *Appl. Phys. A* **2007**, *89*, 85–89. [CrossRef]
14. Schooley, J.F.; Hosler, W.R.; Cohen, M.L. Superconductivity in semiconducting $SrTiO_3$. *Phys. Rev. Lett.* **1964**, *12*, 474–475. [CrossRef]
15. Bednorz, J.G.; Müller, K.A. Possible high-Tc superconductivity in the Ba-La-Cu-0 system. *Z. Phys. B Cond. Matter* **1986**, *64*, 189–193. [CrossRef]
16. Bednorz, J.G.; Müller, K.A. Perovskite-type oxides—The new approach to high-T_c superconductivity. *Nobel Lect. Dec.* **1987**. Available online: https://www.nobelprize.org/uploads/2018/06/bednorz-muller-lecture.pdf (accessed on 1 April 2020). [CrossRef]
17. Tomioka, Y.; Shirakawa, N.; Shibuya, K.; Inoue, I.H. Enhanced superconductivity close to a nonmagnetic quantum critical point in electron-doped strontium titanate. *Nat. Commun.* **2019**, *10*, 738. [CrossRef]
18. Kleemann, W.; Schäfer, F.J.; Müller, K.A.; Bednorz, J.G. Domain state properties of the random-field xy-model system $Sr_{1-x}Ca_xTiO_3$. *Ferroelectrics* **1988**, *80*, 297–300. [CrossRef]
19. Kleemann, W.; Kütz, S.; Rytz, D. Cluster glass and domain state properties of $K_{1-x}Li_xTaO_3$. *Europhys. Lett.* **1987**, *4*, 239–245. [CrossRef]
20. Kleemann, W.; Albertini, A.; Chamberlin, R.V.; Bednorz, J.G. Relaxational dynamics of polar nano-domains in $Sr_{1-x}Ca_xTiO_3$, $x = 0.002$. *Europhys. Lett.* **1997**, *37*, 145–150. [CrossRef]
21. Chamberlin, R.V.; Haines, D.N. Percolation model for relaxation in random systems. *Phys. Rev. Lett.* **1990**, *65*, 2197–2200. [CrossRef]
22. Bedanta, S.; Kleemann, W. Supermagnetism. *J. Phys. D Appl. Phys.* **2009**, *42*, 013001. [CrossRef]
23. Kleemann, W. Relaxor ferroelectrics: Cluster glass ground state via random fields and random bonds. *Phys. Status Solidi B* **2014**, *251*, 1993–2002. [CrossRef]
24. Bianchi, U.; Kleemann, W.; Bednorz, J.G. Raman scattering of ferroelectric $Sr_{1-x}Ca_xTiO_3$, $x = 0.007$. *J. Phys. Condens. Matter* **1994**, *6*, 1229–1238. [CrossRef]
25. Carpenter, M.A.; Howard, C.J.; Knight, K.S.; Zhang, Z. Structural relationships and a phase diagram for $(Ca, Sr)TiO_3$ perovskites. *J. Phys. Condens. Matter* **2006**, *18*, 10725–10749. [CrossRef]
26. Haeni, J.H.; Irvin, P.; Chang, W.; Uecker, R.; Reiche, P.; Li, Y.L.; Choudhury, S.; Tian, W.; Hawley, M.E.; Craigo, B.; et al. Room-temperature ferroelectricity in strained $SrTiO_3$. *Nature* **2004**, *430*, 758–761. [CrossRef]
27. Pertsev, N.A.; Tagantsev, A.K.; Setter, N. Phase transitions and strain-induced ferroelectricity in $SrTiO_3$ epitaxial thin films. *Phys. Rev. B* **2000**, *61*, R825–R829. [CrossRef]
28. Fuchs, D.; Schneider, C.W.; Schneider, R.; Rietschel, H. High dielectric constant and tunability of epitaxial thin film capacitors. *J. Appl. Phys.* **1999**, *85*, 7363–7369. [CrossRef]
29. Stengel, M.; Spaldin, N.A. Origin of the dielectric dead layer in nanoscale capacitors. *Nature* **2006**, *443*, 679–682. [CrossRef]
30. Lee, D.; Lu, H.; Gu, Y.; Choi, S.-Y.; Li, S.-D.; Ryu, S.; Paudel, T.R.; Song, K.; Mikheev, E.; Lee, S.; et al. Emergence of room-temperature ferroelectricity at reduced dimensions. *Science* **2015**, *349*, 1314–1317. [CrossRef]
31. Kang, K.T.; Seo, H.I.; Kwon, O.; Lee, K.; Bae, J.-S.; Chu, M.-W.; Chae, S.C.; Kim, Y.; Choi, W.S. Ferroelec-tricity in $SrTiO_3$ epitaxial thin films via Sr-vacancy-induced tetragonality. *Appl. Surf. Sci.* **2020**, *499*, 143930. [CrossRef]

32. Bréhin, J.; Trier, F.; Vicente-Arche, L.M.; Hemme, P.; Noël, P.; Cosset-Chéneau, M.; Attané, J.-P.; Vila, L.; Sander, A.; Gallais, Y.; et al. Switchable two-dimensional electron gas based on ferroelectric Ca: $SrTiO_3$. *Phys. Rev. Mater.* **2020**, *4*, 041002. [CrossRef]
33. Noël, P.; Trier, F.; Vicente-Arche, L.M.; Bréhin, J.; Vaz, D.C.; Garcia, V.; Fusil, S.; Barthélémy, A.; Vila, L.; Bibes, M.; et al. Non-volatile electric control of spin–charge conversion in a $SrTiO_3$ Rashba system. *Nature* **2020**, *580*, 483–486. [CrossRef]
34. Salmani-Rezaie, S.; Ahadi, K.; Strickland, W.M.; Stemmer, S. Order-disorder ferroelectric transition of strained $SrTiO_3$. *Phys. Rev. Lett.* **2020**, *125*, 087601. [CrossRef]
35. Rischau, C.W.; Lin, X.; Grams, C.P.; Finck, D.; Harms, S.; Engelmayer, J.; Lorenz, T.; Gallais, Y.; Fauqué, B.; Hemberger, J.; et al. A ferroelectric quantum phase transition inside the superconducting dome of $Sr_{1-x}Ca_xTiO_{3-\delta}$. *Nat. Phys.* **2017**, *13*, 643–648. [CrossRef]
36. Wölfle, P.; Balatsky, A.V. Superconductivity at low density near a ferroelectric quantum critical point: Doped $SrTiO_3$. *Phys. Rev. B* **2018**, *98*, 104505. [CrossRef]
37. Takada, Y. Theory of superconductivity in polar semiconductors and its application to n-type semicon-ducting $SrTiO_3$. *J. Phys. Soc. Jpn.* **1980**, *49*, 1267–1275. [CrossRef]
38. Gabay, M.; Triscone, J.-M. Superconductivity: Ferroelectricity woos pairing. *Nat. Phys.* **2017**, *13*, 624–625. [CrossRef]
39. Collignon, C.; Lin, X.; Rischau, C.W.; Fauqué, B.; Behnia, K. Metallicity and superconductivity in doped strontium titanate. *Ann. Rev. Cond. Matt. Phys.* **2019**, *10*, 25–44. [CrossRef]
40. Wang, J.L.; Yang, L.W.; Rischau, C.W.; Xu, Z.K.; Ren, Z.; Lorenz, T.; Hemberger, J.; Lin, X.; Behnia, K. Charge transport in a polar metal. *NPJ Quantum Mater.* **2020**, *4*, 61–68. [CrossRef]
41. Available online: https://en.wikipedia.org/wiki/RKKY_interaction (accessed on 15 April 2020).
42. Glinchuk, M.D.; Kondakova, I.V. Ruderman−Kittel−like interaction of electric dipoles in systems with carriers. *Phys. Stat. Sol.* **1992**, *174*, 193–197. [CrossRef]
43. Schumann, T.; Galletti, L.; Jeong, H.; Ahadi, K.; Strickland, W.M.; Salmani-Rezaie, S.; Stemmer, S. Possible signatures of mixed-parity superconductivity in doped polar $SrTiO_3$ films. *Phys. Rev. B* **2020**, *101*, 100503. [CrossRef]
44. Itahashi, Y.M.; Ideue, T.; Saito, Y.; Shimizu, S.; Ouchi, T.; Nojima, T.; Iwasa, Y. Nonreciprocal transport in gate-induced polar superconductor $SrTiO_3$. *Sci. Adv.* **2020**, *6*, eaay9120. [CrossRef] [PubMed]
45. Binder, K.; Young, A.P. Spin glasses: Experimental facts, theoretical concepts, and open questions. *Rev. Mod. Phys.* **1986**, *58*, 801–976. [CrossRef]
46. Binder, K.; Reger, J.D. Theory of orientational glasses: Models, concepts, simulations. *Adv. Phys.* **1992**, *41*, 547–627. [CrossRef]
47. Shvartsman, V.V.; Bedanta, S.; Borisov, P.; Kleemann, W.; Tkach, A.; Vilarinho, P.M. $(Sr,Mn)TiO_3$: A magnetoelectric multiglass. *Phys. Rev. Lett.* **2008**, *101*, 165704. [CrossRef] [PubMed]
48. Laguta, V.V.; Kondakova, I.V.; Bykov, I.P.; Glinchuk, M.D.; Tkach, A.; Vilarinho, P.M.; Jastrabik, L. Electron spin resonance investigation of Mn^{2+} ions and their dynamics in Mn-doped $SrTiO_3$. *Phys. Rev. B* **2007**, *76*, 054104. [CrossRef]
49. Lebedev, A.I.; Sluchinskaya, I.A.; Erko, A.; Kozlovskii, V.F. Direct evidence for off-centering of Mn impurity in $SrTiO_3$. *JETP Lett.* **2009**, *89*, 457–460. [CrossRef]
50. Levin, I.; Krayzman, V.; Woicik, J.C.; Tkach, A.; Vilarinho, P.M. X-ray absorption fine structure studies of Mn coordination in doped perovskite $SrTiO_3$. *Appl. Phys. Lett.* **2010**, *96*, 052904. [CrossRef]
51. Kleemann, W.; Bedanta, S.; Borisov, P.; Shvartsman, V.V.; Miga, S.; Dec, J.; Tkach, A.; Vilarinho, P.M. Multiglass order and magnetoelectricity in Mn2+ doped incipient ferroelectrics. *Eur. Phys. J. B* **2009**, *71*, 407–410. [CrossRef]
52. Tkach, A.; Vilarinho, P.M.; Kholkin, A.L. Polar behavior in Mn-doped $SrTiO_3$ ceramics. *Appl. Phys. Lett.* **2005**, *86*, 172902. [CrossRef]
53. Tkach, A.; Vilarinho, P.M.; Kleemann, W.; Shvartsman, V.V.; Borisov, P.; Bedanta, S. Comment on "The origin of magnetism in Mn-doped $SrTiO_3$". *Adv. Funct. Mater.* **2013**, *23*, 2229–2230. [CrossRef]
54. Tkach, A.; Okhay, O.; Wu, A.; Vilarinho, P.M.; Bedanta, S.; Shvartsman, V.V.; Borisov, P. Magnetic anomaly and dielectric tunability of $(Sr,Mn)TiO_3$ thin films. *Ferroelectrics* **2012**, *426*, 274–281. [CrossRef]
55. Ribeiro, R.A.P.; Andrés, J.; Longo, F.; Lazaro, S.R. Magnetism and multiferroic properties at $MnTiO_3$ surfaces: A DFT study. *Appl. Surf. Sci.* **2018**, *452*, 463–472. [CrossRef]
56. Shvartsman, V.V.; Bedanta, S.; Borisov, P.; Kleemann, W.; Tkach, A.; Vilarinho, P.M. Spin cluster glass and magnetoelectricity in Mn-doped $KTaO_3$. *J. Appl. Phys.* **2010**, *107*, 103926. [CrossRef]

57. Valant, M.; Kolodiazhnyi, T.; Axelsson, A.-K.; Babu, G.S.; Alford, N.M. Spin ordering in Mn-doped $KTaO_3$? *Chem. Mater.* **2010**, *22*, 1952–1954. [CrossRef]
58. Venturini, E.L.; Samara, G.A.; Laguta, V.V.; Glinchuk, M.D.; Kondakova, I.V. Dipolar centers in incipient ferroelectrics: Mn and Fe in $KTaO_3$. *Phys. Rev. B* **2005**, *71*, 094111. [CrossRef]
59. Golovina, I.S.; Shanina, B.D.; Geifman, I.N.; Andriiko, A.A.; Chernenko, L.V. Specific features of the EPR spectra of $KTaO_3$:Mn nanopowders. *Phys. Sol. State* **2012**, *54*, 551–558. [CrossRef]
60. Golovina, I.S.; Lemishko, S.V.; Morozovska, A.N. Percolation magnetism in ferroelectric nanoparticles. *Nanoscale Res. Lett.* **2017**, *12*, 382. [CrossRef]
61. Singh, K.; Maignan, A.; Simon, C.; Hardy, V.; Pachoud, E.; Martin, C. The spin glass Delafossite $CuFe_{0.5}V_{0.5}O_2$: A dipolar glass? *J. Phys. Condens. Matter* **2011**, *23*, 126005. [CrossRef]
62. Choudhury, D.; Mandal, P.; Mathieu, R.; Hazarika, A.; Rajan, S.; Sundaresan, A.; Waghmare, U.V.; Knut, R.; Karis, O.; Nordblad, P.; et al. Near-room-temperature colossal magnetodielectricity and multiglass properties in partially disordered La_2NiMnO_6. *Phys. Rev. Lett.* **2012**, *108*, 127201. [CrossRef] [PubMed]
63. Sharma, S.; Basu, T.; Shahee, A.; Singh, K.; Lalla, N.P.; Sampathkumaran, E.V. Multiglass properties and magnetoelectric coupling in the uniaxial anisotropic spin-cluster-glass Fe_2TiO_5. *Phys. Rev. B* **2014**, *90*, 144426. [CrossRef]
64. Rathore, S.S.; Vitta, S. Effect of divalent Ba cation substitution with Sr on coupled "multiglass" state in the magnetoelectric multiferroic compound $Ba_3NbFe_3Si_2O_{14}$. *Sci. Rep.* **2015**, *5*, 9751. [CrossRef] [PubMed]

4

Concluding Remarks

As noted by Wolfgang Kleemann, co-editor of this book, it is a well-known fact that E. Schrödinger was the first to stress the high probability of ferroelectricity in crystals, analogously to ferromagnetism, mentioned in his habilitation thesis (Erwin Schrödinger. Kinetik der Dielektrika, Sitz-Ber., Math.-Naturwiss. Klassse, Univ. Wien (1912). This suggestion was made following a conjecture of Peter Debye, Phys. Zeitschr. 13 (1912).

Ferroelectric behaviour was discovered by Valasek in 1921 because he was interested in understanding the **Piezoelectric** behaviour of Rochelle Salt, which had been used during the First World War (1914–18) to detect underwater incoming submarines. All along the 20th century the understanding of **Ferroelectricity**, experimental, theory and applications, has grown substantially all the way. Now, and in the 21st century it will remain a rich field of study, as it has been pointed out well in "Nature Materials" (Feb. 2020).

In particular, understanding of the double well potential, which is the key to ferroelectric behaviour, as it is also the key to hydrogen bonding in so many aminoacids and proteins, can be expected to go up in the forth-coming decades. And many of the potential applications, like direct energy conversion from thermal to electric energy, is very likely to be developed properly in the not-too-distant future.

The Editors
23 June 2021

5

Appendix: Obituaries

1. Joseph Valasek

J. Valasek was born on 27 April 1897, in Cleveland, Ohio, and died on 4 October 1993 in Minneapolis at the age of 98. He was the discoverer of ferroelectricity, which he identified in Rochelle Salt, in the 1920s.

He measured its transition temperature (he called it Curie Temperature, T_C) below which its spontaneous polarization was reversible. It may be noted that the chemical structure of Rochelle Salt is very complex and that below T_C = 23°C the spontaneous polarization grows up to a maximum of 0.25 μC/cm^2 and then decreases again at a lower temperature, disappearing at $T_X \approx -20°C$. Valasek investigated Rochelle Salt because it had a substantial **piezoelectric** activity. Both transitions temperatures are strongly pressure dependent.

Phase transitions and cooperative phenomena in solids flowered only in the mid-1950s, after the end of the Second World War, long after Valasek had changed field. When he died, he was Professor Emeritus of Physics at the University of Minneapolis. He was a quiet and modest man and did not seek recognition and honour for his important discovery.

2. C. B. Sawyer

Ferroelectric **hysteresis loops** can be easily observed in the screen of an oscilloscope by inserting a crystal plate (with conducting electrodes at both sides) by means of a circuit, first described by C. B. Sawyer and C.H. Tower (Phys. Rev. **35**, 269, 1930), in which the ferroelectric plate is connected in series to a capacitor C_O and at the other side is connected to V (the voltage source). The horizontal connections to both sides of the capacitor C_O (whose voltage is proportioned to the polarization of the crystal C_X) goes to the horizontal connections of the oscilloscope, and the voltage of the source V is connected to the horizontal contacts to the oscilloscope. At $T < T_C$ (ferroelectric phase) hysteresis loops are observed if the field is higher than the **coercive field** of the crystal; at $T > T_C$ (paraelectric phase), a non-linear "**sigmoid**" curve is observed. At sufficiently large T the nonlinearity approached the typical linearity of a regular paraelectric.

3. P. Scherrer

The ferroelectric activity of KH_2PO_3 (KDP) was first reported by Busch and Scherrer (Naturwiss 23, 737, 1935). At room temperature the crystal structure is $\bar{4}2m$ non-centrosymmetrical (piezoelectric). A phase transition occurs at a relatively low temperature 123ºK (–150ºC) to a ferroelectric phase. Busch measure the dielectric constant $\varepsilon_C(T)$ and found a very large peak at 123ºK. Busch observed an enormous peak going up to $\varepsilon_C(123°C) \simeq 10^5$, while at room temperature value is $\varepsilon_C(300°C) \simeq 50$. $\varepsilon_C(T)$ fulfils very well the Curie-Weiss law: $\frac{1}{\varepsilon} = \frac{4\pi}{C}(T - T_o)$. Measurements of the spontaneous polarization at low temperature show that $P_S(100°K) \simeq 4.6\ \mu C/cm^2$, almost saturated. KDP shows a large peak in the specific heat $C_p(T)$ at $T = 123°K$. Structural characterization of KDP was carried out by Frazer and Pepinski. Slater published a theory of ferroelectricity in KDP in J. Chem. Phys. (1941)

4. H. Megaw

Helen D. Megaw (1907–2002) was born in Dublin, Ireland. Her family moved to Belfast in 1921, just before the partition of Ireland. After a short period in the Methodist College, Belfast, she went to Roedean School, Brighton (1922–1925). One of her aunts was secretary at a College in Cambridge, and Helen´s ambition was to study there. For financial reasons she decided to go to Queen's University, Belfast. In 1925 she had intended to study Mathematics, but she had enjoyed Chemistry at school, and she opted for both Natural Sciences and Mathematics. Her Director of Studies explained het that she had to study three subjects, Chemistry, Physics and

Mathematics. She told her that she was required to study **experimental** subjects. Helen chose Mineralogy. She then specialized in Physics, obtaining a Class II in Part II the year 1930. When Prof. E. Rutherford, at the Cavendish Laboratory, told Helen that there was no opportunity for her to do postgraduate work in the Physics Department she went to Prof. A. Hutchinson, whose Department of Mineralogy had a strong crystallographic tradition.

Helen completed her Ph.D. in 1934. She got then a fellowship which enabled her to spend a year at the Chemistry Institute of Vienna with Prof. H. Mark. This was followed by a year in 1935–1936 working with Prof. Simon, at the Clarendon Laboratory, Oxford. While she taught Physics for seven years, she was able to continue research during the School holidays, returning to Cambridge and working there in a private lab, on the crystallography of diamonds. In 1943, she returned to full-time research as an X-ray crystallographer at Philips Mitcham Works Research Laboratory. There she studied **barium titanate** $BaTiO_3$, the material of a new ceramic capacitor from America, which had been sent to Philips by mistake. This triggered a life-long interest in **perovskites** ABO_3 compounds, many of them ferroelectric, or antiferroelectric.

In her young days she was an indefatigable traveller and a keen winter sports woman. Helen believed that "the combination of research and teaching is vital for the progress of knowledge".

5. W. P. Mason

Warren Perry Mason (September 28, 1900–August 23, 1986) was an American electrical engineer and physicist at Bell Labs.

He graduated from Columbia University, and he published four books and nearly a hundred scientific papers. In addition, he issued over two hundred relevant patents, more than anyone else at Bell Labs at a time in which transistors, using doping semiconductors, revolutionized electronics and opened the way for personal computers.

His work included acoustics, filters, crystals and ceramics, materials science, polymer chemistry, ultrasonics, bonding to semiconductors, internal friction, and viscoelasticity.

Mason founded the field of distributed-element circuits. He was the first to experimentally show viscoelasticity in individual molecules. He found experimental evidence of electron-phonon coupling in solids and made measurements that aided the theories of phonon drag and superconductivity. Many of Mason's inventions in electronics are still widely used by modern circuit designers.

In connection with ferroelectricity, he, in collaboration with Bernd Matthias, produced the first high level microscopic theory of ferroelectricity using Statistical Mechanics, which was applied successfully to the perovskite ferroelectric barium titanate, $BaTiO_3$, and to other perovskites ABO_3, with many practical applications.

6. B. Matthias

Bernd Theodor Matthias (June 8, 1918–October 27, 1980) was a German-born American physicist credited with discoveries of hundreds of elements and alloys of metallic superconductors. It was commonly assumed then among physicists that when Matthias discovered a higher temperature superconductor, theorists became immediately active at developing a theory capable of predicting that high transition temperature. He worked also in ferroelectrics, and collaborated with W.P. Mason in producing the first successful statistical mechanical theory (similar too Weiss theory in ferromagnetism) to explain ferroelectricity in $BaTiO_3$. This theory was later applied successfully to hydrogen bonded ferroelectrics, like KH_2PO_4 and Triglycine sulphate.

Matthias was born in Frankfurt, Germany. He received his Ph.D. in physics from the Federal Institute of Technology in Zurich in 1943. He immigrated to the United States in 1947 and went to work for Bell Laboratories. He taught at the Massachusetts Institute of Technology and the University of Chicago before joining the physics faculty of University of California, San Diego in 1961, where he remained for the rest of his scientific career. Matthias was also a member of the JASON defense advisory group. In 1965 he was elected to both the National Academy of Sciences and the American Academy of Arts and Sciences. Paul C. W.

Chu, director of the Texas Center for Superconductivity, Houston, was Matthias' former student.

Matthias, Miller and Remeika reported the discovery of ferroelectricity in Triglycine Sulphate (TGS) at $T < 47°C$ in Phys. Rev. **104**, 849 (1956).

7. J. C. Slater

John Clarke Slater (December 22, 1900–July 25, 1976) made major contributions to the theory of the electronic structure of atoms, molecules and solids and to microwave electronics. He received a B.S. in Physics from the University of Rochester in 1920 and a Ph.D. in Physics from Harvard in 1923. Then he did post-doctoral work at the universities of Cambridge (UK) and Copenhagen (Denmark). When he returned to the US, he joined the Physics Department at Harvard.

In 1930, Slater was appointed Chairman of the MIT Department of Physics. He wrote 14 books between 1933 and 1968. During World War II, his work on microwave transmission, made in collaboration with Bell Laboratories was of major importance in the development of radar.

In 1950, Slater founded the Solid State and Molecular Theory Group within the Physics Department. In 1951 he left MIT to spend a year at the Brookhaven National Laboratory (Upton LI, New York) USAEC (US Atomic Energy Commission) until he retired from MIT in 1965.

He joined afterwards the Quantum Theory Project of the University of Florida as Research Professor for five years.

In 1964, he was awarded an honorary degree by Florida University. Slater was important at the formulation of the Bohr-Kramers-Slater theory, within which the terms "Slater determinant" and "Slater orbital" design important elements of the theory.

Slater's contributions to the theory of ferroelectricity consists in reasonable corrections approximating the Lorentz's field in non-cubic lattices which result in: $\varepsilon \approx \frac{C}{T-T_0}$

This is the Curie-Weiss law for $C >> T_0$, which is the case in both KDP (Slater: J. Chem. Phys., 1941) and $BaTiO_3$ (Slater: Phys. Rev., 1950).

See for instance, Jona and Shirane, "Ferrolectric Crystals", 1962.

8. G. Shirane

Gen Shirane (May 15, 1924, Nishinomiya, Jan. 16, 2005, New York) was a Japanese – American experimental solid state physicist, well known for his investigations using neutron scattering as a probe to study crystals undergoing phase transitions (ferroelectric, structural magnetic and superconducting). He spent most of his life in Brookhaven National Laboratory, Upton L. I., New York.

He received his BE in (aeronautical) engineering physics in 1944 and his Dr. Sc. in physics in 1947 with a thesis on ferroelectrics. From 1948 to 1952 he was research associate in physics at the Tokyo Institute of Technology, He moved to Pennsylvania State University where he stayed from 1952 to 1955 as research associate. Then from 1956 to 1957 he became associate physicist at BNL. Afterwards Westinghouse Research Laboratories, in Pittsburgh, he was research physicist from 1959 to 1963, and then advisor. At this time, he published in the International Series of Monographs on Solid State Physics (with Franco Jona IBM Research Center, New York) the excellent monograph **"Ferroelectric Crystals**", a Pergamon Press Book, with the MacMillan Company, New York, 1962. In 1963 he moved permanently to Brookhaven, where the High Flux Beam Reactor, the best of the world then and for many years, was under construction. There he developed the triple-axis neutron spectrometer, improving the neutron spectrometer previously constructed by B. Brockhouse in Canada, optimizing the signal to noise ratio. His expertise led to his publication, with S. Shapiro and J. Tranquada, of the **definitive monograph** on the subject. He worked on spin-wave dispersion and critical phenomena in ferromagnets (Cr), on soft mode investigations in structural ($SrTiO_3$) and ferroelectric transitions, and in electron phonon coupling in superconductors (Nb_3Sn).

In the 1960s and 1990s, Shirane and co-workers confirmed the **soft mode** theories of **P. W. Anderson** and **W. Cochran** beyond the original soft mode theory. He trained many young physicists, some of them from the UAM laboratory, Madrid (Spain) including Beatriz Noheda, in the late 1990s and early 2000s, with whom G. Shirane and D. Cox discovered a significant new **monoclinic phase** in PZT (mixed perovskite $PbZr_{1-x}Ti_xO_3$ solid solution) totally unexpected. Shirane has an extraordinarily high h-index of 103, which means 103 publications with more than 103 quotations, in the scientific literature, and a total of nearly 40.000 citations.

He came to Spain for the EMF 5 held at Benalmadena (Málaga) in 1983 and to the IMF 10, held at the U. Complutense (Madrid) in 2001.

9. W. Cochran

William Cochran (30 July 1922–28 August 2003) was born in Scotland and was educated at Borough High School in Edinburgh. He studied physics at the University of Edinburgh and completed his Ph.D. under Arnold Beevers in X-ray crystallography of sucrose, using isomorphous replacement. He then moved to the University of Cambridge to work with Lawrence Bragg, and obtained tenure in 1951. He realised that isomorphous replacement was the key to solving protein structures. With **Francis Crick**, he invented methods for deducing **helical patterns** from crystallographic data, which ultimately led to the solution of the **DNA structure** by Watson and Crick, published in "Nature" in 1953, which resulted in the award of the **Nobel Prize** for Medicine in 1958.

Cochran went to Canada to study neutron diffraction with Bertram **Brockhouse** (a future Nobel Prize in Physics) and used the solid-state theory of lattice dynamics to explain the phenomenon of ferroelectricity in crystals. This idea was also advanced around the same time by the American physicist Philip Anderson (another future Nobel Prize in Physics). Cochran's basic idea was that upon cooling from higher to lower temperature, symmetry breaking can take place, going the crystal lattice from a **paraelectric** (symmetric) phase to a **ferroelectric** (asymmetric) phase.

Cochran returned to Edinburgh in 1964 and became Department Head in 1975 of the single Department made up from the Natural Philosophy and the Mathematics Physics Departments. He was elected Fellows of the Royal Society (FRS) in 1962, and won the Howard N. Potts Medal in 1984. Jona and Shirane Monograph "Ferroelectric Crystals" (MacMillan, NY, 1962) concludes that Cochran's theory provides a clear picture of the balance between **long-** and **short-range** forces, leading to $\omega_T(T) \rightarrow 0$ (soft mode at $T \rightarrow T_0$) when $\varepsilon \propto \frac{C}{T - T_0}$ as required by the Lyddane-Sachs-Teller relation.

10. J. Stankovski

Prof. Jan Stankovski, born in Poznan (Poland) on Jan. 1934, was a pioneer of Electron Paramagnetic Resonance (EPR) Spectroscopy in Poland. He used EPR to investigate many solid-state materials such as ferroelectrics, ferromagnets and high temperature superconductors, as well as proton glasses and fullerenes. He died in Poznan on Sept. 4, 2009. He graduated in physics from the Adam Mickiewicz University, Poznan in 1956. His mentor was Prof. Arkendiusz Piekra who inspired in him a passion for physics which made him look to research in physics as a human adventure towards human knowledge in general. His Master and Ph D dissertations were both about ferroelectrics. His first research paper was published in 1958 on the investigation of nonlinearity in ferroelectric Rochelle Salt. His last paper was published in 2010, under the title "Size driven ferroelectric effects in TGS induced by high pressures".

Prof. Stankovski was a cofounder of the Institute of Molecular Physics of the Polish Academy of Science. In 1964 he and his team put forward the first ammonia laser made in Poland. He created in 1977 the Division of Low Temperature Physics necessary to investigate superconductivity, including so called high temperature superconductors like $YBa_2CuO_{7-\delta}$.

He was very interested in popularizing physics among young people. He organized in 1985 a "Summer with Helium" Scientific workshop in Odolanov, with young students from all over Poland. Many of them became later outstanding professional physicists.

Professor Stankovski was a Full Member of the Polish Academy of Science, a Member of the Academy of Science of Slovenia, a Vice-President of the Groupement AMPERE (1996–2002), and a Fellow of the Institute of Physics, London.

He organized a Conference on Radio and Microwave Spectroscopy from 1976 to 2009, attended by many scientists from around the world interested in EPR, NMR, NQR Spectroscopies as well as in ferroelectrics.

He was also a member of the Editorial Boards of "Ferroelectrics", "Applied Magnetic Resonance", "Superconductivity Review" and the "Bulletin of Magnetic Resonance"

11. R. Blinc

Robert Blinc (October 30, 1933–September 26, 2011) was a prominent Slovene physicist from the time when Yugoslavia was a single country under Tito. He completed his undergraduate studies in 1958 at the Faculty of Natural Sciences in Ljubljana and received a Ph.D. a year later. Then he went to the Massachusetts Institute of Technology as a post doc. When he returned to Ljubljana, he continued his work at the Jožef Stefan Institute and became Professor in 1970 and Dean since 2004.

He was one of pioneers of Nuclear Magnetic Resonance for investigations of phase transitions and liquid crystals. His first paper on the subject was published in early 1958. He was President of the AMPERE Groupement during 1990–1996, and member of the Slovenian Academy of Sciences and Arts, serving as Vice President from October 2, 1980, to May 6, 1999.

He was also a member of the European Academy of Sciences and Arts, and the Society of Mathematicians, Physicists and Astronomers of Slovenia. He died in Ljubljana in 2011 after a life of professional dedication to research and teaching of Physics. Last but not least, he was President of the European Advisory Committee of Ferroelectricity when Prof. K.A. Müller stepped down, after receiving the Physics Nobel Prize for his discovery with Bednorz of the high temperature superconductors.

12. L. E. Cross

Professor L. Eric Cross, Obituary L. Eric Cross (1923–2016) IEEE, passed away at his home in University Park, Pennsylvania, USA on December 28, 2016. He was born in Morkey, Yorkshire, England in 1923. He studied physics at Leeds University, UK, graduating with BSc and Ph.D. Degree in 1952. His thesis had a very simple title "Ferroelectric Phenomenon". Cross began his research career studying dielectric and ferroelectric materials at the British Electrical Research Association.

In the early 1960s, he accepted an offer from Professor R. Roy to take up an academic position at Pennsylvania State University, and successively served as Associated Professor, Professor and Assistant Director of the Materials Research Laboratory (MRL).

In 1972, he was appointed Associate Director of the Materials Research Laboratory, and became its Director in 1983, remaining there till he formally retired. Retirement was only in effect from his administrative duties: He continued as active as ever until a few years before his death. His important work on Sodium Niobate, co-authored by B.J. Nicholson was published in 1955 in the "Philosophical Magazine, Series 7. His last book "Domains in Ferroic Crystals and Thin Films", co-authored by A. Tagantsev and Jan Fousek, (recently deceased), was published by Springer in 2010. Many of his contributions received a large number of citations. They were devoted to perovskites like $BaTiO_3$, PLZT – Lead Zirconium

Titanate, and Lead lanthanum Zirconate Titanate, etc. The properties investigated included ferroelectric, piezoelectric, electrostrictive, pyroelectric, dielectric and electroscopic behaviour. He was involved in the applications of ferroelectrics to **sensors** and **actuators**. Under his leadership the MRL at Penn State University became the leading research center for Ferroelectricity and Related Phenomena in the USA and one of the leading centers in the world.

During his long career of 65 years, Professor Eric Cross received many awards and honors, including the **Von Hippel** Award of the Materials Research Society.

Eric and his wife Lucilla enjoyed a happy family life with their six children, Peter, Matthew, Daniel, Rebeca, Rachel and Elizabeth. Last, but not least, he did write "**History of Ferroelectricity**", in collaboration with his good friend R. E. Newnham.

13. J. P. Scott

James Floyd Scott was born in Beverly, New Jersey, on May 1942, and died in Cambridge, UK in April 2020.

As noted by Neil Mathur in the obituary of Jim Scott in Physics Today, July 2020, his knowledge of ferroelectric materials was encyclopaedic, and he did combine it with a great personal creativity.

He had an incredible gift for story-telling, which can be seen in the transcript of his **2018 IEEE interview** (https://ethw.org/Oral-History:James_F._Scott)

Jim Scott aimed to integrate ferroelectrics with semiconductors, and tried to demonstrate that **fatigue-free** ferroelectric switching was possible and practical.

In the 1970s he visited researchers in ferroelectricity in the USSR and in China, when those countries were behind the iron curtain for Westerners. In Russia, while visiting Pyotr Kapitsa, the Physics Nobel Prize winner in 1978, he met Galya, a beautiful blond Russian lady whom he married in 1982. Arkady Levanyuk, a Russian theoretical physicist invited by J. A. Gonzalo to be a visiting Professor in early 1980s at UAM, Madrid, told him that Jim's objective was then to end the Cold War between the USA and the USSR by promoting marriages between American young men and Russian young ladies and vice versa.

Jim obtained his BA from Harvard University in 1963, and his PhD from the Ohio State University in 1966. There after he went to Bell Labs where he devoted five years to do **Raman Scattering** experimental work and to explain the soft mode behavior in Strontium Titanate. In 1970, he began to work in Ferroelectrics with W. Cochran at the University of Edinburg in Scotland. Coming back to the US he spent 20 years at the University of Colorado Boulder. Jim Scott cofounded Ramtron International Corp. in 1984, and two years later Symetrix Corp. in Colorado Springs.

Within the next few years, Symetrix was collaborating with the Matsushita Electric Corp in Japan to exploit the idea that bismuth oxide planes can act as oxygen reservoirs for fatigue-free switching. Thereafter he exploited that idea with Sony in Japan before he moved to Australia in 1991. As the new millennium downed ferroelectric random-access memories (FERAMs) began to enjoy commercial applications in Sony's PlayStation 2, and later in Japanese railway fare cards.

Jim's work on ferroelectric materials ended, as it had begun, in Scotland, at the University of St. Andrews, before returning to Cambridge, due to poor health, in April 2020.

14. P. W. Anderson

Philip Warren Anderson (December 13, 1923–March 29, 2020) was an American theoretical physicist and Nobel Prize winner in Physics (1977) who made very substantial contributions to the theories of ferroelectricity, antiferromagnetism, symmetry breaking (1962), particle physics, and high-temperature superconductivity.

He was born in Indianapolis, Indiana, and grew up in Urbana, Illinois. He graduated from University Laboratory High School in Urbana in 1940, and then he went to Harvard University for undergraduate and graduate work, with a wartime in between at the Naval Research Laboratory. Very young he had a close association with particle-nuclear physicist H. Pierre Noyes, molecular physicist Henry Silsbee, and J. H. van Vleck, with whom many years later he would receive the Nobel Prize in Physics.

From 1949 to 1984, he was employed by Bell Laboratories in New Jersey, where he worked on a wide variety of problems in condensed matter physics. He conjectured that extended states can be localized by the presence of disorder in a system (Anderson's localization), he invented the Anderson Hamiltonian to describe the site-wise interaction of electrons in a transition metal; he proposed symmetry breaking within particle physics, playing a role in the Standard Model, developed a theory behind the Higgs mechanism to generate mass in some elementary particles created the pseudospin approach to the Bardeen-Cooper-Schrieffer theory of superconductivity, and made substantial contributions to the theory of superfluidity of He^3, and helped to originate the subject of spin-glasses.

From 1967 to 1975, Anderson was a professor of theoretical physics at Cambridge University. He received the Nobel Prize in 1977 for his investigations on the electronic structure of magnetic and disordered systems, related to electronic switching and memory devices in computers. He was invited by Mortimer Key, M. Gomez and J.A. Gonzalo to participate in the International Conference on Ferroelectricity and Superconductivity held at the Caribe Hilton of San Juan of Puerto Rico in December 1975, but politely declined while accepting to be sponsor as long as he would not be committed himself to come.

In the mid 1980's, after the discovery of the high T_C Cuprates by Müller and Bednorz, he published theoretical work to explain this unexpected phenomenon, He retired from Bell Labs in 1984, and became Professor Emeritus of Physics at Princeton University.

In "Ferroelectric Crystals" (Jona and Shirane), p. 383, it is mentioned that ferroelectricity in crystals can be treated as a problem in lattice dynamics, as pointed out by W. Cochran and P.W. Anderson who did it in a preprint of a paper presented at the Moscow Conference on Dielectrics, December 1958.

15. V. L. Ginzburg

Vitaly Lazarevich Ginzburg, (October 4, 1916–November 8, 2009) was a Soviet and Russian theoretical physicist, who received the Nobel Prize for Physics in 2003.

He graduated from the Physics Faculty of Moscow State University in 1938. Among his main achievements are the Landau–Ginzburg phenomenological theory for phase transitions, which is applicable to ferroelectric phase transitions, and the theory of electromagnetic wave propagation in plasmas. He contributed to discredit the Russian biologist Trofim Lysenko, who opposed openly to Gregor Mendel's genetic theory. His doctoral advisor was Prof. Igor Tamm.

Ginzburg was the Editor-in-Chief of the scientific journal "Uspekhi Fizicheskikh Nauk", and was head of the Academic Department of Physics and Astrophysics. He was a secular Jew, but after the downfall of Communism in Russia, became active in Jewish life, and supported the state of Israel.

He worked at the P. N. Lebedev Physical Institute of Soviet and Russian Academy of Sciences.

In August 2001, he came to Madrid to be an Invited Lecturer at the 10th International Meeting of Ferroelectricity, where he spoke about "Phase Transitions in Ferroelectrics: Some Historical and Other Remarks". While he was in Madrid, he asked to be accompanied to a "bullfight", which the local organizers did, advised by Prof. Arkadi Levanyuk, Ukranian Professor of Physics at the UAM, Madrid, who had done his Ph.D. Thesis with Vitaly Ginzburg.

Printed in the United States
by Baker & Taylor Publisher Services

Printed in the United States
by Baker & Taylor Publisher Services